Emerging Nanomaterials and Their Impact on Society in the 21st Century

Edited by

N.B. Singh[1], Md. Abu Bin Hasan Susan[2] and Ratiram Gomaji Chaudhary[3]

[1]Department of Chemistry and Biochemistry, SBSR, Sharda University, Greater Noida, India

[2]Department of Chemistry, University of Dhaka

Dhaka 1000, Bangladesh

[3]Associate Professor and Head, Department of Chemistry, Seth Kesarimal Porwal College of Arts, Science and Commerce, Kamptee, India

Published by **Materials Research Forum LLC**
Millersville, PA 17551, USA

Published as part of the book series
Materials Research Foundations
Volume 135 (2023)
ISSN 2471-8890 (Print)
ISSN 2471-8904 (Online)

Print ISBN 978-1-64490-216-5
eBook ISBN 978-1-64490-217-2

Distributed worldwide by

Materials Research Forum LLC
105 Springdale Lane
Millersville, PA 17551
USA
https://www.mrforum.com

Manufactured in the United States of America
10 9 8 7 6 5 4 3 2 1

Table of Contents

Nanomaterials: Overview and Historical Perspectives
Mridula Guin and N.B. Singh 1

Nanomaterials in the Lubricant Industry
Ashima Srivastava, Pratibha Singh, Preeti Jain, Kirti Srivastava, Sanjeev Sharma 23

Carbon Nanomaterials and Their Applications
Neksumi Musa, Sushmita Banerjee, N.B. Singh, Usman Lawal Usman........................ 40

Functionalized Carbon Nanomaterials: Fabrication, Properties, and Applications
P.R. Bhilkar, R.S. Madankar, T.S. Shrirame, R.D. Utane, A.K. Potbhare, S. Yerpude, S.R. Thakare, Sarita Rai, A.A. Abdala, Ratiram G. Chaudhary 72

Smart Nanomaterials and Their Applications
Md. Kawsar Alam, Rejwana Karim, A. J. Saleh Ahammad... 100

Emerging Nanomaterials in Drug Delivery and Therapy
Mahmuda Nargis, Raju Ahmed, Abu Bin Ihsan 125

Hybrid Nanomaterials: Historical Developments, Classification and Biomedical Applications
Jarin Ibedita Edi, Abdul Muhaymin, Md. Abu Bin Hasan Susan................................. 152

Carbon Nanomaterials for Efficient Perovskite Solar Cells
T. Swetha, Surya Prakash Singh 178

Nanoemulsions: Preparation, Properties and Applications
Preeti Gupta, Vinita, S.S. Das.. 200

Effect of Nanomaterials on the Properties of Binding Materials in the Construction Industry
N.B. Singh, B. Middendorf, Richa Tomar 226

Nanoparticles Incorporated Soy Protein Isolate for Emerging Applications in Medical and Biomedical Sectors
Priya Rani, Priyaragini Singh, Abhay Pandit, Rakesh Kumar...................................... 265

Emerging Nanomaterials in Healthcare
Ephraim Felix Marondedze, Lukman O. Olasunkanmi, Atheesha Singh, Penny Poomani Govender.. 284

Toxicological Effects of Nanomaterials on the Aquatic Biota
Usman Lawal Usman, Sushmita Banerjee, Nakshatra Bahadur Singh, Neksumi Musa304

Nanomaterials and Safety Concerns to End Users
A. Ale, S. Municoy, J. Cazenave, M.F. Desimone 341

Preface

It is rightly said that one who controls materials, controls technology. The 21st century is the century of materials and particularly of nanomaterials. Nanomaterials have a tremendous role in the development of science and technology. Almost all branches of science and technology use nanomaterials in one way or the other. A large number of research papers, review articles and books have been published on the topic. However, a comprehensive literature on the topic at one place is rarely found. To facilitate the timely widespread utilization of this new theme on nanoscience and technology, it is important to have comprehensive material on different aspects of nanomaterials in the form of a book.

The present book on Emerging nanomaterials and their impact on society in the 21st century contains 14 chapters on different topics written in a simple language by experts from different parts of the globe. The present book emphasizes on synthesis, classification, historical background, characterizations, and applications of nanomaterials in various sectors like nanosensors, healthcare, solar cells, energy storage, hydrogels, nanocatalysts, sport industry, automobile sector, construction industry, lubricant industry, defence and security, textiles, food sciences, agriculture, biomedical sectors etc.

The book will be useful to students, researchers, academics, scientists, and technocrats. We would like to acknowledge and thank diligent work of the authors in developing the chapters. We also thank each and every one who contributed to improving the present book.

N. B. Singh
Md. Abu Bin Hasan Susan
Ratiram Gomaji Chaudhary

Emerging Nanomaterials and Their Impact on Society in the 21st Century   Materials Research Forum LLC
Materials Research Foundations 135 (2023) 1-22   https://doi.org/10.21741/9781644902172-1

Chapter 1

Nanomaterials: Overview and Historical Perspectives

Mridula Guin[1*] and N.B. Singh[1,2]

[1]Department of Chemistry and Biochemistry, Sharda University, Greater Noida, India

[2]Research Development Cell, Sharda University, Greater Noida, India

* mridula.guin@sharda.ac.in

Abstract

Much earlier than the start of the "nanoera", people were using various nanosized materials but were not aware of nanotechnology. More than a thousand years BC, people used natural fabrics with much smaller size (1-20 nms). The British museum boasts about Licurg's bowl showing different colours under different conditions, since it contains nanosize gold and silver. However, the nano word was not known. The concept of nanoscience came for the first time by the lecture of Prof. R. Feynman in 1959 at the session of the American Physical Society, saying that *"there is plenty of room at the bottom"*. The word Nanotechnology was introduced for the first time by N. Taniguchi in Tokyo in 1974. The idea of Feynman was developed by E. Drexler in 1986 through his book. Now nanoscience and nanotechnology have become one of the most important areas of science and technology dealing with materials of the size less than 100 nm in size. At the moment, there is no sector, where nanotechnology is not being used. Now it is said to be the era of nanoscience and technology.

Keyword

Nanomaterial, Nanotechnology, Nanoscience, Nanosize, Lycurgus Cup

Contents

Nanomaterials: Overview and Historical Perspectives 1

1. Introduction 2

2. Terms associated in nanotechnology 3

3. Historical development of nanotechnology 4

4. Modern era of nanotechnology 11

5. Types of nanomaterials 15

6.   **Applications of nanomaterials**..**16**

7.   **Concluding remarks** **17**

References **18**

1.   Introduction

Nanomaterials encompasses a large variety of materials having dimension of 1 nm to 100 nm (Fig.1) [1]. The matter at nano dimensions displays distinctive properties compared to their bulk materials [2]. Nanoscale materials are chemically more reactive because of their large specific surface area. In nanomaterial, the laws of classical physics are ruled out and quantum confinement leads to unique optical, magnetic and electrical effect. Although nanomaterials are known from ancient times. Nano science and Nano technology are in the forefront of the 21st century. Manipulation of matter at nano dimension and controlling them for novel applications fulfills the humans imaginations in several directions. The amalgamation of various scientific disciplines such as chemistry physics, electronics, biology, engineering etc. expedite the application of nanomaterials in computer chips, energy sector, health care and medical diagnosis, safety equipment, space exploration etc. Over the next few years, nanotechnology will bring more breakthroughs by deeply impacting economic development in the society.

The existence of nanoparticles can be traced back to the initial phases of human history. But it has evolved magnificently during the industrial revolution. Nobel Laureate Richard Zsigmondy is the first person to coin the term 'nanometer'. He used a microscope to measure the size of gold particles from colloid solution.

The steppingstone of modern nanotechnology is laid by Richard Feynman during the lecture presented at American Physical Society meeting in 1965. There he introduced the idea of nanotechnology by the concept "There's Plenty of Room at the Bottom" [3]. Feynman hypothesis leads to novel ideas of manipulating matter at the atomistic level. He is widely regarded as the father of nanotechnology. The revolution of nanotechlogy started 15 years after the famous lecture of Feynman. In 1980, fullerene was discovered by the trio scientist Kroto, Smalley and Curl [4]. This further leads to increased interest in the field of nanotechnology. Invention of carbon nanotubes by Japanese scientist Iijima is another development in the history of nanomaterials [5].

Thus, controlling the materials in the nano dimension has brought us technology of stainless textile, scratch resistant surfaces, anti-ageing cream and many other benefits. In this chapter historical perspective of nanomaterials will be discussed and revolution of nanotechnology products in the current generation will be highlighted.

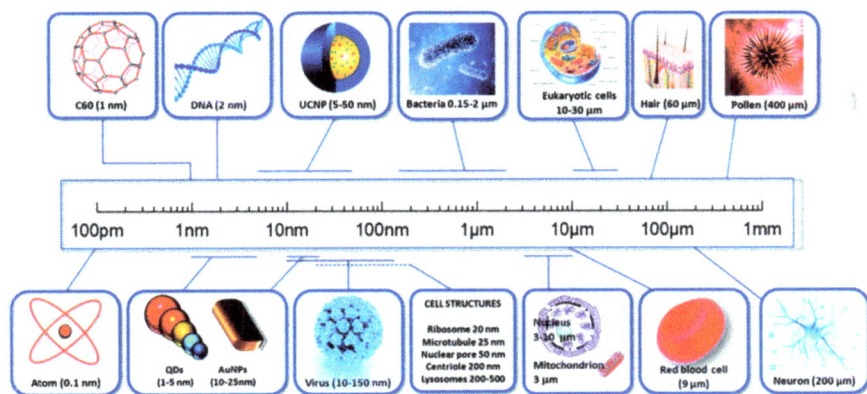

Figure 1. A basic concept on length scale showing size of nanomaterials and their comparison to biological components and the definition of 'micro' and 'nano' sizes [1].

2.   Terms associated in nanotechnology

The prefix nano is associated with a large number of terms in the nanoworld. It is originated from the Greek word 'nannos' which means very small. The definition of nanomaterial given by the organization for standardization (ISO) dictates that it is a material with any external dimension in the nanoscale or internal structure in the nanoscale [6-7]. Accordingly, they defined nanoparticle as an object with size 1 to 100 nm in nanoscale in any of the three external dimension [8]. Various terms associated with the term nano are given in Table1 [9].

Table 1. Various terms associated with the term "nano" [9].

Term	Size and explanation
Nanoscale	Size range between 1 to 100 nm
Nanoscience	Nanoscience involves investigation and understanding of matter at the nanoscale where the properties of the materials are completely different as compared to bulk materials.
Nanotechnology	It is the technology involving in developing materials, devices or systems by controlling matter at nanoscale.
Nanomaterial	Any material having at least one dimension in the nanoscale range of 1 to 100 nm.

Nano object	It is a material with one, two or three external dimensions in the nanoscale range
Nanoparticle	A material for which all three dimensions are in nanoscale range
Nanofibre	It is a nano object for which any two dimensions are in nanoscale range while the third dimension is significantly larger in size.
Nanostructure	The material with discrete functional parts in which the internal or surface structure is at least one of them in nanoscale region.
Nanocomposite	Multicomponent having multiphase domains in which at least one of the phases has a minimum of one dimension in nanoscale region.
Nanorod	The materials in which the shortest and longest axes are lengthwise different. The width of nanorods is between 1-100 nm and aspect ratio more than one.
Nanowire	Nanowire is similar to nanorod, only difference being their aspect ratio is very high.
Nanosphere	These are nanoparticle with aspect ratio of 1.
Nanotube	Nanotubes are basically hollow nanofibers
Nanostructured material	These are structured materials built by elements, molecules, crystallitesor clusters in the nanoscale region.
Engineered nanomaterials	Anthropogenically created materials having one or more dimension in the range of 1-100 nm
Nonmanufacturing	It refers to the production of materials at nanoscale through top down and bottom-up approach

3. Historical development of nanotechnology

Although it may appear that nanotechnology is a modern technology but its history is very old. The earliest use of nanoparticles can be traced back to the ancient time in 4th century AD. The Romans have shown an incredible example of nanotechnology in the glass industry known as Lycurgus cup (Fig.2) [10] which is 1600 years old and still kept in the British Museum for exhibition. It is also called dichroic glass because it exhibits two separate colors in different lighting conditions. When light falls on the front side of the cup, it appears green while it appears red when it gets illuminated from back side.

Research performed by the scientists unveils that the Lycurgus cup is made with soda-lime glass mixed with 1% gold and silver along with 0.5% manganese [11-12]. The fascinating colour change of this glass is due to the presence of colloidal gold. Advanced method of transmission electron microscopy confirmed the existence of 50-100 nm size of the gold and silver particles. The result of X-ray analysis showed that the silver and gold are in the ratio of about 7:3, additionally containing 10% copper dispersed in the glass matrix. In

2007, it was found that the dichroic nature of the glass arises because of plasmon stimulation process of metal nanoparticles. The red colour is due to the absorption of gold nanoparticles at 520 nm while the green colour is due to the absorption of silver nanoparticles. This is one of the important examples in the history of synthetic nanotechnology in the Roman period.

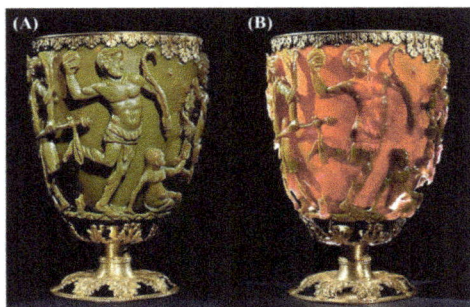

*Figure 2. The Lycurgus cup. The glass appears green in reflected light (**A**) and red-purple in transmitted light (**B**).*

In the late medieval era, the European churches show another example of nanotechnology. The beautiful stained-glass window of the cathedral with dazzling, bright red and yellow colours are made with mixing gold and silver nanoparticles (Fig.3a) [13]. Based on the size of the nanoparticles, the colours of the stained glass exhibit different bright glaze (Fig.4) [14]. Between the 9th-17th centuries, the Islamic world used ceramic glaze and later on it was dispersed in the European community [15]. These ceramics were made from silver, copper and other metal nanoparticles. In the 16th century, renaissance pottery was created by Italians [16]. During the 13th-18th centuries, the Ottoman techniques were applied to build Damascus swords (Fig.3b) [17], saber blades which were strengthened and sharpened using cementite nanowires and carbon nanotubes. Although the reason of these extraordinary properties in these materials were not known to the artists at that time.

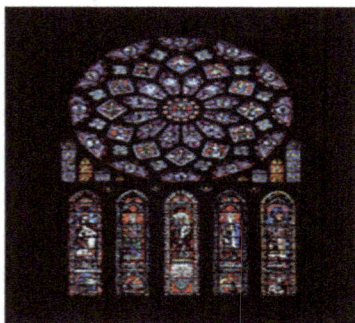

(a) Rose window   (b) Damascus Sword

Fig.3. (a) Rose window on the north facade of Notre Dame Cathedral (b) Damascus Sword known for its sharpness, flexibility and durability.

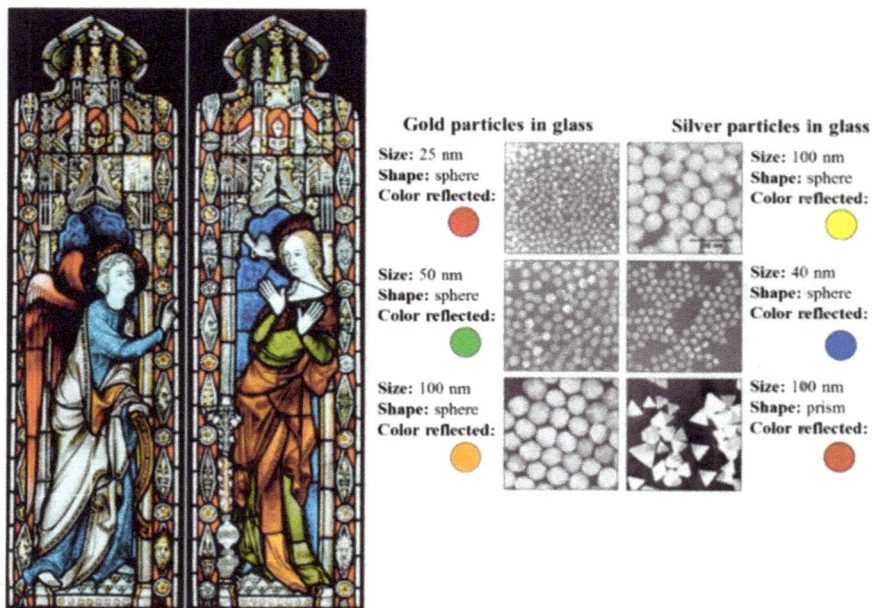

Fig. 4. Effect of nanoparticles on the colors of the stained-glass windows.

Looking back in history, it is evident that nanomaterials were used by humans since ancient times. However, research in nanotechnology begin from mid-19[th]century. Michael Faraday was one of the pioneers in developing nanotechnology. He was working with gold colloids in 1857 to understand their properties and to his surprise, he found exceptional optical and electrical properties of gold nanoparticles [18]. At that time controlling and knowing the size of the nano particles was not possible. In 1925, Richard Zsigmondy (Fig.5a) measured the size of gold nanoparticles and coined the term nanoparticles. He received the Nobel prize for his contribution in the development of nanotechnology. Around 1940, instead of carbon black the use of silica nanoparticles for the reinforcement of rubber started in the USA.

(a) Richard Zsigmondy   (b) Richard Feynman

Fig.5. a) Richard Zsigmondy: Coined the term "Nanoparticles"(b) Richard Feynman: Famous quote "There is plenty of room at the bottom".

The speech of Richard Feynman [Fig.5b] another Nobel Laureate at the meeting of American Physical Society in 1959 talked about the synthesis of atomic and molecular size by special techniques. His famous quote "There is plenty of room at the bottom " put the steppingstone in the progress of nanomaterials. His prediction came into reality as nanoscale machines were developed, where one can manipulate atoms according to their wish. Thus, he is regarded as the father of modern nano technology.

Japanese scientist Norio Taniguchi [Fig.6a] used various methods based on nanotechnology such as atomic layer deposition, focused ion beam etc. for fabricating semiconductors in nano scale [19]. In 1974, the term 'nanotechnology' was first used by him. According to him, nanotechnology consists of processing, separation, consolidation and deformation of materials by one atom or one molecule.

In the field of nanotechnology Eric Drexler [Fig.6b] is another important name. He is the first Ph.D. in molecular nanotechnology in the world. He published two books titled " Engines of creation: The Coming era of the nanotechnology" and "Nanosystems: Molecular, Machinery, manufacturing and computation" where he predicted about the existence of nanorobots and its implications [20-21]. His vision described in those books created a large impact on the development of nanotechnology. He educated people about the advantages and danger of nanotechnology through a nonprofit organization Foresight [22].

(a) Norio Taniguchi   *(b) Eric Drexler*

Fig.6. (a) Norio Taniguchi: Coined the term "Nanotechnology" (b) Eric Drexler: Writer of the book "Engines of creation".

Invention of scanning tunneling microscope (STM) in 1981 by Gerd Binnig and Rohrer (Fig.7a) [23,24] revolutionized the development of nanotechnology. This breakthrough invention enables people to visualize the atomic scale image of 3D structure of the nanomaterial surface and earned them the Nobel prize in 1986. The IBM logo was designed by moving xenon atoms using the STM tip which brings Feynman's vision into reality (Fig.8) [25,26].

Emerging Nanomaterials and Their Impact on Society in the 21st Century   Materials Research Forum LLC
Materials Research Foundations 135 (2023) 1-22   https://doi.org/10.21741/9781644902172-1

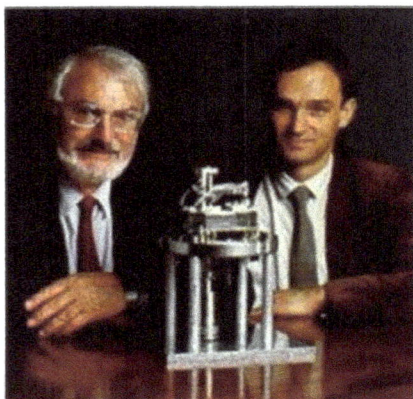

(a) Gerd Binning and Heinrich Rohrer   (b) Sumio Iijima

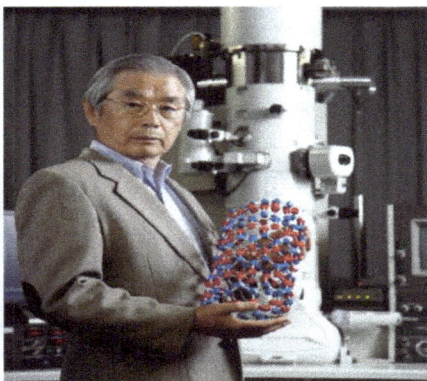

Fig.7. (a) Gerd Binning and Heinrich Rohrer: Inventors of STM (b) Sumio Iijima:
Discoverer of CNTs.

Fig.8. IBM logo crafted by 35 Xe atoms moving into Ni(110) surface using STM [26].

In 1986, another powerful instrument Atomic Force Microscope (AFM) was developed by
Gerd Binnig, Calvin Quate and Christoph Gerber [27,28]. Presently AFM is the most
powerful tool for processing and imaging nanoscale materials at atomic level.

The discovery of C_{60} or fullerene by Richard Smalley, Harold W. Kroto and Robert F. Curl
in 1985 raised the nanotechnology field to a new height [4]. Although C_{60} was first reported
by Japanese scientist Eiji Osawa in 1970, it did not get the publicity until the findings of
trio Smalley, Kroto and Curl.

C_{60} [Fig.9] is also called Buckminster fullerene or buckyballs in the honor of famous architect Buckminster Fuller building the geodetic dome resembling a football. This form of carbons is structurally stronger and lightweight. Because of their heat and electrical permeability, they are widely used for drug delivery and several other applications. Soon after this, Sumio Iijima in 1991 discovered carbon nanotubes [Fig.7b] [5]. Their unique properties and nanoscale dimensions are utilized in many advanced technological applications.

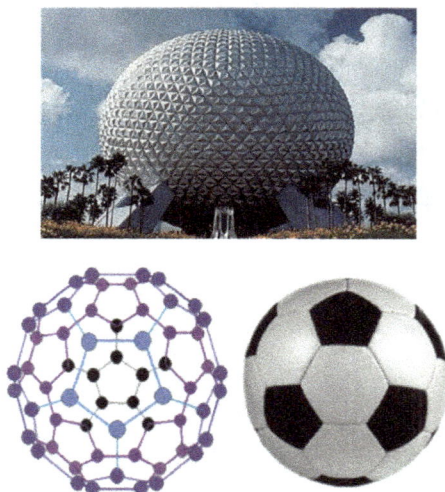

Fig.9. Similarity of C60 with football and Geodetic dome.

Around the same time, nanostructured catalytic materials MCM-41 and MCM-48 were discovered by C.T. Kresge and his colleagues [29,30]. These materials found applications in water remediation, crude oil refinement, drug delivery etc. In 1992, controlled synthesis of quantum dots/ nanocrystals was proposed by Moungi Bawendi from MIT [31]. The research on quantum dots leaps to higher level in the next few years and found applications in several areas.

In the biomedical field, Au nano shells developed by Naomi Halas and his team in 2003, has shown high potential [32,33]. They can be tuned to absorb light according to their size in diagnosing and curing of breast cancer cells. The painful process of biopsies, chemotherapy or surgery can be eliminated in the near future.

In 2006, a nanoscale car was made by Jams Tour and his coworkers [34,35]. This discovery demonstrated nanoscale patterns which is highly useful in the field of optoelectronics and medicine. The 21[st]century is well progressed with advancement in nanotechnology in diverse fields.

Many countries now have nanotechnology programs to make further progress in this field. People fondly address the current time as the Nano-Age because of the sky rocketing development of nanotechnology in recent time. A schematic presentation of the timeline in the progress of nanoscience is shown in Fig.10 [9].

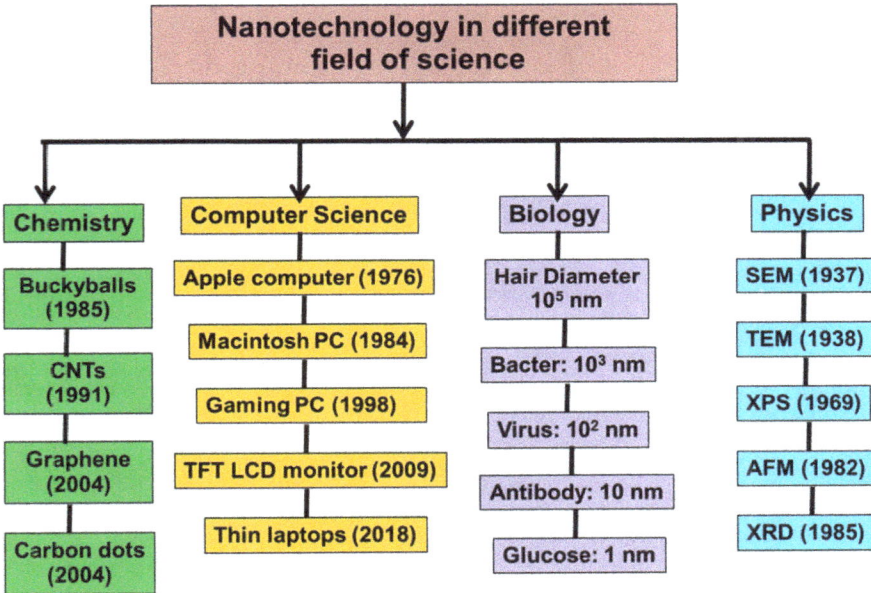

Fig.10. Progresses in Nanoscience [9].

4.   Modern era of nanotechnology

The modern era of nanotechnology started from the time of invention of the Scanning Tunneling Microscope (STM) by Binnig and Rohrer. The sharp tip in STM moving closer to the conductive surface making overlap between the wave function of STM tip atoms and surface atoms. Electrons tunnel through the vacuum from the tunnel tip to the surface under the applied voltage. The first STM image of silicon nanostructure was obtained in 1983 (Fig.11) [36,37].

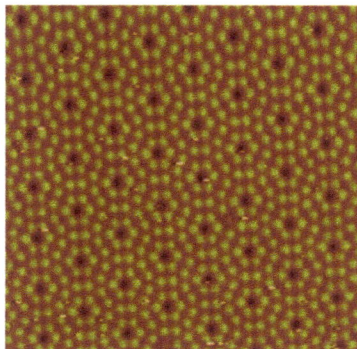

Fig.11. STM image of Si(111) surface [37].

STM is nowadays used routinely to image surfaces and used as a tool to manipulate atoms and molecules to create new product. Along with STM, Atomic Force Microscope (AFM) and Scanning Probe Microscope (SPM) are widely used in modern nanotechnology research.

The discovery of fullerenes or buckyballs lead to the development of new carbon chemistry. Carbon nanotubes are first developed and observed in Transmission Electron Microscopy (TEM) by Iijima add momentum in the progress of nanotechnology. They have found applications in various technological field

The discovery of carbon dots (C-dots) was made accidentally in 2004 by Xu et al. while purifying single walled carbon nanotubes (SWCNTs) [38]. They are highly promising materials in drug delivery, bioimaging and biosensing applications. Because of their good optical and electronic properties, they also found applications in photovoltaic devices, catalysis and sensing [39-42]. The next star that emerged in carbon-based nanomaterial is graphene which have taken the field of nanoscience and nanotechnology by storm. Many disciplines such as computer science, bioengineering etc. are completely taken over by nanotechnology. The size of the computers decreases rapidly while improving the efficiency. A big progress in electrical appliances specially smart phone technology happened due to the advancement in nanotechnology.

In the biomedical field nanotechnology played a big role in terms of biomedicine for diagnosis and curing of many diseases [43]. The discipline of bio-nanotechnology deals with drug delivery, bio-imaging and diagnosis based on nanomaterials. A plethora of nanomaterial products in the medical field are available nowadays. Nanomaterial products used as regenerative medicine, drug delivery, antibacterial activities, biomarkers, nanoelectrodes, nanobiosensors are few examples of nonpharmaceutical [44].

DNA nanotechnology is an interdisciplinary area developed by Paul Rothemund in 2006 [45]. The development of complex and bigger self-assembled DNA in the name of "scaffolded DNA origami" is one important application of nanotechnology. Extensive research in this area on various biopolymers including DNA shows promising results in diagnostic and sensing applications [46]. The efficiency of chemotherapy drugs is improved by applying functional nanomaterials to modulate biological processes in nano-oncology field. This will significantly reduce the toxicity of chemotherapy application and improvement in tumor treatment [47-50].

Nanotechnology is currently used for building a safe, pollution free environment. New generation solar cell, hydrogen storage system, fuel cells are made by novel nanomaterials. Treatment of water and air from toxic elements is one of the most important areas benefitted by nanotechnology.

Currently research is going on to develop nanomedicine computational research. The field of nano informatics is still in its infancy stage and need further research. Machine learning approaches can accurately predict various process such as cellular uptake, bioactivity and cytotoxicity [51-53]. The powerful algorithms are of great help in designing highly efficient nanocarriers. Prediction of quantitative structure activity relationship (QSAR), quantitative structure property relationship (QSPR), Adsorption, Distribution, Metabolism, Excretion and Toxicity (ADMET), data mining, network analysis belong to nano informatics. Design and analysis of nanomaterial become easier through modeling in nano-informatics. Chemotherapy gets considerable upliftment by machine learning approaches to detect tumor cell and resist them. Targeted drug delivery and gene therapy are two latest additions in nano informatics for cancer treatment [54]. Various important events in the development of nanotechnology are presented in Table 2.

Table 2. Timeline of development of nanotechnology [9].

Timeline	Phenomenon
4th Century	Lycurgus cup
500-1450	Stained glass window in cathedrals
1450-1600	Deruta Pottery
1857	Synthesis of colloidal ruby gold NPs by Farday
1908	Light Scattering NPs by Gustav Mie
1928	Near field optical microscope by Edward Synge
1931	Invention of TEM by Max Knoll and Ernst Ruska
1936	Invention of field electron microscope by Erwin Muller
1947	Discovery of semiconductor transistor by Shokley, Brattain, and Bardeen
1951	Invention of field ion microscope by Erwin Muller
1953	Discovery of DNA by Watson and Crick

1956	Molecular engineering by Hippel
1958	Electron tunneling by Leo Esaki
1959	Phenomenal talk "There's plenty of room at the bottom' by Feynman
1960	Zeolites for catalysis by Plank and Rosinski
1963	Invention of ferrofluids by Papell
1965	Development of Moore's law
1970	Prediction of C_{60} by Eiji Osawa
1974	Nanotechnology term by Taniguchi
1974	Molecular electronics by Ratner and Aviram
1977	Discovery of SERS by R.P. Van Duyne
1980	Discovery of self-assembly monolayer by Jacop Sagiv
1981	Invention of STM by Binnig and Rohrer
1981	Molecular engineering by Eric Drexler
1981	Discovery of nanocrystalline quantum dots by Alexey Ekimov
1982	Development of the concept of DNA by Nadrian Seeman
1983	Discovery of colloidal quantum dots by Louis Brus
1985	Discovery of Buckminster fullerene by Smalley, Curl and Kroto
1986	Invention of AFM by Binnig, Gerber and Quate
1987	Single electron tunneling transistor by Averin and Konstantin
1990	Writing of the logo of IBM using Xe atoms by Eigler and Schweizer
1991	Discovery of multiwalled carbon nanotubes by Iijima
1992	Discovery of MCM-41 by Kresge
1993	Discovery of single walled carbon nanotubes by Iijima and Bethune
1996	Self-assembly monolayers of DNA by Chad Mirkin and Robert Letsinger
1997	First nanotechnology company Zyvex founded
1998	Creation of transistor using CNT by Cees Dekker
1999	Dip pen nanolithography by Chad Mirkin
2000	Feedback controlled lithography by Hersam and Lyding

2000	US National Nanotechnology Initiative by Bill Clinton
2001	Molecular nanomachines by Carlo Montemagno
2002	Functionalization of CNT with DNA by Cees Dekker
2003	Nanotechnology Research Development Act by George Bush
2003	Development of gold nanoshells Naomi Halas
2004	Discovery of Graphene by Geim and Novoselov
2004	Discovery of carbon dots by Xu et al
2005	Nanocar using buckyball by James Tour
2006	DNA origami by Paul Rothemund
2007	Artificial molecular machines by Stoddart
2008	Discovery of Green fluorescent protein by Shimomura, Chalfie and Tsien
2009	DNA structure folded into 3D rhomohedral crystals by Nadrian Seeman
2010	Ultrafast lithography for 3D nanoscale textured surface by IBM
2011	Use of STM for electronic and mechanical properties by Leonhard Grill
2016	Design of molecular machines by Sauvage, Stoddart and Feringa
2017	Gravitational waves (Nobel prize in physics, 2017)
2018	Smallest tic-tac-toe game board using DNA
2018	Shrinking objects to nanoscale

5.   Types of nanomaterials

Generally, the nanomaterials (NMs) are classified according to their physical and chemical properties such as shape and dimensionality and phase composition (Fig.12) [55].

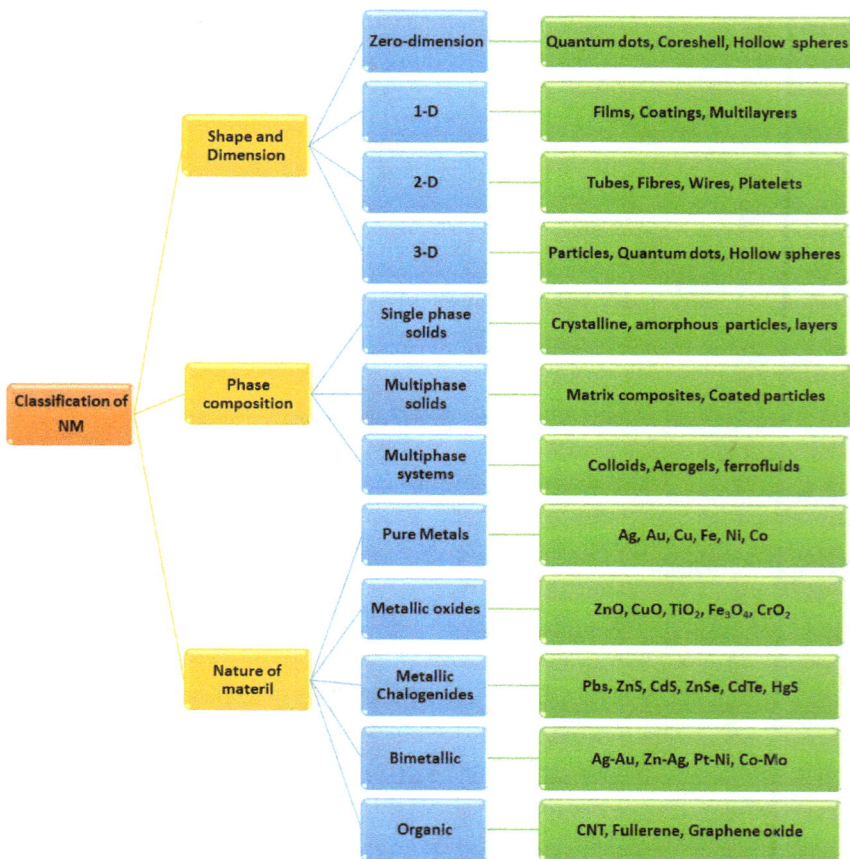

Fig. 12. Classification of nanomaterials [55].

6. Applications of nanomaterials

Nanomaterials are now being used in almost every field as given by Fig. 13.

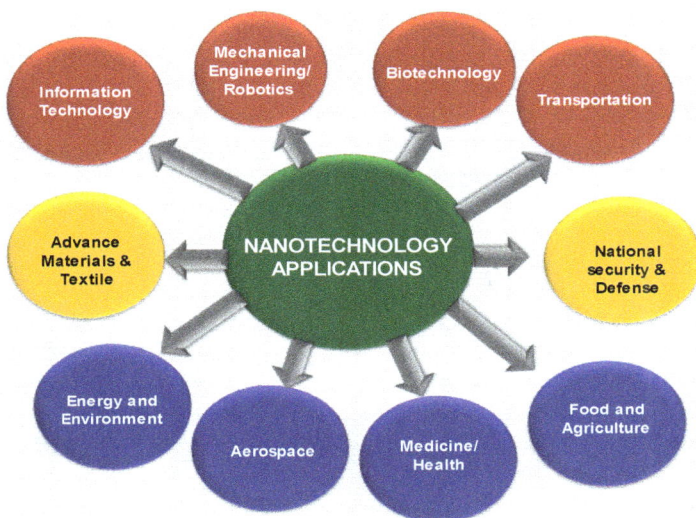

Fig. 13 Applications of NMs in different sectors.

7.   Concluding remarks

Nanomaterials have revolutionized various sectors with their unique properties. Stainless fabrics, lightweight machinery, scratch resistant and dustless surface, colour changing pigments, anti-aging beauty products etc. are possible because of the advancement in nanotechnology. The research on nanomaterials started from the middle of the 19th century. Michael Faraday initiated the development of nanotechnology by investigating the optical and electrical properties of gold nanoparticles. Zsigmondy first measured the size of nanoparticles and introduced the term nanometer. Then Feynman envisioned and predicted many discoveries that may be possible by manipulating matter at atomic and molecular level. Scientist Norio Taniguchi introduced the term nanotechnology and contributed immensely to the field. Eric Dexler in his famous book discussed about nanotechnology with highly efficient products. Invention of STM and AFM are two major developments in the progress of nanotechnology. Discovery of C_{60} and carbon nanotubes are example of remarkable innovation that led to tremendous progress in the field of biomedical, sensing, electronics and fuel cells. These materials are ultra-thin, light weight, durable with high precision contribute extensively to many industrial processes.

References

[1] A. Gnach, T. Lipinski, A. Bednarkiewicz, J. Rybka,J. A. Capobianco, Upconverting nanoparticles: Assessing the toxicity, Chem. Soc. Rev. 44 (2015) 1561-1584. https://doi.org/10.1039/C4CS00177J

[2] S. Kargozar, M. Mozafari, Nanotechnology and nanomedicine: Start small, think big, Mater. Today Proc. 5 (2018) 15492-15500. https://doi.org/10.1016/j.matpr.2018.04.155

[3] R. P. Feynman, There's plenty of room at the bottom, Eng. Sci. 23 (1960) 22-36.

[4]H. W. Kroto, J. R. Heath, S. C. O'Brien, R. F. Curl, R. E. Smalley, C60: Buckminsterfullerene. Nature, 318 (1985) 162-163. https://doi.org/10.1038/318162a0

[5] S. Iijima, Helical microtubules of graphitic carbon. Nature, 354(1991) 56-58. https://doi.org/10.1038/354056a0

[6] Iso.org, 80004-1:2010, I. (2017). ISO/TS 80004-1:2010-Nanotechnologies-Vocabulary-Part 1 Core terms (2017). [online] Iso.org. Available from: https://www.iso.org/standard/51240.html.

[7] ISO/TS 27867, Nanotechnologies-terminology and definitions for nano-objects-nanoparticle, nanofibre and nanoplate, 2008. Available from: https://www.iso.org/standard/44278.html.

[8] Shop.BSI, bsigroup.com. (2017). PAS 71:2011-Nanoparticles. Vocabulary-BSI British Standards Institute. [online] Available from: http://shop.bsigroup.com/ProductDetail/pid=000000000030214797.

[9] S. Bayda, M. Adeel, T. Tuccinardi, M. Cordani, F. Rizzolio, The History of Nanoscience and Nanotechnology: From Chemical-Physical Applications to Nanomedicine, Molecules, 25 (2020) 112. doi:10.3390/molecules25010112 https://doi.org/10.3390/molecules25010112

[10] The British Museum. Available online: www.britishmuseum.org/research/collection_online/collection_object_details.aspx?objobjec=61219&partId=1

[11] D. J. Barber, I. C. Freestone, An investigation of the origin of the colour of the Lycurgus Cup by analytical transmission electron microscopy, Archaeometry, 32 (1990) 33-45. https://doi.org/10.1111/j.1475-4754.1990.tb01079.x

[12] I. Freestone, N. Meeks, M. Sax, C. Higgitt, The Lycurgus Cup-A Roman nanotechnology, Gold Bull. 40 (2007) 270-277. https://doi.org/10.1007/BF03215599

[13] This image is published on https://www.alamy.com/stock-photo/north-rose-window-notre-damecathedral. html and retrieved from Google Images

[14] The New York Times. Available online: www.nytimes.com/imagepages/2005/02/21/science/20050222_NANO1_GRAPHIC.html

[15] T. Pradell, A. Climent-Font, J. Molera, A. Zucchiatti, M. D. Ynsa, P. Roura, D. Crespo, Metallic and nonmetallic shine in luster: An elastic ion backscattering study, J. Appl. Phys. 101 (2007) 103518. https://doi.org/10.1063/1.2734944

[16] C. P. Poole, F. J. Owens, Introduction to Nanotechnology, John Wiley & Sons: New York, NY, USA, 2003.

[17] M. Reibold, P. Paufler, A. A. Levin, W. Kochmann, N. Pätzke, D.C. Meyer, Materials: Carbon nanotubes in an ancient Damascus sabre, Nature, 444 (2006) 286. https://doi.org/10.1038/444286a

[18] M. Faraday, The Bakerian Lecture: Experimental Relations of Gold (and Other Metals) to Light, Philos. Trans. R. Soc. Lond. 147 (1857) 145-181. https://doi.org/10.1098/rstl.1857.0011

[19] N. Taniguchi, C. Arakawa, T. Kobayashi, On the basic concept of nano-technology. In Proceedings of the International Conference on Production Engineering, Tokyo, Japan, 26-29 August 1974.

[20] E. K. Drexler, Engines of Creation: The Coming Era of Nanotechnology; Anchor Press: Garden City, NY, USA, 1986.

[21] E. K. Drexler, C. Peterson, G. Pergamit, Unbounding the Future: The Nanotechnology Revolution; William Morrow and Company, Inc.: New York, NY, USA, 1991.

[22] E. K. Drexler, Molecular engineering: An approach to the development of general capabilities for molecular manipulation, Proc. Natl. Acad. Sci. USA 78 (1981) 5275-5278. https://doi.org/10.1073/pnas.78.9.5275

[23] G. Binnig, H. Rohrer, C. Gerber, E. Weibel, Tunneling through a controllable vacuum gap. Appl. Phys. Lett. 40 (1982) 178. https://doi.org/10.1063/1.92999

[24] G. Binnig, H. Rohrer, C. Gerber, E. Weibel, Surface Studies by Scanning Tunneling Microscopy. Phys. Rev. Lett. 49 (1982) 57-61. https://doi.org/10.1103/PhysRevLett.49.57

[25] D. M. Eigler, E. K. Schweizer, Positioning single atoms with a scanning tunnelling microscope, Nature, 344 (1990) 524-526. https://doi.org/10.1038/344524a0

[26] Institute of Physics Polish Academy of Sciences. Available online: http://info.ifpan.edu.pl/~{}wawro/ subframes/Surfaces.htm

[27] G. Binnig, C. F. Quate, C. Gerber, Atomic Force Microscope, Phys. Rev. Lett. 56 (1986) 930-933. https://doi.org/10.1103/PhysRevLett.56.930

[28] G. Binnig, Atomic Force Microscope and Method for Imaging Surfaces with Atomic Resolution. U.S. Patent 4724318A, 16 October 1990.

[29] C. T. Kresge, M. E. Leonowicz, W. J. Roth, J. C. Vartuli, J. S. Beck, Ordered mesoporous molecular sieves synthesized by a liquid-crystal template mechanism, Nature, 359 (1992) 710-712. https://doi.org/10.1038/359710a0

[30] J. S. Beck, J. C. Vartuli, W. J. Roth, M. E. Leonowicz, C. T. Kresge, K. D. Schmitt, C. T. W. Chu, D. H. Olson, E. W. Sheppard, S. B. McCullen, A new family of mesoporous molecular sieves prepared with liquid crystal templates, J. Am. Chem. Soc. 114 (1992) 10834-10843. https://doi.org/10.1021/ja00053a020

[31] M. Bawendi, P. Carroll, W. Wilson, L. Brus, Luminescence properties of CdSe quan¬tum crystallites: resonance between interior and surface localized states, J. Chem. Phys. 96 (2) (1992) 946-954. https://doi.org/10.1063/1.462114

[32] C. Loo, A. Lin, L. Hirsch, M.-H. Lee, J. Barton, N. Halas, J. West, R. Drezek, Nanoshell-Enabled Photonics-Based Imaging and Therapy of Cancer, Technol. Cancer Res. Treat. 3 (2004) 33-40. https://doi.org/10.1177/153303460400300104

[33] L. R. Hirsch, R. J. Stafford, J. A. Bankson, S. R. Sershen, B. Rivera, R. E. Price, J. D. Hazle, N. J. Halas, J. L. West, Nanoshell-mediated near-infrared thermal therapy of tumors under magnetic resonance guidance, Proc. Natl. Acad. Sci. USA 100 (2003) 13549-13554. https://doi.org/10.1073/pnas.2232479100

[34] Y. Shirai, A. J. Osgood, Y. Zhao, K. F. Kelly, J. M. Tour, Directional Control in Thermally Driven Single-Molecule Nanocars, Nano Lett. 5 (2005) 2330-2334. https://doi.org/10.1021/nl051915k

[35] J.-F. Morin, Y. Shirai, J. M. Tour, EnRoute to a Motorized Nanocar, Org. Lett. 8 (2006) 1713-1716. https://doi.org/10.1021/ol060445d

[36] G. Binnig, H. Rohrer, C. Gerber, E. Weibel, 7x7 Reconstruction on Si(111) Resolved in Real Space. Phys. Rev. Lett. 50 (1983) 120-123. https://doi.org/10.1103/PhysRevLett.50.120

[37] Institute of Physics Polish Academy of Sciences. Available online: http://info.ifpan.edu.pl/~{}wawro/subframes/Surfaces.htm

[38] X. Xu, R. Ray, Y. Gu, H. J. Ploehn, L. Gearheart, K. Raker, W. A. Scrivens, Electrophoretic Analysis and Purification of Fluorescent Single-Walled Carbon Nanotube Fragments, J. Am. Chem. Soc. 126 (2004) 12736-12737. https://doi.org/10.1021/ja040082h

[39] J. C. G. Esteves da Silva, H. M. R. Gonçalves, Analytical and bioanalytical applications of carbon dots, TrAC Trends Anal. Chem. 30 (2011) 1327-1336. https://doi.org/10.1016/j.trac.2011.04.009

[40] P.B. Chouke, K.M. Dadure, A.K. Potbhare, G.S. Bhusari, A. Mondal, K. Chaudhary, V. Singh, M.F. Desimone, R.G. Chaudhary, D.T. Masram Biosynthesized δ-Bi2O3

Nanoparticles from Crinum viviparum Flower Extract for Photocatalytic Dye Degradation and Molecular Docking. ACS Omega. 7 (2022) 20983-20993. https://doi.org/10.1021/acsomega.2c01745

[41] A K. Potbhare, R.G. Chaudhary, P.B. Chouke, A. Rai, A. Abdala, R. Mishra, M. Desimone, Graphene-based materials and their nanocomposites with metal oxides: Biosynthesis, electrochemical, photocatalytic and antimicrobial applications. Mater. Res. Forum. 83 (2020) 79-116. https://doi.org/10.21741/9781644900970-4

[42] L. Cao, X. Wang, M. J. Meziani, F. Lu, H. Wang, P. G. Luo, Y. Lin, B. A. Harruff, L. M. Veca, D. Murray, S.-Y. Xie, Y.-P. Sun, Carbon Dots for Multiphoton Bioimaging, J. Am. Chem. Soc. 129 (2007) 11318-11319. https://doi.org/10.1021/ja073527l

[43] C. Kinnear, T. L. Moore, L. Rodriguez-Lorenzo, B. Rothen-Rutishauser, A. Petri-Fink, Form Follows Function: Nanoparticle Shape and Its Implications for Nanomedicine, Chem. Rev. 117 (2017) 11476-11521. https://doi.org/10.1021/acs.chemrev.7b00194

[44] V. Weissig, T. K. Pettinger, N. Murdock, N. Nanopharmaceuticals (part 1): Products on the market, Int. J. Nanomed. 9 (2014) 4357-4373. https://doi.org/10.2147/IJN.S46900

[45] P. W. K. Rothemund, Folding DNA to create nanoscale shapes and patterns, Nature, 440 (2006) 297-302. https://doi.org/10.1038/nature04586

[46] N. C. Seeman, Nucleic acid junctions and lattices, J. Theor. Biol. 99 (1982) 237-247. https://doi.org/10.1016/0022-5193(82)90002-9

[47] V. Kumar, S. Bayda, M. Hadla, I. Caligiuri, C. Russo Spena, S. Palazzolo, S. Kempter, G. Corona, G. Toffoli, F. Rizzolio, Enhanced Chemotherapeutic Behavior of Open-Caged DNA@Doxorubicin Nanostructures for Cancer Cells, J. Cell. Physiol. 231, (2016) 106-110. https://doi.org/10.1002/jcp.25057

[48] V. Kumar, S. Palazzolo, S. Bayda, G. Corona, G. Toffoli, F. Rizzolio, DNA Nanotechnology for Cancer Therapy, Theranostics, 6 (2016) 710-725. https://doi.org/10.7150/thno.14203

[49] S. Palazzolo, M. Hadla, C. R. Spena, S. Bayda, V. Kumar, F. Lo Re, M. Adeel, I. Caligiuri, F. Romano, G. Corona, Proof-of-Concept Multistage Biomimetic Liposomal DNA Origami Nanosystem for the Remote Loading of Doxorubicin, ACS Med. Chem. Lett. 10 (2019) 517-521. https://doi.org/10.1021/acsmedchemlett.8b00557

[50] S. Palazzolo, M. Hadla, C. R. Spena, I. Caligiuri, R. Rotondo, M. Adeel, V. Kumar, G. Corona, V. Canzonieri, G. Toffoli, An Effective Multi-Stage Liposomal DNA Origami Nanosystem for In Vivo Cancer Therapy, Cancers, 11 (2019) 1997. https://doi.org/10.3390/cancers11121997

[51] P. Y. Lee, K. K. Y. Wong, Nanomedicine: A new frontier in cancer therapeutics. Curr. Drug Deliv. 8 (2011) 245-253. https://doi.org/10.2174/156720111795256110

[52] P.B. Chouke, A.K. Potbhare, N.P. Meshram, M. M. Rai, K.M. Dadure, K. Chaudhary, A.R. Rai, M. Desimone, R. G. Chaudhary, D.T. Masram, Bioinspired NiO nanospheres: Exploring in-vitro toxicity using Bm-17 and L. rohita liver cells, DNA degradation, docking and proposed vacuolization mechanism. ACS Omega. 7 (2022) 6869−6884. https://doi.org/10.1021/acsomega.1c06544

[53] M. Cordani, A. Somoza, Á. Targeting autophagy using metallic nanoparticles: A promising strategy for cancer treatment, Cell. Mol. Life Sci. 76 (2019) 1215-1242. https://doi.org/10.1007/s00018-018-2973-y

[54] N. Sharma, M. Sharma, Q. M. Sajid Jamal, M. A. Kamal, S. Akhtar, Nanoinformatics and biomolecular nanomodeling: A novel move en route for effective cancer treatment, Environ. Sci. Pollut. Res. Int. (2019) 1-15. https://doi.org/10.1007/s11356-019-05152-8

[55] P. Khanna, A. Kaur, D. Goyal, Algae-based metallic nanoparticles: Synthesis, characterization and applications, J. Microbiol. Methods, 163 (2019) 105656. https://doi.org/10.1016/j.mimet.2019.105656

Materials Research Foundations 135 (2023) 23-39   https://doi.org/10.21741/9781644902172-2

Chapter 2

Nanomaterials in the Lubricant Industry

Ashima Srivastava[1], Pratibha Singh[2], Preeti Jain[3*], Kirti Srivastava[1] & Sanjeev Sharma[2]

[1]Department of Chemistry, JSS Academy of Technical Education, Sector 62, Noida, India

[2]Enviro Infra Solutions Pvt. Ltd., Sector 9, Vasundhara, Ghaziabad 201012

[3]Department of Chemistry & Biochemistry, School of Basic Sciences and Research, Sharda University, Greater Noid

*preeti.jain@sharda.ac.in

Abstract

Nanomaterials in lubricants are solid particles having nanometer size (particle size ranging between 1-100 nm) and possessing intriguing tribological properties. When these nanoparticles are incorporated in the lubricating base oil, they have an add-on effect on diminishing friction, wear and tear of the two interacting surfaces in relative motion. Ideally the shape of nano-lubricants is spherical which is encountered in C_{60} fullerenes, carbon onions, inorganic fullerenes of metal oxides, sulphides, etc. On the other hand, they can have a cylindrical (carbon nanotubes), disc or wire shape also. Nanomaterials have arisen as environmentally-benign lubricant additives to elevate the tribological properties of the traditional lubricants utilized in the fields of automobiles and various industries. The presence of nanoparticles in lubricants enhances the mechanistic aspects of thin film (boundary film), thick film and extreme pressure lubrication. When the layer of lubricating oil is disrupted, for example, the solid nanoparticles in it can take up the load without decomposing and provide essential backup. Other advantages include excellent mechanical strength due to their compact size and stability at high temperatures and pressures. They also have the ability to adsorb on the surfaces of metal machines, preventing direct metal-to-metal contact and thereby wear and tear. These benefits include increased equipment usage, improved fuel efficiency, and increased maintenance intervals. As a result, the current chapter examines several types of nanomaterials as lubricant additives, as well as their applications and benefits in various automotive/engine/industrial lubricants.

Keyword

Nanoparticles, Lubricant, Additives, Friction, Wear, Mechanism

Materials Research Foundations 135 (2023) 23-39   https://doi.org/10.21741/9781644902172-2

Contents

Nanomaterials in the Lubricant Industry **23**

1.   Introduction..**24**

2.   Base oils and their classification...**27**

2.1   Liquid lubricants or lubricating oils 28

2.2   Mechanism of lubrication 29

2.3   Nanoparticles as lubricant additives..30

3.   Types of nanoparticles as lubricant additives.....................**32**

3.1   Metal oxides..32

3.2   Metals..33

3.3   Metal sulphides...33

3.4   Carbon-based nanomaterials...33

3.5   Boron-based nanomaterials 33

3.6   Nanocomposites..34

4.   Factors affecting the lubricating properties of nanoparticles as lubricants **34**

4.1   Size of nanoparticle 34

4.2   Shape of nanoparticles...34

4.3   Surface functionalization...35

4.4   Concentration of nanoparticles..35

5.   Application fields of nanolubricants.....................................**36**

6.   Roadblocks to full scale use of nanolubricants...................**36**

Conclusions **37**

References **37**

1.   Introduction

When dry surfaces move relative to one another, asperities (microscopic peaks and valleys) on the surfaces may rub, lock together, and break apart. This leads to friction, abrasive wear and tear of machine parts (Fig 1). Any substance introduced between moving, sliding or rolling surfaces to reduce the friction in order to avoid or reduce wear and tear is known as lubricant. The process of applying the lubricant in between the moving surfaces is called lubrication. Lubrication is indispensably significant for both domestic and industrial processes. It prevents the direct metal to metal contact, hence reduces wear and tear and surface deformation. It prevents the loss of energy so efficiency is increased. It acts as a coolant in engines. Sometimes it acts as seal, such as around the piston ring and prevents the entry of moisture, dust and dirt around it. It reduces the power losses and maintenance costs of machines.

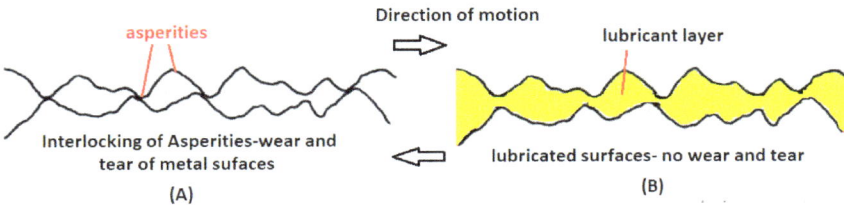

Figure 1. (A) Microscopic view of two metal surfaces under sliding motion in the absence of lubricant, (B) microscopic view of two metal surfaces under sliding motion in the presence of lubricant

Exceptional efforts have been made in recent years to develop environmentally friendly lubricants that improve working productivity, extend the life of machines, and reduce maintenance expenses. For many modern machines, there is no single oil that is the best lubricant. Specific additives are added to these oils in order to increase their properties [1]. These oils have the lubricating qualities that are necessary for certain machines. Some of the additives along with their different functions are given below in Table 1.

Table 1: Additives and their functions

Additive Type	Chemical used	Functions	Applications
Antioxidant phenolic or amino compounds	Aromatic phenolic or amino compounds.	Retard the oxidation of lubricating oils by getting themselves preferentially oxidized.	Added in lubricants used in internal combustion engines
Oiliness-carrier	Vegetable oils like castor oil and fatty oils like stearic acid., palmatic oil	Increasing the oiliness of the lubricating oil	Added in lubricants used in boundary lubrication

Viscosity-index improver	Polymers and copolymers of olefins.	Increases the viscosity index	Added in the machines working under broad range of temperature change
Pour-point depressant	Alkylated naphthalene and phenolic polymers, polymethacrylates.	Lowers the pour point of the oil	Reduced interlocking
Dispersants	High molecular weight amines and amides, polymers	Prevent or retard sludge formation	In lubricating oils used in internal combustion engines .
Metal deactivators	Amines, sulphides or phosphides	Stop the catalytic effect of metals on oxidation rate	Protect the metal surfaces by forming a protective layer over the metal surface/
Antifoaming agents	Glycol, glycerol, silicon polymers	Prevent the formation of stable foams	
Corrosion inhibitors	Organo metallic compounds, zinc dithiophosphates	Protect the metal in contact with the lubricant from corrosion	In lubricating oils and for bearing ets.
Rust inhibitors	Amine phosphates, Alkyd succinic acid, fatty acids.	Protect ferrous metals from rusting	
Extreme Pressure additives	Sulphurized fats, organic phosphorus compounds etc.	React chemically with the metal, forming a surface layer which prevents seizure and welding of the metals.	
Abrasion inhibitors	Tricresyl phosphate	They provide abrasion resistance.	

While these additives have unique properties, they may have at least one of the following drawbacks: reduced oil miscibility; metal part corrosion; lubricant colour obscuration; and an increased level of sulphur and/or phosphorus in the finished product. As a result, lubricant additives that can improve the overall characteristics and hence performance of the lubricant while also being compatible with metallic parts of engines, etc. are needed. [2]. Critical purposes behind the investigation of new sort of additives and optimization of their concentrations are:

- Consistently increasing austerity in working conditions
- developments in equipment technology
- growing lubrication requirements
- increasingly rigorous regulatory requirements in various markets

Before discussing the special category of lubricant additives in detail, the brief account of major types of base oils, their mechanism of lubrication is discussed below.

2.   Base oils and their classification

They can be classified in the following ways (Fig 2)

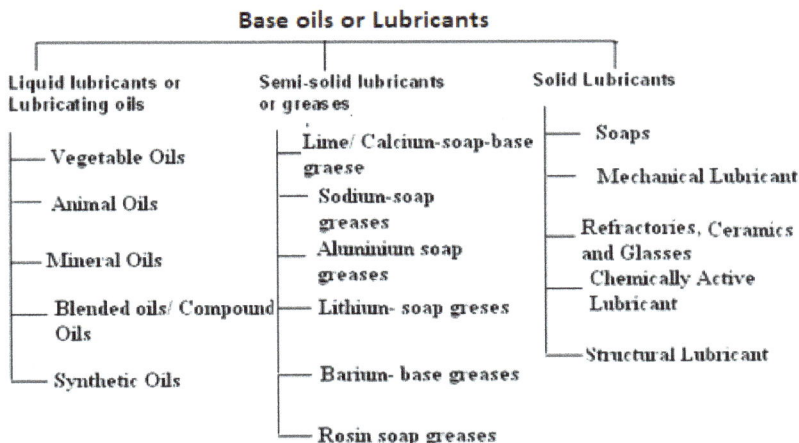

Base oils or Lubricants

Liquid lubricants or Lubricating oils	Semi-solid lubricants or greases	Solid Lubricants
Vegetable Oils	Lime/ Calcium-soap-base graese	Soaps
Animal Oils	Sodium-soap greases	Mechanical Lubricant
Mineral Oils	Aluminium soap greases	Refractories, Ceramics and Glasses
Blended oils/ Compound Oils	Lithium- soap greses	Chemically Active Lubricant
Synthetic Oils	Barium- base greases	Structural Lubricant
	Rosin soap greases	

Figure 2 Classification of Base Oils

2.1   Liquid lubricants or lubricating oils

Lubricating oils work by creating a continuous fluid coating between two moving metallic surfaces, reducing friction and wear. The following characteristics must be present in a good lubricant:

- It must have a high boiling point or a low vapour pressure;
- It must have high thermal stability and oxidation resistance;
- It must also have acceptable viscosity for certain operating conditions;
- It must have a low freezing point; and
- It must be non-corrosive.

Lubricating oils are further sub-classified as:

(i) Animal and Vegetable oils

- These oils are higher fatty acid glycosides with excellent oiliness (by virtue of which the oil sticks to the surface of the metal even under high temperature and high load).
- However, they are expensive, quickly oxidised, and have a tendency to hydrolyze when exposed to wet air or water.
- These oils decompose when heated without being distilled, and so are referred to as "fixed oils." They are utilised as blenders to make petroleum oils more oily.

(ii) Petroleum oils or Mineral oils:

- They are made via fractional distillation of crude petroleum oils, and the hydrocarbon chain length ranges from C12 to C50. They are inexpensive and relatively stable in operation.
- They have a low level of oiliness, which can be boosted by adding higher molecular weight vegetable or animal oils.
- Because crude liquid petroleum oil has a lot of contaminants like wax, it can't be used as is. Asphalt, pigmented chemicals, and other oxidizable contaminants should be avoided. Different processes, including as dewaxing, acid refining, and solvent refining, can be used to eliminate these impurities.

(iii) Blended oils or Additives for lubricating oils:

For many modern machines, there is no single oil that is the best lubricant. To improve the qualities of petroleum oils, specific additives are added. These lubricants are referred to as "mixed oils" or "compounded oils," and they provide the lubricating qualities that are necessary for certain machines.

2.2   Mechanism of lubrication

Lubrication is accomplished mainly by three methods:

- Hydrodynamic or fluid film lubrication

- Thin film or boundary lubrication

- Extreme pressure lubrication

(i) Hydrodynamic or fluid film or Thick-film lubrication

The moving or sliding surfaces are separated from one other by a bulk lubricant coating (at least 1000 Angstrom thick) when the load is low and the speed is high. This bulk lubricant coating keeps the microscopic peaks and valleys from interlocking by preventing direct surface-to-surface contact. As a result, friction is reduced and wear is avoided. Figure 3 depicts the lubrication of a fluid film.

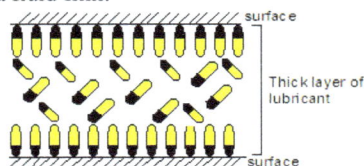

Figure 3 fluid-film lubrication

The internal resistance of the lubricant particles sliding over one another causes slight friction. The thickness and viscosity of the lubricant, as well as the relative velocity and area of the moving/sliding surfaces, define friction in such a system. The coefficient of friction for a fluid film lubricated system could be as low as 0.002 to 0.03.

(ii) Thin film or boundary lubrication

When the lubricant is not viscous enough to provide a film of sufficient thickness to separate the surfaces under severe loads, friction can still be decreased by using the right lubricant that forms a surface film that adheres strongly to the surface. These films are usually only one or two molecules thick, but they provide adequate protection to keep metal from contacting metal (A thin layer of lubricant is adsorbed on the metallic surfaces). Border lubrication is the term for this form of boundary protection. In such cases, the coefficient of friction is usually between 0.05 and 0.15.

Active organic compounds with long chain molecules and active end groups are the best additives in general. These compounds link securely and intricately with one another, generating a layer that forms on the metal's surface and self-binds powerfully. As a result, a thin coating is formed that is difficult to penetrate. When two surfaces, each covered by a boundary layer, collide, they tend to glide along their outermost surfaces, with the actual faces of the surfaces contacting each other only occasionally. Liquids are seldom effective border lubricants. Solids with long chains, high inter-chain attraction, low shear resistance, and a high temperature tolerance provide the ideal boundary lubricants. Obviously, the boundary lubricant must be able to retain a strong bond to the surfaces even at high temperatures and load pressures.

The oiliness of the lubricant determines the effectiveness of boundary lubrication. The ability of a lubricant to stick to a surface is referred to as oiliness. The oiliness of vegetable oils and their fatty acids is higher. Oleic acid ($C_{17}H_{33}COOH$), stearic acid ($C_{17}H_{35}COOH$), and so on. These oils' polar carbonyl group reacts with metal surfaces to generate a continuous thin lubricating coating. Hydrocarbon chain of the fatty acid gets oriented outwards in a perpendicular direction as shown in Fig 4:

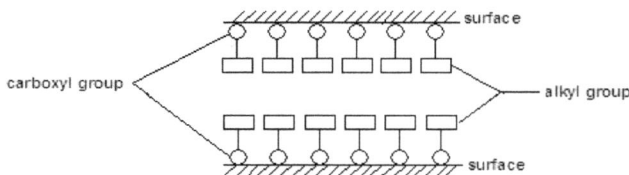

Figure 4 Boundary lubrication

(iii) Extreme pressure lubrication

When a system is subjected to high speeds and loads, more heat is created between the moving surfaces. As a result, the liquid lubricant loses its ability to stick and decomposes or evaporates. To solve these challenges, lubricants are combined with chemicals known as Extreme Pressure additives. Organic compounds containing sulphur, phosphorus, or chlorine are commonly used as additives. Such additions include chlorinated esters, sulphurized oils, and tricresyl phosphate. These compounds react with metallic surfaces to generate solid coatings with high melting temperatures, such as chloride, sulphide, or phosphide. These films are capable of withstanding extremely high loads and temperatures.

2.3   Nanoparticles as lubricant additives

In the field of using nanomaterials based on carbon, metals, metal chalcogenides, metal oxides (single or mixed), metal sulphides, metal borates and metal carbonates, rare earth compounds, and other materials to improve the properties and thus performance of lubricants, significant progress has been made [3-5]. The size, shape, and concentration of nanoparticles have an impact on their ability to reduce friction and wear. They disperse in base oil more easily than micron-sized additions. The various benefits of nanolubricants (Lubricants with nanoparticle additives) are given in Fig 5.

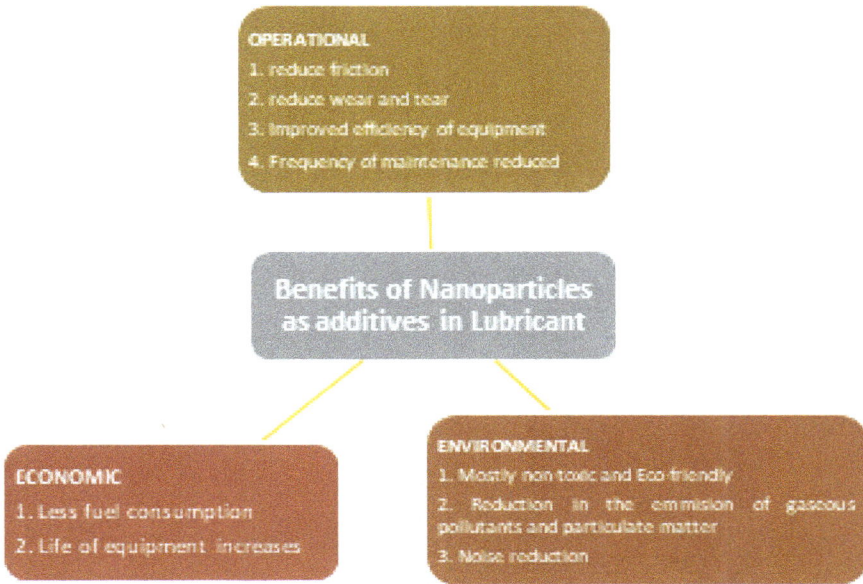

Figure 5. Benefits of nanolubricants

The following are some of the nanoparticle-assisted lubricating methods [6-9]. (Fig. 6):

- Ball-bearing effect (small spherical nanoparticles rolling between moving surfaces) • Nanoparticles most likely interact with sliding surfaces to generate a surface protective coating

- Filling valleys with nanoparticles to prevent the asperities of two moving surfaces from colliding (Mending effect)

- Shearing of trapped nanoparticles at the interface • Sintering and repair effect: Nanoparticles compact on the friction surface as a result of high pressure and temperature.

- Nanoparticles dispersed in the base oil change the viscosity and other tribological properties of the oil, improving lubrication.

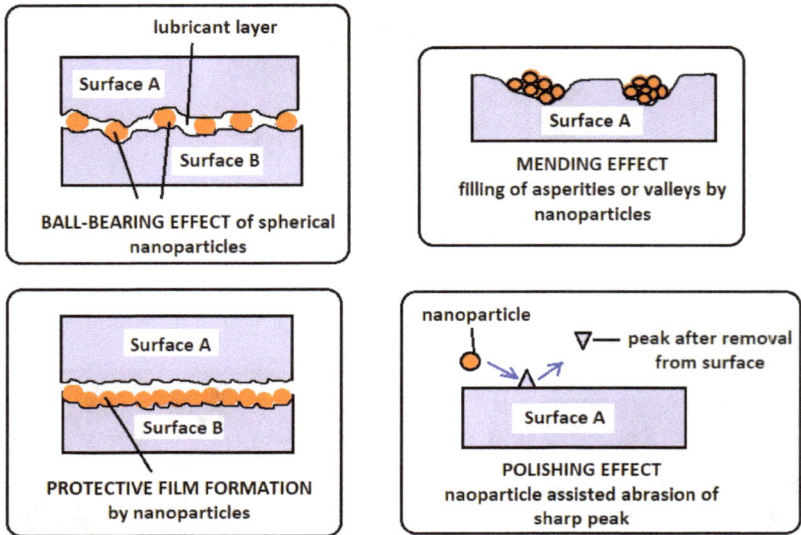

Figure 6. mechanisms of nanoparticle- assisted lubrication

3.   Types of nanoparticles as lubricant additives

3.1   Metal oxides

They function best in vacuum or dry environments since they have the lowest friction. The lubricating action is appropriate due to the weak inter-layer attraction force. They can also be used in low-temperature applications. They produce an adsorbed layer on the friction surface under low load conditions, but under high load, the oxides may form ceramic films that reduce friction. TiO_2, CuO, Fe_3O_4, ZnO, Co_3O_4, SiO_2, and Al_2O_3 are good examples of additives that have shown to reduce friction during lubrication and preserve the materials [10-11]. For example, accumulating CuO nanoparticles in the form of a film or layer on moving, rolling, or sliding surfaces reduces shearing stress and so improves the base oil's lubricating efficacy. The viscosity of lubricant with TiO_2 nanoparticles has increased. By using TiO_2 nanoparticle as a lubricant, the bearing's load carrying ability has improved significantly. The addition of Fe_3O_4 nanoparticles to paraffin oil increases its load-bearing ability as well as its wear resistance [12]. Metal oxide nanoparticles have a number of drawbacks, one of which is that they breakdown quickly in humid and oxidizing environments. They're also unsuitable for usage in high-temperature environments due to oxidation. They also have a limited stability of dispersion.

3.2   Metals

Because of their superior friction-reducing, anti-wear, and self-repairing capabilities, metallic nanoparticles with low shear stress, high plasticity, and low melting point have been employed as friction modifiers. Cu-containing nano-lubricants have notable features due to their small particle size, low melting temperature, and expected ductility [13]. They, like mixed oxides, generate surface coatings on friction surfaces, which increases lubrication. When metallic nanoparticles were utilized in pairs, they provided superior lubrication than when they were used separately [14-15]. Copper nanoparticle-containing emulsions for cold rolling of steel strips have been observed to have higher performance and lubrication [15]. Nevertheless, if sulphur or chlorine is present in the operating surroundings, metal oxide nanoparticles tend to undergo fast corrosive decay.

3.3   Metal sulphides

Molybednum disulphide (MoS_2) has a layered sandwich-like structure, a low coefficient of friction, and is thermally stable up to 300 degrees Celsius in air. It is a versatile solid lubricant in bulk scale due to these qualities. MoS^2 nanoparticles, which have a hollow inorganic fullerene-like structure, are employed as a lubricant additive (friction-modifier). The tribological performance has improved dramatically [16]. The characteristics of tungsten disulphide (WS_2) nanoparticles are likewise similar.

3.4   Carbon-based nanomaterials

Carbon quantum dots (CQD), Nano diamond, fullerene C_{60}, carbon nanotubes, graphene, and graphite are among the most common. The majority of them have a lamellar structure. They also have exceptional chemical stability, wear resistance, lubricity, heat resistance, and durability [17]. They have a low coefficient of friction and come in large quantities at a low cost. Carbon nanotube walls have been shown to quickly open and change into graphene-like layers when subjected to strong loads. This inter-layer sliding effect contributes to base oil's lubricating capabilities. Fullerenes have superior lubricating characteristics than nanodiamonds due to a similar phenomenon. In a study on nanodiamonds as a lubricant additive, it was found that when compared to traditional lubricating oils, there was a 70% reduction in the wear of agricultural tractor engines. [18]. The coefficient of friction ranges between 0.1-0.2. But they have less load-bearing ability compared to other categories of nano-lubricants. Almost all of them fail to provide low friction and wear in dry, inert of vacuum conditions.

3.5   Boron-based nanomaterials

They have good lubricating qualities at high temperatures and have a high chemical stability. They don't even oxidize at temperatures above 1000 °C. Because of its capacity to generate surface boundary films that slide over one another easily, nano-sized boric acid lowers the coefficient of friction and prevents wear and tear on moving surfaces. Up to 700 °C, the friction coefficients of boron nanoparticles (in the presence of air) are 0.2–0.3. They can be used with metalworking oils. Boron additions are expected to remove the main sources of environmentally hazardous emissions and wastes by substituting sulphur and phosphorus [19]. They lose lubricating power when working at high vacuum levels. In a damp atmosphere, they are prone to rusting. Impurities like boron oxide have a detrimental impact on the lubricant properties of boron nitrides.

3.6   Nanocomposites

Composites typically outperform individual NPs due to the combined influence of multiple NP types. The performance of nanocomposites including magnetite and poly(dodecyl acrylate) in three parameters of mineral-based oil, namely anti-wear tendency, viscosity index, and pour point [20], was improved. Silver/graphene nanocomposite disperses well in base oil without clumping, effectively preventing aggregation and ensuring good dispersion stability. This was attributed to graphene's layered structure and Ag nanoparticles' spherical form. The layered structure provides self-lubrication, while the spherical nano-Ag particles provide a ball-bearing and mending effect, resulting in outstanding lubricant performance. Moreover, they do not contain any corrosion and pollution causing harmful elements [21].

4.   Factors affecting the lubricating properties of nanoparticles as lubricants

4.1   Size of nanoparticle

The size of a nanoparticle is an important factor that directly affects the tribological action of lubricants. Nanoparticles with a smaller size can easily penetrate sliding or moving surfaces. It has long been known that shrinking the grain size of nanomaterials increases their hardness (grain boundary strengthening for particles smaller than 100 nm). In contrast, for sizes less than 10nm, a reduction in size makes them softer (Inverse Hall-Petch effect). It is worth noting that nanoparticles with hardness greater than the hardness of rubbing rolling of sliding surfaces are not desired, as this would result in surface abrasion (results from large size of nanoparticles). Nanoparticles should spread uniformly in the base oil for optimal lubrication and long-term stability. As per the Stokes' law, smaller the size, better is the dispersion. Consequently, nanoparticles neither agglomerate nor settle down and function effectively [4].

4.2   Shape of nanoparticles

The shape of nanoparticles has a significant impact on the performance of nanolubricants. Spherical, onion, granular, sheet, and tube are all examples of spherical shapes. The tribological performance of nanoparticles with spherical and onion forms is exceptional. Spherical nanoparticles have an extraordinarily high load bearing ability due to the ball-bearing effect [22]. The shape of an onion is spherical on the outside and lamellar on the

Emerging Nanomaterials and Their Impact on Society in the 21st Century   Materials Research Forum LLC
Materials Research Foundations 135 (2023) 23-39   https://doi.org/10.21741/9781644902172-2

interior. Even if the outward spherical structure is ripped off under severe pressure, the lamellar structure will continue to lubricate.

4.3   Surface functionalization

Nanoparticles tend to agglomerate because of their high surface area and surface energy, resulting in precipitation and poor lubrication. Aggregation must be avoided in order to increase the stability of nanoparticle dispersion in base oil, which can be performed by surface modification or functionalization of nanoparticles [23] (Fig.7). Surfactants or polymers can be used to accomplish this. They use two methods to keep the surfaces stable:

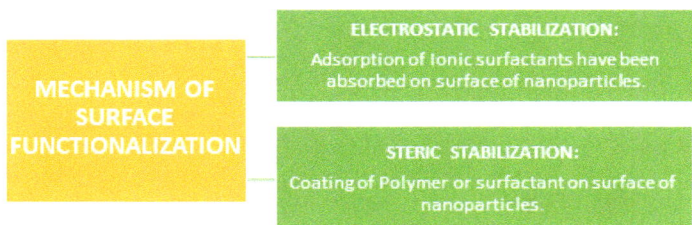

Figure 7. Mechanism of Surface functionalization

A zeta potential of greater than 30 mV is thought to give adequate stability for electrostatically stabilized nano dispersions. It is critical to choose a surfactant that is compatible with the base oil in order to obtain maximum stability. Water-soluble surfactants are recommended for polar base fluids, but oil-soluble surfactants are better for non-polar medium. For oil base fluids, non-ionic surfactants with low HLB values (hydrophilic-lipophilic balance values) are desirable.

Oleic acid, stearic acid, ascorbic acid, polyethylene glycol, alkanethiols, lauric acid, alkenyl succinimide, and sorbital monooleate are all examples of surface modifiers [4, 24]

4.4   Concentration of nanoparticles

To achieve the predicted improvement in lubricating properties of base oil, an optimal concentration of nanoparticles is required (Fig.8). The formation of a continuous, stable, and protective layer is prevented at concentrations less than the optimum amount, but clustering of nanoparticles causes the lubricant to become abrasive above the critical limit at higher concentrations. As a result, the goal of adding nanoparticles isn't achieved at all [25].

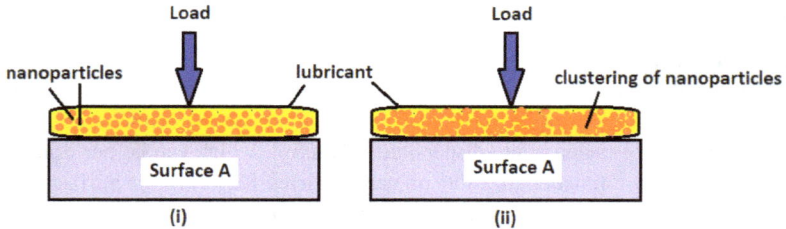

*Figure 8 (i) Effective Lubrication with optimum concentration of nanoparticles
(ii) Ineffective Lubrication with excess concentration of nanoparticles*

5. Application fields of nanolubricants

Globally there is a need for lubricants in in various sectors which are energy efficient, cost-Lubricants that are energy efficient, cost-effective, and sustainable are needed in a variety of industries around the world. Various types of lubricants are used depending on the working parameters (temperature, pressure, load, and speed of the machinery involved in the procedure). Nanoparticles are being used as additives in automotive and industrial oils for the first time. Nanoparticles have also been added to greases and metalworking oils to increase their tribological qualities. Motor oils, gear oils, compressor oils, heavy duty motor oils, turbine oils, hydraulic oils, and other lubricants require high-performance lubrication, which can be done with nanoparticles as additives [26].

6. Roadblocks to full scale use of nanolubricants

The various technical and trade-based roadblocks to the full-scale market use of nanolubricants are depicted in Fig 9.

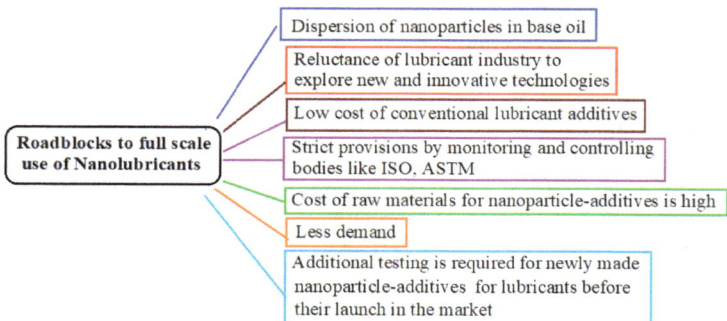

Figure 9. Roadblocks to full scale of nanolubricants

Conclusions

Nanolubrication technology is a promising subject of science that has been rapidly developing in recent years. This chapter discusses the use of nanoparticles as lubricant additives to improve the anti-friction/anti-wear qualities (tribological performance) of base oils, particularly in harsh working situations. It demonstrates the several types of nano-additives, their advantages over conventional additives, and their uses in diverse industries. The various lubrication mechanisms caused by the presence of nanoparticles in the oil are also examined. They explain the interactions between nanoparticles, base oil, and rubbing/rolling or sliding surfaces, as well as confirming nanoparticles' beneficial role in lubrication. Only by studying lubrication mechanisms at the molecular level under operating conditions can energy-efficient lubricants be developed. Advanced research inputs are required to make nano-lubricants a "commercialized product" in the lubricant business and to overcome certain constraints of nano-lubricants.

References

[1] Wu, Y. Y., W. C. Tsui, and T. C. Liu. "Experimental analysis of tribological properties of lubricating oils with nanoparticle additives." Wear 262, no. 7-8 (2007): 819-825. https://doi.org/10.1016/j.wear.2006.08.021

[2] Battez, A. Hernández, Rubén González, J. L. Viesca, J. E. Fernández, JM Díaz Fernández, A. Machado, R. Chou, and J. Riba. "CuO, ZrO2 and ZnO nanoparticles as antiwear additive in oil lubricants." Wear 265, no. 3-4 (2008): 422-428. https://doi.org/10.1016/j.wear.2007.11.013

[3] Chen, C. S., X. H. Chen, L. S. Xu, Z. Yang, and W. H. Li. "Modification of multi-walled carbon nanotubes with fatty acid and their tribological properties as lubricant additive." Carbon 43, no. 8 (2005): 1660-1666. https://doi.org/10.1016/j.carbon.2005.01.044

[4] Uflyand, Igor E., Vladimir A. Zhinzhilo, and Victoria E. Burlakova. "Metal-containing nanomaterials as lubricant additives: State-of-the-art and future development." Friction 7, no. 2 (2019): 93-116. https://doi.org/10.1007/s40544-019-0261-y

[5] Mosleh, Mohsen, Neway D. Atnafu, John H. Belk, and Orval M. Nobles. "Modification of sheet metal forming fluids with dispersed nanoparticles for improved lubrication." Wear 267, no. 5-8 (2009): 1220-1225. https://doi.org/10.1016/j.wear.2008.12.074

[6] Alves, Salete Martins, Bráulio Silva Barros, Marinalva Ferreira Trajano, Kandice Suane Barros Ribeiro, and E. J. T. I. Moura. "Tribological behavior of vegetable oil-based lubricants with nanoparticles of oxides in boundary lubrication conditions." Tribology international 65 (2013): 28-36. https://doi.org/10.1016/j.triboint.2013.03.027

[7] Mosleh, Mohsen, Neway D. Atnafu, John H. Belk, and Orval M. Nobles. "Modification of sheet metal forming fluids with dispersed nanoparticles for improved lubrication." Wear 267, no. 5-8 (2009): 1220-1225. https://doi.org/10.1016/j.wear.2008.12.074

[8] Chou, Chau-Chang, and Szu-Hsien Lee. "Tribological behavior of nanodiamond-dispersed lubricants on carbon steels and aluminum alloy." Wear 269, no. 11-12 (2010): 757-762. https://doi.org/10.1016/j.wear.2010.08.001

[9] Cizaire, L., B. Vacher, T. Le Mogne, J. M. Martin, L. Rapoport, A. Margolin, and Reshef Tenne. "Mechanisms of ultra-low friction by hollow inorganic fullerene-like MoS2 nanoparticles." Surface and coatings technology 160, no. 2-3 (2002): 282-287. https://doi.org/10.1016/S0257-8972(02)00420-6

[10] Kedzierski, Mark A. "Effect of concentration on R134a/Al2O3 nanolubricant mixture boiling on a reentrant cavity surface." International Journal of Refrigeration 49 (2015): 36-48. https://doi.org/10.1016/j.ijrefrig.2014.09.012

[11] Chaudhry, A.U., Abdala, A.A., Lonkar, S.P., Chudhary, R.G., Mabrouk, A., "Thermal, electrical, and mechanical properties of highly filled HDPE/graphite nanoplatelets composites ", Materials Today: Proceedings, 29 (2020), 704-708. https://doi.org/10.1016/j.matpr.2020.04.168

[12] Huang, Wei, Xiaolei Wang, Guoliang Ma, and Cong Shen. "Study on the synthesis and tribological property of Fe3O4 based magnetic fluids." Tribology Letters 33, no. 3 (2009): 187-192. https://doi.org/10.1007/s11249-008-9407-1

[13] Li, Yanhong, TianTian Liu, Yujuan Zhang, Pingyu Zhang, and Shengmao Zhang. "Study on the tribological behaviors of copper nanoparticles in three kinds of commercially available lubricants." Industrial lubrication and tribology (2018). https://doi.org/10.1108/ILT-05-2017-0143

[14] Padgurskas, Juozas, Raimundas Rukuiza, Igoris Prosyčevas, and Raimondas Kreivaitis. "Tribological properties of lubricant additives of Fe, Cu and Co nanoparticles." Tribology International 60 (2013): 224-232. https://doi.org/10.1016/j.triboint.2012.10.024

15] Chaudhary, R.G., Potbhare, A.K., Chouke, P.B., Rai, A.R., Mishra, R.K., Desimone, M.F., A Abdala, A.A. "Graphene-Based Nanomaterials and their Nanocomposites with Metal Oxides: Biosynthesis, Electrochemical, Photocatalytic and Antimicrobial Applications". Magnetic Oxides and Composites II, 83 (2020)79-116. https://doi.org/10.21741/9781644900970-4

[16] Yadgarov, Lena, Vincenzo Petrone, Rita Rosentsveig, Yishay Feldman, Reshef Tenne, and Adolfo Senatore. "Tribological studies of rhenium doped fullerene-like MoS2 nanoparticles in boundary, mixed and elasto-hydrodynamic lubrication conditions." Wear 297, no. 1-2 (2013): 1103-1110. https://doi.org/10.1016/j.wear.2012.11.084

[17] Mishra, R.K., Verma, K., Chaudhary, R.G., Lambat, T., Joseph, K. "An efficient fabrication of polypropylene hybrid nanocomposites using carbon nanotubes and PET fibrils", Materials Today: Proceedings, 29 (2020), 794-800. https://doi.org/10.1016/j.matpr.2020.04.753

[18] Nasiri-Khuzani, G., M. Asoodar, M. Rahnama, and H. Sharifnasab. "Evaluation of engine parts wear using nano lubrication oil in agricultural tractors." Global Journal of Science Frontier Research 12, no. 8-D (2012).

[19] Canter, Neil. "Boron nanotechnology-based lubricant additive." Tribology & Lubrication Technology 65, no. 8 (2009): 12.

[20] Dey, Koushik, Gobinda Karmakar, Mahua Upadhyay, and Pranab Ghosh. "Polyacrylate-magnetite nanocomposite as a potential multifunctional additive for lube oil." Scientific reports 10, no. 1 (2020): 1-10. https://doi.org/10.1038/s41598-020-76246-4

[21] Wang, Li, Peiwei Gong, Wei Li, Ting Luo, and Bingqiang Cao. "Mono-dispersed Ag/Graphene nanocomposite as lubricant additive to reduce friction and wear." Tribology International 146 (2020): 106228. https://doi.org/10.1016/j.triboint.2020.106228

[22] Luo, Ting, Xiaowei Wei, Haiyan Zhao, Guangyong Cai, and Xiaoyu Zheng. "Tribology properties of Al2O3/TiO2 nanocomposites as lubricant additives." Ceramics International 40, no. 7 (2014): 10103-10109. https://doi.org/10.1016/j.ceramint.2014.03.181

[23] Uflyand, Igor E., Vladimir A. Zhinzhilo, and Victoria E. Burlakova. "Metal-containing nanomaterials as lubricant additives: State-of-the-art and future development." Friction 7, no. 2 (2019): 93-116. https://doi.org/10.1007/s40544-019-0261-y

[24] Wu, Lili, Yujuan Zhang, Guangbin Yang, Shengmao Zhang, Laigui Yu, and Pingyu Zhang. "Tribological properties of oleic acid-modified zinc oxide nanoparticles as the lubricant additive in poly-alpha olefin and diisooctyl sebacate base oils." Rsc Advances 6, no. 74 (2016): 69836-69844. https://doi.org/10.1039/C6RA10042B

[25] Ghaednia, Hamed, Hasan Babaei, Robert L. Jackson, Michael J. Bozack, and J. M. Khodadadi. "The effect of nanoparticles on thin film elasto-hydrodynamic lubrication." Applied physics letters 103, no. 26 (2013): 263111. https://doi.org/10.1063/1.4858485

[26] Ali, Mohamed Kamal Ahmed, Hou Xianjun, Liqiang Mai, Cai Qingping, Richard Fiifi Turkson, and Chen Bicheng. "Improving the tribological characteristics of piston ring assembly in automotive engines using Al2O3 and TiO2 nanomaterials as nano-lubricant additives." Tribology International 103 (2016): 540-554. https://doi.org/10.1016/j.triboint.2016.08.011

Emerging Nanomaterials and Their Impact on Society in the 21st Century   Materials Research Forum LLC
Materials Research Foundations 135 (2023) 40-71   https://doi.org/10.21741/9781644902172-3

Chapter 3

Carbon Nanomaterials and Their Applications

Neksumi Musa[1,2,*], Sushmita Banerjee[1], N.B. Singh[3], and Usman Lawal Usman[1]

[1]Department of Environmental Sciences, SBSR, Sharda University, Greater Noida, India

[2]Department of Science Laboratory Technology, Adamawa State Polytechnic, Nigeria

[3]Department of Chemistry and Biochemistry, SBSR, Sharda University, Greater Noda, India

*jodaneks@yahoo.com

Abstract

Modern science and technology centred on carbon-based nanomaterials are changing at a fast pace, with the potential to replace or complement current systems. Carbon-based materials that can be produced and characterised at the nanoscale have become a cornerstone in nanotechnology. The morphologies and topographies of these carbon compounds can be quite diverse. As the well-known families of fullerenes and carbon nanotubes demonstrate, they can have hollow or filled frameworks and can assume a variety of forms. In this chapter, the synthesis, types, applications (water purification, biomedical, gas sensors, and transport materials in plant breeding) and some impacts of nano-carbon materials such as nano-spheres, nano-tubes, nano-fibres, and helices are discussed.

Keywords

Allotropes, Applications, Carbon, Nanomaterials, Nanoparticles

Contents

Carbon Nanomaterials and Their Applications...40

1.   Introduction..42

2.   Properties of nano-carbons...44

3   Fullerenes...45

      3.1   Types of fullerenes 46

      3.2   Synthesis of fullerenes...46

      3.2.1 The arc discharge technique 46

4.   The carbon nanotubes (CNTs) 47

      4.1   Types of carbon nanotubes (CNTS)..47

4.1.1 Single-walled carbon nanotubes (SWCNTS)....................................47

4.1.2 Multiple-walled carbon nanotubes (MW-CNTS)..............................47

5.   Synthesis of carbon nanotubes (CNTs)..48

5.1   Arc discharge or plasma-based synthesis technique 48

5.2   Laser ablation technique 48

5.3   Plasma-enhanced chemical vapour deposition (PE-CVD) technique...48

5.4   Thermal synthesis technique 49

5.5   Chemical vapour deposition (CVD) Technique 49

6.   Nano-diamonds 50

6.1   Synthesis of nanodiamonds 50

7.   Applications of carbon nanomaterials in water treatment...................51

7.1   Fullerenes and water purification 51

7.2   Carbon nanotubes and water purification...51

7.3   Nanoporous activated carbon and water purification........................52

7.4   Graphene and water purification 53

8.   Carbon nanomaterials in diagnosis and therapy: clinical applications..54

8.1   Biomedical applications of nanodiamonds..54

8.2   Chemo-resistant cancers and nanodiamond drug delivery...............55

8.3   Optimized magnetic resonance imaging using nanodiamonds 56

8.4   Carbon nanotubes with functionalized surfaces for improved drug delivery...56

8.5   Tissue engineering and regeneration using functionalized carbon nanotubes 56

8.6   Ex-vivo stem cell development using functionalized carbon nanotubes 56

8.7   Deployment of carbon nanotubes in photoacoustic imaging 57

8.8   Carbon nanofibers for electrochemical sensors and biosensors 57

9.   Graphene and graphene oxide in bioanalytical sciences 57

10.   Electrochemical sensors based on carbon nanoparticles....................58

10.1   Carbon nanotube-based electrochemical sensors 58

10.2   Amperometric transducers made of carbon nanotubes 58

10.3   DNA sensors based on carbon nanotubes 59

10.4   Gas sensors using carbon nanotubes 59

11.   Carbon nanoparticles and their applications in the plant system 60

11.1   Seed germination, seedling growth, plant development, and
       phytotoxicity effects of CNMs 60

11.2   Effects of fullerenes (C_{60}) and fullerols ($C_{60}(OH)n$) on
       plants 60

11.3   Effects of graphene on plants 62

11.4   Effects of carbon nanoparticles on plants..62

11.5   Effects of graphene oxide on plants 63

11.6   Effects of mesoporous carbon nanoparticles on plants 63

11.7   Effects of fluorescent carbon dots and carbon nanodots on
       plants 64

Conclusion and future prospects 64

References 65

1.   Introduction

Nanomaterials are materials with a minimum of one exterior dimension in the size range of 1 to 100 nanometres, whereas nanoparticles are objects with three external dimensions at a specific nanoscale [1]. Carbon is a fascinating element that has been regarded as an essential component of all living things. Carbon's capacity to link to itself to create oligomers and polymers permits it to play such a significant part in biological processes. This feature may also be employed to create the diverse structures that make carbon such a valuable asset due to its exceptional features such as outstanding mechanical characteristics, variable morphologies, high thermal conductivity, and high corrosion resistance [2]. Due to their exceptional electrochemistry, the one-dimensional (1D) carbon nanotubes, two-dimensional graphene (2D), three-dimensional nano-diamonds (3D), and even the zero-dimensional fullerenes (0D) are all carbon nanomaterials with potential for usage in related fields [3].

C_{60} is the most stable form of fullerenes which is a zero-dimensional nanomaterial. It has a truncated icosahedron shape which contains twelve (12) pentagons and twenty (20) hexagons. Due to their availability for chemical modification and high electron mobility, from optics to medical science and research, fullerenes have been utilized from optics to medical science and research [4].

Graphene and carbon nanotubes are all sp^2 hybridized carbon allotropes. Due to their remarkable features such as electrochemical stability, high charge carrier mobility, and mechanical flexibility have gotten a lot of interest [5]. Hence, chemical separation, super-lubricity, electronics, and catalyst support are only a few of the applications of graphene and carbon nanotubes.

In contrast to the 2D graphene, 1D carbon nanotubes, and 0D fullerenes, the 3D nano-diamonds possess both the sp3 and sp2 carbon atoms, with the reconstruction of the sp^2 playing a key part in its stability [6]. The Nano-diamonds have appealing surface areas, adjustable surface morphologies, and high hardness, hence, their novel applications in cosmetics and biomedicine [7].

Carbon nanomaterials (CNMs) have attracted a lot of interest in recent decades due to their exceptional features. Figure 1 depicts the types of different carbon nanomaterials available

The existing three (3) allotropes of carbon as shown in Figure 2 (fullerene carbon nanospheres first prepared by Krotto et al in 1985, carbon nanotubes (CNT) having a cylindrical shape, and graphene obtained from graphite) contains an extended hexagonal lattice of sp^2-bonded carbon atoms [6]. Hence, they are characterized by remarkable chemical, mechanical, and electrical features that offer tremendous possibilities in applications such as carbon fibres, energy storage, conversion devices, biosensors, and catalysts. Carbon nanomaterials have proven to be a versatile class of nanomaterials used in a wide range of consumer and industrial applications, from nano-diamonds and carbon dots to carbon nanotubes and graphene [8]. Carbon nanomaterials are being used to improve a variety of products, including electronics, lubricants, composites, and sporting equipment [8].

Carbon nanoparticles have a wide range of applications due to their unique chemical characteristics and the variety of carbon nanostructures that may be created. Variations in sp, sp^2, and sp^3 hybridizations are responsible for the variety of carbon nanostructures, as well as their structural and chemical characteristics [9]. The unique hexagonal lattices of carbon nanomaterials are mediated by sp^2 hybridization, in which carbon atoms join together with the aid of oxygen atoms. There are two single bonds and one double bond in a double bond. To make the classic, trigonal planar fashion is applied. A 2D hexagonal lattice holds a large number of carbon nanomaterials [10]. Sp^3 hybridization, in which carbon atoms have four single bonds in a tetrahedral configuration, gives diamonds their classic cubic crystal formations. Other carbon forms, such as octahedral and spherical clusters, result from transitions from sp^3 to sp^2 carbons. In carbon nanomaterials, the ratio of $sp/sp^2/sp^3$ hybridizations influences the creation of flat 2D nanomaterials (graphene), hollow 1D nanomaterials (carbon nanotubes), and closed 0D nanomaterials known as nano-diamonds. Furthermore, this ratio also governs additional features of carbon nanomaterials [10]. These include electrical, magnetic, structural strength, and chemical characteristics. All these unique properties contribute to the distinct benefits of various carbon nanomaterials for a wide variety of applications [11].

The particle composition and surface characteristics of carbon nanomaterials are linked to their behaviour and characterization. Furthermore, strengthening the surface characteristics of these carbon nanostructures might improve their chemical reactivity and properties. These nanoparticles can be used alone or in combination with other materials like membranes or structured media [12].

The structural chemistry of carbon is determined by the fact that it has four electrons available for bonding. These four electrons known as sp^3 electrons are utilized to establish four bonds with other atoms in the standard valence bond model [13]. Carbon to carbon single bonds is created in the simplest situation where bonding solely exists between carbon atoms, and the traditional diamond structure is formed. On the other hand, carbon's capacity to make numerous bonds amongst elements is one of its most distinguishing characteristics. Carbon may also form double and triple bonds by linking to another carbon atom in this fashion [14]. The chemical and physical characteristics of single, double, and triple interacting carbon to carbon bonds vary, and the ability to regulate the synthesis of structures comprising these units has led to an exploitation of carbon chemistry. One of the important events that have led to the present nanotechnology revolution has been the ability to manufacture all-carbon nanomaterials, particularly ones with networks of carbon-to-carbon double bonds [12].

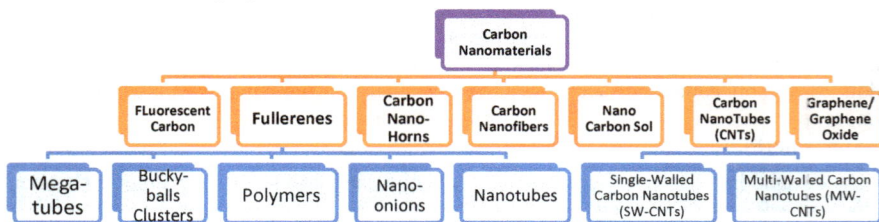

Fig. 1: Types of carbon nanomaterials

2.   Properties of nano-carbons

Nano carbons have a variety of unique characteristics [6]. These include:

- Having an approximate young's modulus of 1 TPA for a single-walled carbon nano-tube, the tubular carbon is the strongest material synthesized yet.

- The hardest substance known is carbon in the form of a diamond or single-walled carbon nano-tube. Carbon's thermal conductivity is varied. The thermal conductivity of single-walled carbon nanotubes along the tube axis is ten times that of copper (ballistic), yet they are an insulator when seen perpendicular to the axis.

- Carbon materials' surfaces can be functionalized. This can result in a new generation of reagents that can be utilized in novel applications, such as composite materials.

- Carbon is a light element (it is around six times lighter than iron), thus constructions formed of it are usually light.

- Depending on the structure and carbon-to-carbon linkage, carbon can operate either as an insulator, semiconductor, or conductor.

- According to optical investigations of carbon, single-walled carbon nano-tubes exhibit absorbance of 0.98 a.u. − 0.99 a.u. over a wide range of wavelengths, making them a near-perfect black body.

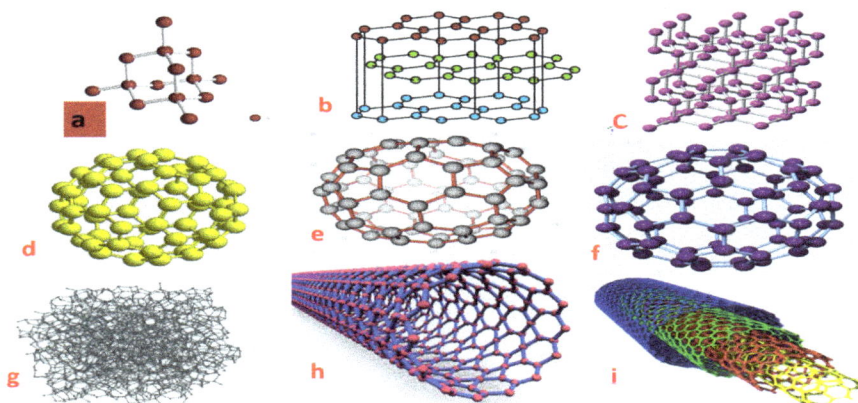

Figure 2: Eight Allotropes of Carbon: a) diamond, b) graphite, c) lonsdaleite, d) C_{60}(Buckminsterfullerene), e) C_{540} (fullerene), f) C_{70} (fullerene), g) amorphous carbon, h) single-walled carbon nanotube, i) multi-walled carbon nanotube

3   Fullerenes

Fullerenes which are commonly called the Bucky-balls (C_{60}, and C_{70}) have structures akin to that of a football discovered. They are generally characterized by weak intermolecular bonding, spherical structural form, and strong intra-molecular bonding, hence, they have attracted a lot of attention concerning their lubricating performances [15]. The chemical, as well as the colourful physical features of the molecules of C_{60}, is determined by their distinctive structure. Concerning their physical properties, the fullerenes (C_{60}) are sphere-shaped JI bonded molecules and non-polar [16]. At room temperature and pressure, they are solids and are readily soluble in organic solvents. In both ground and excited states, the C_{60} are good oxidizing agents. Based on investigations of fullerenes and their derivatives, it was discovered that they have the potential for use in superconductors, molecular magnets, biomedicine, electric light sensitivity devices (ELSD), catalysis, luminescent materials, etc. Molecular modification is the present focus of the chemical properties of the fullerenes (C_{60}) [17].

3.1   Types of fullerenes

Based on variations in their structures, the types of fullerenes are given as follows [18]:

- **Bucky-ball Clusters:** the C_{60} (buckminsterfullerene) is the most common member of the bucky-ball while the C_{20} is the smallest member in the clusters. It was discovered that the C_{60} has symmetries such as rotational, reflection symmetries, etc., hence, it is considered to be the most symmetric molecule. The C_{60} has a structural form akin to the geodesic dome [19].

- **Nanotubes:** Having a very dimension with either single or double boundaries like a tube, the nanotubes are hollow from within [20].

- **Mega-tubes:** when compared to nanotubes, the diameter of the mega-tubes is larger. They are characterized by different wall sizes.

- **Nano-onions:** Having a bucky-ball core between its layers and multiple structural carbon layers, the nano-onions are spherical [21].

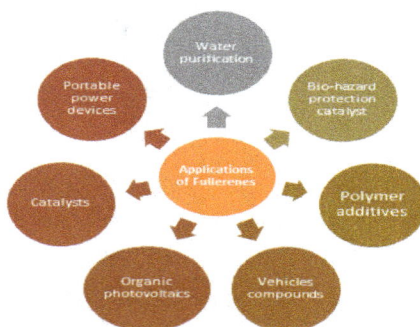

Figure 3: Applications of fullerenes

3.2   Synthesis of fullerenes

3.2.1 The arc discharge technique

The arc discharge is a synthesis technique carried out at high temperatures. Parts of the equipment used are; a water-cooled trap, a furnace for heating, carbon electrodes, a high voltage pulsed-power input, and a quartz tube. At a temperature range of about 25 to 1000^0C, buffer gas is allowed to pass through the quartz tube during the annealing process of the carbon clusters at a pressure of around 500 Torr and a flow rate of 300 $cm^3 s^{-1}$. High-

Emerging Nanomaterials and Their Impact on Society in the 21st Century   Materials Research Forum LLC
Materials Research Foundations 135 (2023) 40-71   https://doi.org/10.21741/9781644902172-3

Performance Liquid Chromatography (HPLC) is used to detect the presence of fullerenes even at low concentrations [22].

4.   The carbon nanotubes (CNTs)

An exceptional feature of carbon is its ability to form carbon nanofibers and carbon nanotubes. These nano-tubes and nano-fibres of carbon usually have a length of 10nm to a few centimetres and a diameter that falls within the range of 1 nm to 100 nm. The first carbon nanotubes were prepared with the aid of arc-discharge evaporation in 1991 [23]. The CNTs when compared to the sp^2 hybridized crystal structure of the graphite, have a cylindrical sp2 carbon just like the graphite. This enables it to easily form ideal rotational and linear nano bearings in the process of nano-rotating [20].

4.1   Types of carbon nanotubes (CNTS)

There are two (2) types of CNTs. These are given as follows:

I.   Single-walled carbon nanotubes (SWCNTs) [24]

II.   Multiple-walled carbon nanotubes (MW-CNTs) [25]

4.1.1   Single-walled carbon nanotubes (SWCNTS)

Depending on the synthesis temperature, the SW-CNTs are made up of an average of 0.4 to 2.0 nm single cylindrical carbon. The diameter of the CNTs is directly proportional to the synthesis temperature. The shape of the CNTs could either be chiral, armchair, helical or in a zigzag array. They have excellent potential for bio-conjugation and drug loading due to their ultra-high surface area [24].

4.1.2   Multiple-walled carbon nanotubes (MW-CNTS)

The MW-CNTs are characterized by several co-axial cylinders having an inner diameter of 1 to 3 nm and an outer diameter within the range of 2 to 100 nm. Consisting of a length of one to quite a few micrometres, every single cylinder is made up of a graphene sheet that envelopes a hollow core. The sp2 hybridization in MW-CNTs is such that a delocalized electron cloud is created along the wall, which is accountable for interactions among adjoining tubular layers, resulting in a less flexible material with more structural defects [25].

Depending on the arrays of the graphite layer, the structure of MW-CNTs can be classified into two. These are:

I.   The first is called the Russian-doll model. This structure is characterized by the arrangement of layers of graphene sheets in a concentric configuration.

II.   C second structure is made up of a graphene sheet rolled up around it and characterized by a structure akin to a parchment [26].

In designing the MW-CNTs, nanoparticles are deposited on the edges or walls of the MW-CNTs. This is bonded or governed by physical bonds and with prospective applications in biosensors, magnetic data storage devices, biomedical, electronic gadgets, and catalysis. Techniques such as metal precursor chemical decomposition, precipitation, and high-temperature hydrolysis are employed.

5.   Synthesis of carbon nanotubes (CNTs)

The following are techniques used for the synthesis of CNTs: Arc discharge or Plasma-based synthesis technique, Laser ablation technique, Plasma-enhanced chemical vapour deposition (PE-CVD) technique, Thermal synthesis technique, and Chemical vapour deposition (CVD) technique [27]. Figure 3 gives a depiction of the different synthesis techniques for carbon nanotubes.

5.1   Arc discharge or plasma-based synthesis technique

This synthesis technique is associated with the use of a high temperature of about $1800\ ^{0}C$. When compared to other synthesis techniques, this process usually results in the development of carbon nanotubes with fewer defects structurally. The first MW-CNTs in 1991 and the subsequent synthesis of the SW-CNTs in 1993 were achieved using this technique. Hence, it has proven to be the most conventional and facile technique for the synthesis of CNTs. A very high temperature is generated when an electric arc is formed between two electrodes of graphite. This temperature is high and adequately enough for carbon sublimation. After cooling and condensation of the carbon vapours, either SW-CNTs or MW-CNTs can be produced. On a general note, when a catalyst is introduced through the packing of a metal such as Nickel, Cobalt, or Iron, the SW-CNTs are produced. Whereas, the MW-CNTs are produced without the addition of catalysts [28].

Because the morphology of carbon nanotubes is greatly affected by the different atmospheres, the percentage yield of the MW-CNTs is dependent upon the pressure of the gas in the reaction vessel.

5.2   Laser ablation technique

This technique is analogous to the plasma-based method of synthesis whereby; high temperature is applied. Here, a pulsed and continuous laser is used in vaporizing targeted graphite situated in filled argon or helium oven. In both the Laser ablation and Plasma-based synthesis techniques, the conditions of reaction which serve as reaction mechanism indicators are the same. By continuous wave-carbon dioxide laser ablation at with no target heating, SW-CNTs were synthesized. It was discovered that the laser power is directly proportional to the average diameter of the SW-CNTs by CO_2 laser [28].

5.3   Plasma-enhanced chemical vapour deposition (PE-CVD) technique

Plasma-enhanced chemical vapour deposition (PE-CVD) is a general term that comprises numerous varying methods of synthesis. These synthesis systems have been utilized in the synthesis of both MW-CNTs and SW-CNTs. By and large, this method can either be directly or remotely carried out. Some SW-CNTs and MW-CNTs field emitter towers can

Emerging Nanomaterials and Their Impact on Society in the 21st Century   Materials Research Forum LLC
Materials Research Foundations 135 (2023) 40-71   https://doi.org/10.21741/9781644902172-3

be produced using the Direct Plasma-enhanced chemical vapour deposition (PE-CVD) technique. Also, both SW-CNTs and MW-CNTs can be synthesized using the Remote Plasma-enhanced chemical vapour deposition (PE-CVD) technique [27].

Unlike Laser ablation, arc discharge, and Solar furnace which use a solid graphite source, there is no need for a graphite source because the carbon which is used in the Plasma-enhanced chemical vapour deposition (PE-CVD) synthesis technique is gotten from feed-stock gases like methane (CH_4) and carbon monoxide (CO). The feed-stock gases are broken down into reactive carbon species (RCS) like C_2 and CH with the aid of a plasma equipped with argon. This is done to aid the formation of the products at low temperatures and pressure.

5.4   Thermal synthesis technique

Thermal synthesis is a general term that represents various chemical deposition techniques such as Flame synthesis, Chemical vapour deposition technique, and CO_2 synthesis technique. In contrast to the Laser ablation and the Arc discharge techniques which are plasma-based processes, the thermal synthesis technique utilizes only the hot reaction zone (~1200 °C) and thermal energy. As reported in most cases, depending on the feedstock of carbon in use, metal catalysts such as nickel, cobalt, iron, and carbon feedstock bring forth carbon nanotubes.

5.5   Chemical vapour deposition (CVD) Technique

This process is carried out by heating the gaseous carbon which contains the molecules. This is done by putting carbon in the gaseous phase and using a coil that is heated or plasma. The molecule is cracked up by heat into a reactive atomic carbon. Transition metal catalysts such as nickel, iron, and cobalt are the most frequently used catalysts and are erstwhile, treated or doped with other metals.

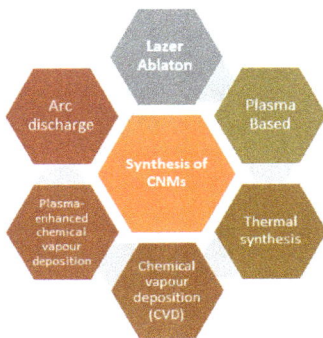

Fig 4: Synthesis of carbon nanoparticles

Emerging Nanomaterials and Their Impact on Society in the 21st Century   Materials Research Forum LLC
Materials Research Foundations 135 (2023) 40-71   https://doi.org/10.21741/9781644902172-3

Hydrocarbons such as ethane, xylene, methane, acetylene, ethanol, and isobutane are the recommended source of carbon in this technique. If the source of carbon is gaseous, the development efficiency of the carbon nanotubes is dependent upon factors such as the free radicals and reactive species, together with the concentration and reactivity of the intermediates in the gaseous phase. This process is a result of the decomposition of hydrocarbon [29].

6.   Nano-diamonds

Nanodiamonds are diamond nanoparticles with a size of less than one micrometre. For various uses, there are multiple techniques for producing nanodiamond particles [30]. Nanodiamonds and their derivatives are being investigated as promising materials in biological and electrical applications because of their low cost and high surface functionalization potential. Nanomaterials' size matters in a variety of applications, and nanodiamonds are no exception [31].

Doping procedures in Nanodiamonds cause a wide variety of, mechanical, magneto-optical, biocompatibility, and electrical, characteristics.

6.1   Synthesis of nanodiamonds

Nanodiamond particle preparation may be done in a variety of ways, but only two types of nanodiamonds are commercially available. The detonation method, which includes the explosion of carbon-containing materials, is commonly used to create nanodiamond particles with diameters of 2–10 nm [32]. The grinding of micron-sized diamond powders at high temperatures and a static high-pressure environment is another frequent approach for producing nanodiamonds. The architectural patterns of the nanodiamond particles created by these two methods differ, as do the scientific applications [33]. The size of the particles, their shape, and the condition and number of N atom defects in the nanodiamonds' cores are the most significant distinctions between the two procedures. This material can be made on a commercial scale since detonated nanodiamonds are in the 2–5 nm size range. Nanodiamond de-aggregation through milling can create colloidal solutions of primary nanodiamond particles with a size range of 2–10 nm, according to current improvements in detonation. In a fraction of a microsecond, detonated nanodiamonds are created on the front side of the detonation wave. In the diamond phase, they are mostly made of carbon. The presence of a range of functional active groups on the surface of the particles is an important feature of the detonation configuration. Throughout the detonation synthesis process, these sorts of groups are formed. The heat capacity of the cooling platform in the detonation chamber determines how many nanodiamonds are created. The monocrystalline nanodiamond particles generated by the high pressure-high temperature technique can range in size from 5 to 20 nanometers. Blocky forms and sizes are possible for nanoparticles [34].

Nanodiamonds are made from a variety of precursors, including CH_4/H_2 gaseous combinations with inert gases. However, the utilisation of oxygen and nitrogen has been documented, but due to the insertion of many faults, this is neither cost-effective nor efficient. The grain size, shape, microstructure, crystalline grade, and development rate of nanodiamonds have all been shown to be affected by changing the amount of oxygen and nitrogen included. As a result, increasing the inclusion ratio improves the nanodiamonds' crystal properties. As precursors for nanodiamonds, carbon nanotubes have been used. For instance, in a hydrogen microwave discharge, multiwall carbon nanotubes can be utilised as a precursor to depositing diamonds. Similarly, nanodiamonds were made from a pure carbon sheet aided by silicon, with the formation of the nanodiamonds controlled by varying the heating temperature and Si %. Carbide-organic systems, graphite, fullerenes, trityl/hexogen, and 1,1,1-trichloroethane polycarbonate are some of the other nanodiamond precursors [35]. Ion Bombardment Technique, Microwave-Assisted CVD, Hydrothermal Method, Pulsed Laser Ablation, Electrochemical Synthesis, Detonation method and Ultrasound Synthesis techniques are generally used for the synthesis of Nanodiamond [36].

7.   Applications of carbon nanomaterials in water treatment

Carbon nanomaterials have the potential to be used in water and wastewater treatment. The adsorption of contaminants such as organic and inorganic pollutants from water and wastewater using various types of carbon as adsorbents is given below:

7.1   Fullerenes and water purification

Fullerenes have proven to be good adsorbents for the adsorption of organic compounds and heavy metals from polluted water. Fullerenes were used to remove hydrophobic organic chemicals (HOC) such as phenanthrene, pyrene and naphthalene from an aqueous solution. Factors such as surface area, micropore volume and volume ratios of mesopores to micropore, and molecular size of the pollutant affect the fullerenes' adsorption capacity. Specifically, due to its hydrophobic surface, the C_{60} fullerene is a potentially excellent adsorbent for the remediation of several organic compounds from water [13].

7.2   Carbon nanotubes and water purification

Carbon nanotubes have been demonstrated to be an excellent adsorbent for removing pollutants from water, resulting in a clean environment. CNTs may adsorb pollutants on their own, but they can also mix with other materials to produce composites with greater adsorption capacity. Usually, the surface of the carbon nanotube is functionalized to explore and maximize the adsorption properties of the adsorbent [37]. This process improves the interaction between the adsorbent and the pollutant via enhanced dispersion in the adsorption system. Because of the impurities which are introduced during the synthesis of carbon nanotubes, the as-synthesized carbon nanotubes (CNTs) do not always give optimum performance. Therefore, the CNTs can be functionalized in the presence of acid and alkali solutions to improve their performance. The occurrence of all these contaminants on the carbon nanotube's surface would cause a change in their essential features. As a result, contaminants will be removed from the carbon nanotube using an acid or alkaline solution to prevent the characteristics from being altered. Based on that fact, the surface of the carbon nanotubes will be altered as a result of the functionalization procedure, which involves attaching an extra functional group to the surface of carbon nanotubes [38]. By replacing weak Van der Waals interactions with strong electrostatic forces, an extra negatively charged functional group at the open end or sidewall of the carbon nanotubes would boost solubility in any type of solvent and reinforce the surface of the carbon nanotubes. Acids such as sulphuric acid (H_2SO_4), potassium permanganate ($KMnO_4$), and nitric acid (HNO_3) contain oxygen and are frequently used as alkaline and acidic medium for the functionalization of carbon nanotubes. Functional groups such as -OH, -COOH and -C=O are usually introduced into the carbon nanotubes. On that basis, carbon nanotubes have been effectively coated with other carbon-based compounds to boost their efficacy in removing pollutants from water. At optimized reaction conditions of 0.2g adsorbent (functionalized CNTs), pH 5.0, 90 °C, and a temperature of 100 minutes, the result for the removal of Cu^+ from aqueous solution showed that CNTs which are functionalized using HNO_3 and H_2SO_4 display a higher adsorption capacity as compared to the non-functionalized CNTs [39]. As a result, carbon nanotubes (CNTs) were chosen as an adsorbent to give a more promising and efficient substitute. $Pb^{2+}>Ni^{2+}>Zn^{2+}>Cu^{2+}>Cd^{2+}$ was the order of metal ion adsorption to different kinds of carbon nanotubes (CNTs), with surface-functionalized CNTs via NaOCl, HNO_3, and $KMnO_4$ having a greater adsorption capacity than non-functionalized carbon nanotubes (CNTs) [40]. The interaction of the ions situated on the surface of CNTs, such as the functional groups present after the functionalization process, and the metal ions define the adsorption mechanism. Therefore, both non-functionalized and functionalized carbon nanotubes are so widely used in the treatment of divalent metal ions in water and wastewater. As a result, carbon nanotubes (CNTs) were chosen as an adsorbent to give a more promising and efficient alternative.

7.3   Nanoporous activated carbon and water purification

Due to the vast supply of agricultural waste, the manufacturing of adsorbents based on biomaterials such as agricultural waste has increased significantly in recent years. Aside from being plentiful, these agricultural sources are inexpensive to procure, and the expense of producing an adsorbent using this agricultural waste is low [41]. These biomaterials, on the other hand, are extremely efficient at low metal concentrations and are reusable. These biomaterials will subsequently be crushed and converted into a carbon-based adsorbent for water purification. Examples of these biomaterials are bagasse, sawdust, activated carbon cloth, crab shells, and olive stone waste. Although, all these carbon-based adsorbents had a good adsorption capacity for removing pollutants from water, their efficiency was insufficient. As a result, by raising the specific surface area or decreasing the particle size, the adsorption efficiency of this activated carbon may be improved. In other words, by functionalizing this activated carbon, it may be transformed into nanoporous activated carbon, which can remove pollutants from water with great efficiency [42].

In the removal of heavy metals (Cd^{2+}, Ca^{2+}, Hg^{2+}, and Cu^{2+}) using nanoporous activated carbon, two sets of oxygen and nitrogen functionalized activated carbon were synthesized.

This was achieved by changing the nitrogen content during synthesis. According to the adsorption data, the maximum adsorption capacity of oxidized nanoporous carbon was larger than that of its unoxidized counterparts. This is because these carbons have a large number of acidic oxygen functional groups connected to their surfaces [43]. The nanoporous activated carbon is also efficient in the adsorption of organic pollutants such as toluene, ethylbenzene, benzene, and p-xylene. Also, since individuals were also found to be at risk from the occurrence of chlorinated compounds more than the permitted level in the water, granular activated carbon was successfully used in the efficient adsorption of chlorinated organic compounds such as tetrachloroethylene and trichloroethylene from water. If adsorption achieves a high capacity of adsorption (Q), and a low equilibrium concentration (C_e), it is deemed successful and highly efficient. When compared to the adsorption data of granular activated carbon, functionalized nanoporous carbon has better adsorption capabilities. As a result of this comparison, the overall benefit of using nano-sized carbon materials would yield a greater adsorption capacity of pollutants from water owing to the large surface area. This is because a material with a large surface area allows for greater surface contact with the adsorbent and adsorbate, resulting in a larger adsorption efficiency [44].

7.4   Graphene and water purification

Because of its unique two-dimensional structure and remarkable thermal, electrical and mechanical capabilities, graphene, which is made up of a few atomic layered graphites, was also used to investigate the adsorption of pollutants from water. The adsorbent for the removal of Cd^{2+} and Co^{2+} from vast volumes of aqueous solutions was graphene oxide nanosheets produced from graphite using the modified Hummers technique. The sorption of metal ions on graphene oxide nanosheets was substantially reliant on pH and slightly dependent on ionic strength, as seen by the effects of ionic strength, pH, and humic acid on $Co2+$ and Cd^{2+}. The presence of humic acid inhibited the adsorption of the two heavy metals on graphene oxide nanosheets because of this finding. As a result, if graphene oxide nanosheets could be mass-produced at quite a cheaper cost, these may become the most widely used material for the purification of water [45].

The use of potassium hexafluorophosphate solution as the electrolyte allowed for the fabrication of functionalized graphene in a facile and low-cost way. Cd^{2+} and Pb^{2+} ions were removed from the aqueous solution using the newly designed adsorbent. Also, the adsorption of Pb^{2+} from the water was effectively achieved using graphene nanosheets produced by vacuum-assisted low-temperature exfoliation. Based on the results, the thermal processing on the graphene nanosheets increased the sorption of Pb^{2+} ions depleted in the oxygen complexes concentration in the graphene nanosheets. The Lewis basicity of graphene nanosheets is increased by thermal processing under a high vacuum. This favours the concurrent sorption of Pb^{2+} ions and protons onto the graphene nanosheets [46]. Using a chemical process and a magnetite particle size of 10 nm, magnetic graphene hybrids were produced. At ambient temperature, these composites were discovered to be superparamagnetic and can be extracted using magnetization. Arsenic adsorption ability was shown to be greater in these superparamagnetic composite materials (As V and As III). A decrease in the agglomeration of bare magnetite resulted in an improvement in the adsorption efficiency of these magnetic graphene composites [47].

In addition to the remediation of heavy metals, numerous studies have shown that graphene may be used for the sorption of dyes. The Hummers–Offerman technique was used to oxidise graphite, which was then utilised to efficiently adsorb dye from water. The surface area of graphite oxide did not change significantly, but the structure did grow, and numerous functional groups were observed to be connected to the surface, which subsequently increased graphite oxide's sorption capacity. The number of dyes such as malachite green and methylene blue adsorbed on the graphene oxide was substantially higher than on graphite [46].

Other carbon nanomaterials such as nitrogen-dopped carbon magnetic nanoparticles were used for the remediation of heavy metals [48]. Polypyrrole nanoparticles were first made, and then they were employed as a carbon precursor to making nitrogen-doped carbon magnetic nanoparticles which were subsequently used for the remediation of Cr^{3+}. The adsorption of Cr 3 + onto nitrogen-doped carbon magnetic nanoparticles was discovered to be spontaneous and endothermic, with the adsorption process being influenced by valence forces. Because of their huge surface area and nitrogen content, nitrogen-dopped carbon magnetic nanoparticles were discovered to exhibit a strong adsorption affinity for Cr^{3+} [49].

8.   Carbon nanomaterials in diagnosis and therapy: clinical applications

As depicted in Figure 5, carbon nanoparticles have the potential to change how patients are treated and diagnosed in healthcare settings. Carbon-based nanoparticles, like carbon nanotubes, nanodiamonds, graphene oxides, carbon nanofibers, and graphene, are potential nanomaterials for a wide range of therapeutic uses, including treating chemo-resistant cancer, enhancing tissue regeneration, stem cell banking, and many others [50].

8.1   Biomedical applications of nanodiamonds

Nanodiamonds are a type of carbon nanoparticle that has a great deal of potential for therapeutic applications as agents in imaging and nanomedical drug delivery. Fluorescent nanodiamonds or small detonation nanodiamonds are the most frequent nanodiamonds used in therapeutic systems. Distinctive facet-dependent surface electrostatic potentials, a chemically inert core that promotes biocompatibility, and a surface adorned with adjustable functional groups are all advantages of nanodiamonds as a vehicle for drug delivery and imaging. Nanodiamonds have the potential to revolutionize medicine, whether employed on their own as a drug delivery system or in conjunction with other components [51].

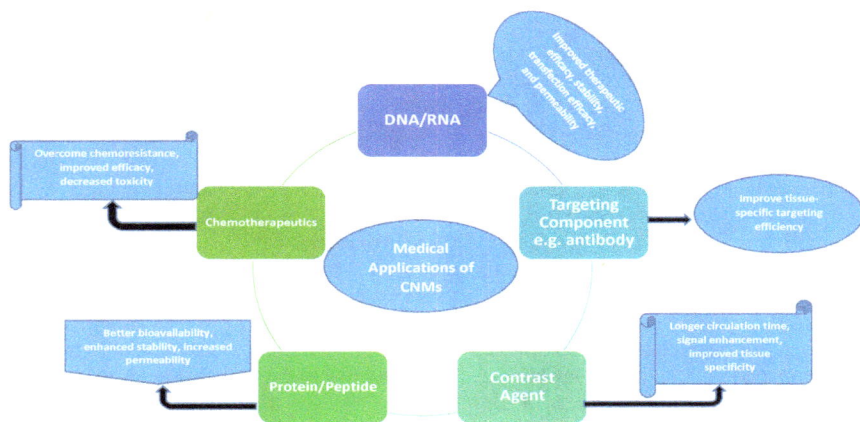

Figure 5: Medical applications of carbon nanoparticles

8.2   Chemo-resistant cancers and nanodiamond drug delivery

Enhanced administration of chemotherapy medications, especially for chemo-resistant tumours, is one of the most potential applications of nanodiamonds. Anthracyclines are a kind of chemotherapy medicine that functions by complexation with DNA and inhibiting topoisomerase II, which kills cancer cells. Although anthracyclines are useful in treating a variety of malignancies, ABC transporter-mediated chemo-resistance is a prevalent mechanism of anthracycline resistance since many ABC transporters may efflux anthracyclines. Although nanodiamonds have been used to administer a variety of anthracyclines and other chemotherapeutic medicines, epirubicin and doxorubicin were used in most of the most extensive research work reported. Unlike the standard unmodified doxorubicin, the modified doxorubicin-nanodiamond complex prevented the occurrence of active efflux of doxorubicin, based on the fact that the complex is retained much longer in ABC transporter overexpressed cells. In mice models of breast and liver cancer, this greater cellular retention resulted in increased death of chemo-resistant cancerous cells, as well as improved survival and smaller tumours [52]. It was discovered that by surface functionalization of nanodiamonds–anthracycline complexes with targeted constituents, the potency of nanodiamonds–anthracycline complexes may be increased yet, maintaining the minimal side effects which is a feature of nanodiamonds–anthracycline combinations. Nanodiamonds are a potent tool for enhancing the delivery of drugs in cancer, especially when it comes to anthracycline effectiveness and safety [51].

8.3   Optimized magnetic resonance imaging using nanodiamonds

Magnetic resonance imaging (MRI) is a useful clinical technique for detecting, diagnosing, and monitoring the trajectory of illnesses that are difficult to identify visually or using biological liquid biomarkers, as with many other malignancies. The exceptional electrostatic surface promising features of nanodiamonds impart the potential to create improved ND-based magnetic resonance imaging contrast agents capable of providing significantly larger imaging responses at concentrations lower than regular contrast agents, and the use of fluorescence signals could be quickly translated into biomedical research domains. The capacity of nanodiamonds to increase the per ion relaxivity of MRI contrast agents implies that nanodiamonds might be used in the clinic to sustain and enhance MRI resolution enhancement by contrast agents at lower doses, resulting in decreased toxicological concerns [53]. In comparison to the synthesis of bidirectional anthracycline-nanodiamonds complexes, Gd^{3+} to nanodiamonds conjugation was accomplished by functionalizing both the Gd^{3+} chelate and the nanodiamonds to covalently bond both constituents together. In comparison to existing therapeutic Gd^{3+} chelates, the nanodiamonds-gadolinium complex demonstrated increased cellular retention and enhanced MR imaging of triple-negative breast cancer tumours in experimental animal models. nanodiamonds-gadolinium have no substantial harmful effects in vitro, according to cytotoxicity assays [54].

8.4   Carbon nanotubes with functionalized surfaces for improved drug delivery

Covalent functionalization and physical confinement can both be used to functionalize and bind drugs onto and within carbon nanotubes due to the vacuous interior and functional groups on the external surface [50].

8.5   Tissue engineering and regeneration using functionalized carbon nanotubes

Carbon-based nanomaterials such as carbon nanotubes have peculiar features that make them useful as frameworks for tissue engineering and regeneration. The capacity of triggering precursor cells at the defect site to differentiate into exact connective tissue species based on specific growth regulators and the surfaces on which they develop is critical to the outcome of this procedure [50]. Because of their immense differentiation potential, stem cells perform a key function in this area. Therefore, to differentiate into specific cell lineage, they still require biochemical inputs. Carbon nanotubes were shown to stimulate normal cellular development and speed up stem cell differentiation into distinct lineages when used to treat skeletal muscle injuries [55].

8.6   Ex-vivo stem cell development using functionalized carbon nanotubes

Because of their great mechanical strength, carbon nanotubes have been shown to have viability as a scaffold material in tissue engineering procedures. Carbon nanotubes have excellent electrical conductivity, making them ideal for use as a neural development scaffold. carbon nanotubes can be functionalized with poly (ethylene glycol), amines, or carboxyls to improve aqueous dispersion, minimize agglomeration, and promote biocompatibility to make them more suitable for biomedical applications. For example, the Ex-vivo development of freeze-thawed, unsorted HSPC from human UBC mononuclear cells can be assisted by carboxylated single-walled carbon nanotubes in the dispersed form [56].

8.7   Deployment of carbon nanotubes in photoacoustic imaging

Carbon nanoparticles were also studied for use as contrast materials in photoacoustic imaging, or photoacoustic tomography, in addition to MRI. Carbon nanotubes were widely investigated as contrast media for photoacoustic tomography and have been shown to enhance photoacoustic tomography imaging, especially in cancer treatments. The design and assessment of cyclic Gly/Arg/Asp (peptide)- single-wall carbon nanotube complex to target and detect tumours in vivo in a tumours model was one of the early examples of enhancements to photoacoustic tomography by using carbon nanotubes as contrast media [57].

8.8   Carbon nanofibers for electrochemical sensors and biosensors

Due to manufacturing processes like electrospinning followed by carbonization, CNFs may be manufactured to significantly greater lengths than CNTs. Also, CNFs made this way can have significantly larger hollow internal dimensions. The electrical conductivity qualities of CNFs are determined by their wider width and the polymer-carbon composite produced during production, making them appealing for electrochemical devices. CNFs offer new opportunities in neurological areas such as electrode fabrication in biosensors that monitor electroactive neurotransmitters because they are flexible, biocompatible, and electrically conductive [58].

9.   Graphene and graphene oxide in bioanalytical sciences

Because of their distinctive rheological and mechanical characteristics, graphene-based nanomaterials have shown potential in enhancing drug delivery, tissue bioengineering, as well as other biomedical applications. Through solution exfoliation and dispersion processes, these nanomaterials are frequently synthesized in a flake-like shape. Graphene nanoparticles are used in spectroscopy (Surface-enhanced laser desorption ionization spectroscopy) as platforms for the extraction of biomolecules (for signal improvement). When contrasted to graphene oxide, graphene performs better as a framework for surface-enhanced laser desorption ionization spectroscopy. On a general note, biomolecules break considerably more readily on graphene oxide than on graphene due to graphene's superior thermal conductivity, which facilitates effective thermal dissipation and hence lessens fragmentation. Because of this advantage, a very effective graphene/TiO2 extraction and detection system for phospho-peptides in cancer cells was developed [50].

10.   Electrochemical sensors based on carbon nanoparticles

Carbon nanotubes, diamond-like carbon, graphene, and crystalline diamond are all carbon nanoparticles with outstanding electrochemical characteristics, which has led to their extensive use. The possibilities of these nanomaterials in sensing systems are undeniable since the unique carbon-derived nanostructures have characteristics that are unimaginable in conventional materials. Consequently, they can function with better selectivity and sensitivity not just in extreme settings, but also throughout a wider thermal and dynamic range.

10.1   Carbon nanotube-based electrochemical sensors

Due to the distinctive structure of the carbon nanotubes such as mechanical deformations, the ability to form very thin wires which simultaneously acquire the conductivity of graphene and the toughness of diamond, and high surface-to-volume ratios, these make them good for use in the development of nanoscale sensors that seem to be sensitive to mechanical, physical, and chemical surroundings. The curvature of graphene sheets affects the electrical characteristics of carbon nanotubes [59].

Chemicals that are either electron donors or acceptors like NH_3, O_2, and NO_2, when dopped with single-wall carbon nanotubes either increase or reduce their conductivity by providing the nanotubes with more holes or charge carriers. Not only do carbon nanotubes improves the electrochemical reactivity of vital biomolecules, but they also enhance electron-based transfer reactions of proteins as well. The capacity of modified electrodes with carbon nanotubes to reduce surface contamination has been established, as in the event of primary oxidation of NADH due to high over-potentials which lead to oxidation products fouling the electrode surface. Nucleic acid is an example of a vital biomolecule that is concentrated by carbon nanotubes which help in improving the sensitivity and selectivity of the probe. It is critical to properly functionalize and immobilize carbon nanotubes to use them in electrochemical sensing applications. Metallic nanoparticles were also used for the electrochemical functionalization of the carbon nanoparticles and the subsequent utilization of the resultant CNTs-metal nanocomposite for catalysis and sensing has increased recently [59].

10.2   Amperometric transducers made of carbon nanotubes

Surfactants are a preferred method for preserving the structure and characteristics of carbon nanotubes since they disrupt the strong Van der Waal attractive interactions between carbon nanotubes and so improve their solubility. Alternative treatments, such as covalent alteration of the surface, are significantly less effective. Sodium dodecyl sulphate is now the most extensively employed surfactant in this procedure, even though different DNA, polymers, and detergents have all shown possibilities as surface-active agents. For the fabrication of thin films at the electrode junction, sodium dodecyl sulphate was employed to provide appropriate homogenous dispersions of carbon nanotubes [60].

The use of carbon nanotubes as nanoprobes is another application. Carbon nanoprobes can be used as atomic force microscopes or scanning tunnelling microscope tips in this application. There are several advantages of using carbon nanotube tips, these are:

- They have high ratio characteristics that permit them to investigate deep fissures and trenches.
- tiny diameters by description
- readily configurable to make effective and functional probes
- elastic capacity to buckle, limiting the strain produced by the atomic force microscope probe and reducing distortion and harm to biological and organic substances

Functionalized carbon nanotubes as atomic force microscope tips have provided additional possibilities in chemistry and biology for molecular identification and chemically sensitive imaging.

10.3   DNA sensors based on carbon nanotubes

Carbon nanotubes are worthy choices to be used in DNA and RNA sensing because of their capacity to create bonds among their conjugated systems and nucleobases. Carbon nanotubes also magnify the DNA/RNA sensing signal owing to their intrinsic electrical properties. Because techniques of amplification like the incorporation of Nanoparticles or enzymes that facilitate the transfer of electrons are often required to augment the generally weak DNA/RNA sensing signal, CNTs' intrinsic conductivity is crucial.

Glassy carbon electrode electrodes modified with multi-walled carbon nanotubes are used to detect and quantify albumin, amino acids, and glucose. Furthermore, by using a single data point attachment to attach DNA to a solid surface, the immobilization of DNA by carbon nanotubes is achieved. This implies that the DNA sensing system's detection limit was enhanced and reduced production time.

Multi-walled carbon nanotubes were used to modify an underlying glassy carbon electrode whereby a well-known biomarker for oxidative DNA damage sensors that had a good electrochemical performance to the oxidation of 8-OHdG was fabricated [61].

10.4   Gas sensors using carbon nanotubes

Numerous uses in a variety of fields, such as industrial, ecological, and health-related analyses, necessitate the use of gas sensors. Gas detection has traditionally been accomplished using large, cumbersome instruments. Small-scale sensors are indeed a great alternative to these earlier forms because they are much less costly. Nevertheless, in needing to achieve acceptability, their expected outcomes must be comparable to that of existing analytical tools. As a sensing variable, nanoparticles have particular benefits in terms of selectivity and sensitivity [60].

Several carbon-based nanodevices have been used for the detection of nitrogen and ammonia (NO_2 and NH_3) gases. At low analyte concentrations, the major sensing mechanism has been the association between the molecule and the carbon nanotube at the CNT-metal interface. Other examples of carbon-based gas sensors are; Platinum-decorated multi-walled carbon nanotubes with nickel oxide functionalisation for the detection of hydrogen gas, multi-walled carbon nanotubes blended composite with cuprous oxide–functionalized–platinum-coated cuprous oxide, an SW-CNTs/Copper-based sensors with improved H_2S gas response, as a sensor module for NO_2, reduced GO-CNT/SnO2 hybrids were used, and PANI/MW-CNTs-based nanocomposite for the detection of trace amounts of NH_3 gas [60].

11. Carbon nanoparticles and their applications in the plant system

Carbon-based nanomaterials (CNMs) have been widely employed for diverse applications in various parts of the plant system as a result of the spectacular advancement in the field of nanotechnology [62].

11.1 Seed germination, seedling growth, plant development, and phytotoxicity effects of CNMs

Because of their potential for application in plant development control, various forms of CNMs (fullerenes, fullerols, CNPs, and CNTs) have recently acquired interest. Depending on the CNM types, concentration, exposure duration, plant species, and growth circumstances in which they are administered, most study studies have shown both helpful and detrimental effects on plants. These outcomes are considered below.

11.2 Effects of fullerenes (C_{60}) and fullerols ($C_{60}(OH)n$) on plants

The fullerene molecule is composed of sixty carbon atoms and is known as C_{60}. Fullerols have fantastic features like high water solubility and biocompatibility, that can be readily modified by adjusting the amount of Hydroxyl group in the component. Due to their diverse biological effects, fullerols NPs have a wide range of applications in biomedical studies. Complementing the germination media of A. thaliana with $C_{70}(C(COOH)_2)$, a water-soluble fullerene malonic acid derivative, produced root tip distortion and hindered root extension. It was discovered that the impact is dose-dependent [63]. This was linked to a disruption of the auxin transport network in root tips, irregularities in cell cycles in the meristematic zone of the root, and decreased internal reactive oxygen species of the cell. The seed vegetative growth was unaffected by the fullerene malonic acid derivative, which was ascribed to the shielding functions of the seed coat in A. thaliana. It is essential to mention that not all functionalized fullerenes can enhance plant development. The carboxy-fullerenes, for example, have a phytotoxic effect which is linked with the production of oxidative stress and cell wall spoilage in N. tobacum in vitro cell cultures. Pre-treating O. Sativa seeds with fullerene-Natural Organic Matter affected CNM absorption, transmission, and translocation. Natural Organic Matter (NOM) was used to increase the movement of C_{70} accumulations the C_{70} was predominantly found mostly in the vascular system of the O. sativa plants, with essentially none in the tissues [62].

Emerging Nanomaterials and Their Impact on Society in the 21st Century   Materials Research Forum LLC
Materials Research Foundations 135 (2023) 40-71   https://doi.org/10.21741/9781644902172-3

There were no impacts or negative effects of C_{60} on plant growth in a vast number of study studies centred on the impact of C_{60} on terrestrial and marine plants. The incorporation of fullerene nanoparticles into the growth media improved the plant's absorption of industrial effluents (trichloroethylene) in a Populus deltoides fragment produced under hydroponic conditions. Lemna gibba's chlorophyll deposition and proliferation were shown to be inhibited by fullerenes. Also, there was a reduction in biomass growth when grassland plants were cultivated in soil amended using fullerene, which was utilised to immobilise pesticide residues. C_{60} compounded the total pesticide concentration in a crop by 30, 45, and % in C. pepo, G. max, and S. Lycopersicum, correspondingly, in a vermiculite substrate, according to another study. Despite this, fullerenes were found somewhere at the surface and in the root tissue. Fullerene C60 absorption and buildup in the root tissue, as well as translocation of lower quantities in the stems and leaves of Raphanus sativus cultivated hydroponically, were seen in hydroponically grown Raphanus sativus. When seedlings of Tectona grandis were given C_{60}, they responded better to removing nitrogen from high amounts of nitrogen-containing effluent [62].

Plant development was aided by fullerols, which increased biomass output by up to 54%. Grain treatments containing fullerols increased the size, quantity, weight, and ultimate yield of Momordica charantia fruits, as well as the quantity of soluble fibre, charantin, carotenoid, and quercetin in M. charantia fruits. B. napus germination percentage, photosynthesis, and ultimate development were all improved after being exposed to fullerol. Seedling growth, photosynthesis indices, ABA buildup, and above-ground plant biomass were all improved in water-stressed crops, as were amounts of non-enzymatic and enzymatic antioxidants, and guaiacol peroxidase, which reduced oxidative damages [62]. Endogenous impacts of polyhydroxy fullerene-fullerol with seedlings priming or foliar application on leaves, on the other hand, were efficient in promoting plant development and improving drought resistance in B. napus L. Polyhydroxy fullerene shows promise as a root development stimulant in barley seedlings, based on its good growth findings. The development of cell walls in the developmental regions was aided via using modest amounts of PHF. Nevertheless, due to the sheer potential for Polyhydroxy fullerene to cause injury owing to its significant build-up in the root tissues, its usage has to be limited. Polyhydroxy fullerene appears to have a stress-relieving effect by activating the antioxidant system [64].

On O. Sativa panicles, the impacts of C_{60} were examined to those of Copper, Cadmium, and Lead nanoparticles. C_{60} concentrations were discovered to be considerably greater in the roots and shoots of O. Sativa than in metallic nanoparticles. Nevertheless, metal ions (Copper, Cadmium, and Lead) were deposited in a considerably greater quantity in the panicles than in C_{60}. Fullerol were shown to be able to permeate root growth tissues and leaves, in which the nanoparticles bond to water in separate cellular compartments. The fullerol water-absorbing action revealed that they may be useful in a variety of crops.

The application of fullerol nanoparticles has been shown to enhance above-ground biomass, sprouting rate, water stress, fruit size, ultimate yield, and increased fruit. The amount of ABA and photosynthetic variables, including the levels of enzymatic and non-enzymatic antioxidants, were all raised in the investigated plants when a suitable amount of FNPs was given. Even though low levels of fullerenes were found to improve root development in various plants, C_{60} fullerenes had a detrimental influence on plant development. additionally, adding the right quantity of fullerenes to the phytoremediation mechanism can boost the plants' nitrification and organic pollutant removal capacity. As a result, fullerenes appear to be a promising choice for use in a phytoremediation process [62].

11.3   Effects of graphene on plants

Graphene is a member of the 2D carbon nanomaterial family, and it has tremendous promise in a wide spectrum of uses because of its unique properties. Seeds of O. Sativa grown on MS media enriched with graphene had a richer stem and root systems than reference seeds. Also, tomato seeds treated with graphene powder showed that there was an increase in the rate of germination of the seeds as compared to the untreated seeds. These results have shown that graphene has a positive effect on the sprouting of seeds, implying that graphene could aid the uptake of water via piercing and rupturing the husks of seedlings. This resulted in improved seedling growth and a significant prevalence of seed germination [65].

On the other hand, when the concentration of graphene was raised, it had an inhibiting impact on the O. Sativa seed sprouting rate. According to another research, graphene had a negative impact on T. aestivum growth and the development of plants. Graphene's phytotoxicity on plants may result in a combination of photosynthetic suppression, root hair loss, oxidative stress and nutrient homeostasis imbalance. Nanoparticles of Sulfonated graphene induced a hormesis effect in the seeds of Z. mays, resulting in increased plant height. The build-up of polychlorinated biphenyls in the roots of Z. mays seedlings was shown to be greatly increased by nanoparticles of sulfonated graphene but decreased in the leaves and stems.

In conclusion, the exposed plants displayed concentration-dependent impacts after being exposed to graphene. Although the capacity of graphene to carry water increases the sprouting rate, larger quantities have been found to have deleterious consequences, and the phytotoxicity of plants is influenced by the period of their contact with graphene [62].

11.4   Effects of carbon nanoparticles on plants

The Electrical, thermal conductivity, and mechanical characteristics of carbon nanoparticles are exceptional. Because the carbon nanoparticles are made entirely of carbon, they have excellent stability, minimal toxicity, an environmentally friendly nature, and conductance [62]. When varying amounts of carbon nanoparticles were applied to N. tabacumL. plants, they grew faster at distinct steps than when normal fertilisers were used. This implies that carbon nanoparticles and their use on the components of N. tabacumplant had higher levels of potassium and Nitrogen. Also, the effect of water-soluble carbon nanoparticles on Triticum aestivum was studied at different doses. The water-soluble carbon nanoparticles were extracted from naturally produced raw carbon nanoparticles found in charcoal and had an average of 20% carboxylic groups affixed. In another study using Arabidopsis thaliana, the impacts of carbon nanoparticles on photomorphogenesis and flowering time were studied. By changing both phytochrome B and photoperiod-dependent channels, the carbon nanoparticles were absorbed by the plant, deposited in leaf tissues, and triggered early blooming [66].

The use of carbon nanoparticles boosted the growth of plants and stimulated the uptake and deposition of both micro and macro elements, boosting fertiliser effectiveness and crop viability. Hence, water-soluble carbon nanoparticles may be a better solution than manure or fertiliser because of the crops' gradual and regulated uptake of nutrients for improved digestion [66].

11.5   Effects of graphene oxide on plants

Using water-soluble graphene nanomaterials in crops can result in increased reactive oxygen species release (ROS). In the seedling phases of vegetables such as red spinach, cabbage, lettuce and tomatoes, reactive oxygen species emission has been reported as a key factor in developmental delays. The administration of graphene oxide caused necrotic lesions and cellular damage, which was linked to the build-up of oxidative stress due to increased reactive oxygen species release production. Plant growth parameters such as shoot weight, area of the leaf, root weight, biomass underground and above the ground, and reduction in the number of leaves were all affected by the graphene oxide. Also, the absorption of graphene oxide by the roots of V. faba L. seedlings showed a favourable impact when graphene oxide was added to the nutrient solution. This was accompanied by considerable vegetative growth, as seen by lower H_2O_2 levels and lipid and protein oxidation. At high graphene oxide concentrations, a substantial reduction in plant development indices was found in *V. faba L.*, along with induced oxidative stress and an elevation in the leakage of electrolytes, both of which were deleterious to the development of the seeds [67].

Nanoparticles of graphene oxide may have a favourable influence on germination rate, seedling development, and root development in a wide range of crop varieties at low concentrations. For most plants, however, higher levels of graphene oxide have negative consequences. Also, graphene oxide nanosheets with an average thickness of 1–2 nm have been shown to have greater favourable impacts on plant growth and development in terms of mechanism [68].

11.6   Effects of mesoporous carbon nanoparticles on plants

Mesoporous carbon nanoparticles produced using metal-organic frameworks offer a wide range of applications. This includes catalysis, energy storage, drug administration, and photodynamic treatment. This could be attributed to their remarkable features such as large surface area, stable mesoporous structure, and outstanding catalytic activity [69].

According to new reports, mesoporous carbon nanoparticles had damaging consequences on O. sativa by altering the concentrations of key phytohormones. The particle size of the nanoparticle affected the generated damage caused by mesoporous carbon nanoparticle administration. These observations indicate the possible danger of mesoporous carbon nanoparticles to crop species, as well as the health risks to humans and livestock [68].

11.7   Effects of fluorescent carbon dots and carbon nanodots on plants

Carbon dots are small carbon nanoparticles with consistent surface modification due to a variety of functions, making them distinctive fluorescent carbonic nanomaterials. They possess several distinguishing characteristics, including ease of manipulation, excellent water solubility, low cytotoxicity, chemical stability, and excellent biocompatibility. Carbon nanodots have long been employed in biological imaging as well as other biosensor techniques, but their impacts on crops have been studied very sparingly. Following ten days of exposure to water-soluble carbon nanodots, the root development of wheat was 10 times that of the control. Water-soluble fluorescent nanodots were inserted into N. tabacum leaves directly. As a result, the rate of photosynthesis of N. tabacum leaves increased by 18%, assisting the fluorescent nanodots in converting ultraviolet to blue light. In another report, treatment with various doses of carbon nanodots induced favourable physical effects in V. radiata plants while causing no phytotoxicity. The fluorescent CDs were conveyed by the circulatory system from the roots to the stems and eventually to the leaves through the apoplastic route, according to confocal microscopy findings [69].

These investigations revealed that water-soluble carbon nanodots may readily pass natural membranes in juvenile plant seedlings, allowing them to transfer additional water and nutrients ahead and grow and mature faster. Carbon nanodots may be used in plant cell and tissue imaging as a result of these studies. However, because the underlying basis of the CNDs' oxidative stress has yet to be examined, further research into their potential effects on other plant development, product safety, and the environment is still needed [70].

Conclusion and future prospects

Because of their remarkable features such as thermal, structural, electrical and mechanical capabilities, substantial initiatives have been undertaken to use carbon nanomaterials in specific applications. As the requirement for safe drinking water has risen dramatically in recent decades, the efforts taken to meet that need have also become more difficult. Due to their high feedstock capability, reduced manufacturing cost, and greater performance, a wide range of carbon-based nanomaterials appear to be one of the most viable options for removing any form of impurities from the water supply. The use of these nanoparticles for water purification has been tested in the laboratory and has shown to be successful. The primary problem here would be to apply this laboratory-scale approach to real-world applications, as nanomaterials' suitability for commercialization differs greatly. A step-by-step approach to this difficulty might be used. Carbon nanoparticles have greatly aided the creation and advancement of electrochemical sensors and biosensors due to their distinctive features. In comparison to traditional non-nanostructured electrochemical systems, both innovative and improved carbon-based probes frequently exhibit improved analytical efficiency. Electrochemical approaches employing carbon-enhanced electrodes in sensing and biosensing instruments are shown potential for use in real-world analytical testing. Carbon nanotubes especially diamonds have been used as components in numerous analytes for sensing (electrochemical). Also, plant scientists and researchers have been paying close attention to various carbon nanomaterials, such as fullerenes, fullerols, single-walled carbon nanotubes, and multi-walled carbon nanotubes for their potential uses in controlling plant development. The beneficial impacts of multi-walled carbon nanotubes on legume plants may offer a solution to rhizobial infestation concerns and poor nitrogenase performance. Solving these challenges might boost bean crop yields. Likewise, Carbon dots might also be used as a fantastic tool for visualising host tissue and obtaining vital phytotoxic data and a type of fertiliser in farming, opening up new nanobiotechnology possibilities. However, there is presently a paucity of knowledge about the molecular mechanisms that underlie the negative and positive impacts of carbon nanomaterials in plants. As a result, it's impossible to say which of the carbon nanomaterials' unique features are linked to the production of genes and proteins required for plant development. Therefore, future studies must concentrate on carbon nanomaterials-induced crop poisoning, carbon nanomaterials transport within the plant organs, and the impact of carbon nanomaterials' Physico-chemical characteristics in the root system. Critical developments in the production and uses of carbon-based nanomaterials, as well as the link between this knowledge and other nanostructures, have been described before. Current work in these domains is aimed at delivering scientific and technological breakthroughs to dimensions where they may be used in manufacturing and applications. Simultaneously with synthesis, modelling and computational studies have progressed toward a coherent conceptual basis for the formation of carbon nanomaterials, and the current study is linking the most recent technological development and understanding acquired from experimental studies to further steer this modelling approach ahead.

References

[1] G. Zhan, J. Huang, M. Du, I. Abdul-Rauf, Y. Ma, and Q. Li, "Green synthesis of Au-Pd bimetallic nanoparticles: Single-step bioreduction method with plant extract," Mater. Lett. 65 (2011) 2989-2991. https://doi.org/10.1016/j.matlet.2011.06.079

[2] T. Rasheed, M. Bilal, F. Nabeel, M. Adeel, and H. M. N. Iqbal, "Environmentally-related contaminants of high concern: Potential sources and analytical modalities for detection, quantification, and treatment," Environment International. 122 (2019) 52-66. https://doi.org/10.1016/j.envint.2018.11.038

[3] M. Q. Fan, F. Xu, and L. X. Sun, "Studies on hydrogen generation characteristics of hydrolysis of the ball milling Al-based materials in pure water," Int. J. Hydrogen Energy, (2007) https://doi.org/10.1016/j.ijhydene.2006.12.020

[4] M. S. Bakshi, "How Surfactants Control Crystal Growth of Nanomaterials," Cryst. Growth Des., vol. 16, no. 2, pp. 1104-1133, 2016, doi: 10.1021/acs.cgd.5b01465. https://doi.org/10.1021/acs.cgd.5b01465

[5] K. Mukhopadhyay, A. Ghosh, S. K. Das, B. Show, P. Sasikumar, and U. Chand Ghosh, "Synthesis and characterisation of cerium(IV)-incorporated hydrous iron(III) oxide as an adsorbent for fluoride removal from water," RSC Adv.7 (2017) 26037-26051 https://doi.org/10.1039/C7RA00265C

[6] W. Zhai, N. Srikanth, L. B. Kong, and K. Zhou, "Carbon nanomaterials in tribology," Carbon N. Y. 119 (2017) 150-171 https://doi.org/10.1016/j.carbon.2017.04.027

[7] V. L. Pandey, S. Mahendra Dev, and U. Jayachandran, "Impact of agricultural interventions on the nutritional status in South Asia: A review," Food Policy. 2016. https://doi.org/10.1016/j.foodpol.2016.05.002

[8] M. Asase, E. K. Yanful, M. Mensah, J. Stanford, and S. Amponsah, "Comparison of municipal solid waste management systems in Canada and Ghana: A case study of the cities of London, Ontario, and Kumasi, Ghana," Waste Management. 2009. https://doi.org/10.1016/j.wasman.2009.06.019

[9] K. RI, "No TitleEΛΕNH," UU Kesehat. 2009, no. 1, 2009.

[10] A. M. Pereyra, M. R. Gonzalez, V. G. Rosato, and E. I. Basaldella, "A-type zeolite containing Ag+/Zn2+ as an inorganic antifungal for waterborne coating formulations," Prog. Org. Coatings, 2014 https://doi.org/10.1016/j.porgcoat.2013.09.008

[11] M. Yadav, S. Chatterji, S. K. Gupta, and G. Watal, "Innovare Academic Sciences PRELIMINARY PHYTOCHEMICAL SCREENING OF SIX MEDICINAL PLANTS USED IN TRADITIONAL MEDICINE," vol. 6, no. 5, 2014.

[12] Y. Long et al., "Flowering time quantitative trait loci analysis of oilseed Brassica in multiple environments and genomewide alignment with Arabidopsis," Genetics, 177 (2007) 2433-2444 https://doi.org/10.1534/genetics.107.080705

[13] R. K. Thines, N. M. Mubarak, S. Nizamuddin, J. N. Sahu, E. C. Abdullah, and P. Ganesan, "Application potential of carbon nanomaterials in water and wastewater treatment: A review," J. Taiwan Inst. Chem. Eng 72 (2017) 116-133 https://doi.org/10.1016/j.jtice.2017.01.018

[14] A. Keshav Krishna and K. Rama Mohan, "Distribution, correlation, ecological and health risk assessment of heavy metal contamination in surface soils around an industrial area, Hyderabad, India," Environ. Earth Sci., 2016, https://doi.org/10.1007/s12665-015-5151-7

[15] U. Kumar, M. Kumar, S. Sankar, and R. Kishore, "Removal of As (V) from aqueous solution by Ce-Fe bimetal mixed oxide Journal of Environmental Chemical Engineering Removal of As (V) from aqueous solution by Ce-Fe bimetal mixed oxide," Biochem. Pharmacol. 4 (2016) 2892-2899 https://doi.org/10.1016/j.jece.2016.05.041

[16] S. Sachan and S. Chaurasia, "Bio-medical waste and its management," Indian J. Environ. Prot. 26 (2006) 255-259

[17] M. F. Ourique, P. V. F. Sousa, A. F. Oliveira, and R. P. Lopes, "Comparative study of the direct black removal by Fe, Cu, and Fe/Cu nanoparticles," Environ. Sci. Pollut. Res., 235 (2018) 749-756

[18] J. Yadav, "Fullerene : Properties, Synthesis and Application," vol. 6 (2017) 1-6.

[19] H. W. Kroto, J. R. Heath, S. C. O'Brien, R. F. Curl, and R. E. Smalley, "Kroto Et Al 1985," Nature, vol. 318, pp. 162-163. https://doi.org/10.1038/318162a0

[20] 杨洋等 et al., "Research on the Complete Set of Polysaccharide Extracting Equipment from Edible Fungi," Nanoscale, 3(1) (2022) 95-120

[21] T. Sugai, H. Omote, S. Bandow, N. Tanaka, and H. Shinohara, "Production of fullerenes and single-wall carbon nanotubes by high-temperature pulsed arc discharge," J. Chem. Phys. 112, (2000) 6000-6005 https://doi.org/10.1063/1.481172

[22] Z. M. Dang, M. S. Zheng, and J. W. Zha, "1D/2D Carbon Nanomaterial-Polymer Dielectric Composites with High Permittivity for Power Energy Storage Applications," Small, 12, (2016) 1688-1701, https://doi.org/10.1002/smll.201503193

[23] P. Pandey and M. Dahiya, "Carbon nanotubes: Types, methods of preparation and applications," Int. J. Pharm. Sci. Res., 6 (2016) 15-21.

[24] H. Jin, D. A. Heller, and M. S. Strano, "Single-particle tracking of endocytosis and exocytosis of single-walled carbon nanotubes in NIH-3T3 cells," Nano Lett., 8 (2008) 1577-1585 https://doi.org/10.1021/nl072969s

[25] R. P. Feazell, N. Nakayama-Ratchford, H. Dai, and S. J. Lippard, "Soluble single-walled carbon nanotubes as longboat delivery systems for platinum(IV) anticancer drug design," J. Am. Chem. Soc., 129 (2007) 8438-8439 https://doi.org/10.1021/ja073231f

[26] L. Thi Mai Hoa, "Characterization of multi-walled carbon nanotubes functionalized by a mixture of HNO3/H2SO4," Diam. Relat. Mater., 89 (2018) 43-51 https://doi.org/10.1016/j.diamond.2018.08.008

[27] T. Arunkumar, R. Karthikeyan, R. Ram Subramani, K. Viswanathan, and M. Anish, "Synthesis and characterisation of multi-walled carbon nanotubes (MWCNTs)," Int. J. Ambient Energy, 41 (2020) 452-456 https://doi.org/10.1080/01430750.2018.1472657

[28] M. Endo, T. Hayashi, Y. A. Kim, and H. Muramatsu, "Development and application of carbon nanotubes," Japanese J. Appl. Physics, Part 1 Regul. Pap. Short Notes Rev. Pap., 45, (2006) 4883-4892 https://doi.org/10.1143/JJAP.45.4883

[29] A. Naqi, N. Abbas, N. Zahra, A. Hussain, and S. Q. Shabbir, "Effect of multi-walled carbon nanotubes (MWCNTs) on the strength development of cementitious materials," J. Mater. Res. Technol. 8, (2019) 1203-1211 https://doi.org/10.1016/j.jmrt.2018.09.006

[30] T. Basu, K. Gupta, and U. C. Ghosh, "Performances of As(V) adsorption of calcined (250°C) synthetic iron(III)-aluminum(III) mixed oxide in the presence of some

groundwater occurring ions," Chem. Eng. J., 2012
https://doi.org/10.1016/j.cej.2011.12.083

[31] S. R. Mishra and M. Ahmaruzzaman, "Cerium oxide and its nanocomposites:
Structure, synthesis, and wastewater treatment applications," Mater. Today Commun.,
28 (2021) 102562 https://doi.org/10.1016/j.mtcomm.2021.102562

[32] T. P. Wagner, "Shared responsibility for managing electronic waste: A case study of
Maine, USA," Waste Manag., 2009 https://doi.org/10.1016/j.wasman.2009.06.015

[33] R. Soni, "Urban Solid Waste Management In India," SSRN Electron. J., (2019)
https://doi.org/10.2139/ssrn.3463653

[34] Q. Qin, L. He, C. Li, and Y. Zang, "Research on Springback Defects in Incremental
Forming of Cu-Al Bimetal," Zhongguo Jixie Gongcheng/China Mech. Eng., (2021) 2-
46

[35] L. Basso, M. Cazzanelli, M. Orlandi, and A. Miotello, "Nanodiamonds: Synthesis
and application in sensing, catalysis, and the possible connection with some processes
occurring in space," Appl. Sci., 10 (2020) 1-28 https://doi.org/10.3390/app10124094

[36] Y. H. Tee, E. Grulke, and D. Bhattacharyya, "Role of Ni/Fe nanoparticle
composition on the degradation of trichloroethylene from water," Ind. Eng. Chem.
Res., 2005 https://doi.org/10.1021/ie050086a

[37] H. He, L. A. Pham-Huy, P. Dramou, D. Xiao, P. Zuo, and C. Pham-Huy, "Carbon
nanotubes: Applications in pharmacy and medicine," Biomed Res. Int.7 (2013)
https://doi.org/10.1155/2013/578290

[38] F. H. Hussein, B. S. Hasan, M. B. Mageed, Z. H. Nafaee, and G. J. Mohammed,
"Pharmaceutical Application of Carbon Nanotubes Synthesized by Flame Fragments
Deposition Method," J. Environ. Anal. Chem. 4 (2017) 1-2
https://doi.org/10.4172/2380-2391.1000e115

[39] S. P. Gautam, "Carbon Nanotubes: Exploring Intrinsic Medicinal Activities and
Biomedical Applications," Open Access J. Oncol. Med. 3 (2019) 230-232
https://doi.org/10.32474/OAJOM.2019.03.000152

[40] M. Thiruvengadam et al., "Recent insights and multifactorial applications of carbon
nanotubes," Micromachines., 12 (2021) 243-270 https://doi.org/10.3390/mi12121502

[41] B. Adsorption and L. K. Shrestha, "High Surface Area Nanoporous Activated
Carbons Materials," 2022.

[42] A. K. Prajapati and M. K. Mondal, "Hazardous As(III) removal using nanoporous
activated carbon of waste garlic stem as adsorbent: Kinetic and mass transfer
mechanisms," Korean J. Chem. Eng., 36 (2019) 1900-1914
https://doi.org/10.1007/s11814-019-0376-x

[43] H. Xu et al., "Nanoporous activated carbon derived from rice husk for high-performance supercapacitor," J. Nanomater., 4 (2014) 1-7 https://doi.org/10.1155/2014/714010

[44] S. Pathan, N. Pandita, and N. Kishore, "Acid functionalized-nanoporous carbon/MnO2 composite for removal of arsenic from the aqueous medium," Arab. J. Chem., 12 (2019) 5200-5211 https://doi.org/10.1016/j.arabjc.2016.12.011

[45] I. Ali, O. M. L. Alharbi, A. Tkachev, E. Galunin, A. Burakov, and V. A. Grachev, "Water treatment by new-generation graphene materials: hope for bright future," Environ. Sci. Pollut. Res., 25 (8), (2018) 7315-7329 https://doi.org/10.1007/s11356-018-1315-9

[46] P. B. Pawar, S. Saxena, D. K. Badhe, R. P. Chaudhary, and S. Shukla, "3D oxidized graphene frameworks for efficient nano sieving," Sci. Rep., 6 (2016) 345-370 https://doi.org/10.1038/srep21150

[47] M. R. Gandhi, S. Vasudevan, A. Shibayama, and M. Yamada, "Graphene and Graphene-Based Composites: A Rising Star in Water Purification - A Comprehensive Overview," ChemistrySelect, vol. 1(15), (2016) 4358-4385, https://doi.org/10.1002/slct.201600693

[48] Z. Wang, D. Shen, C. Wu, and S. Gu, "State-of-the-art on the production and application of carbon nanomaterials from biomass," Green Chem., vol. 20, no. 22, pp. 5031-5057, 2018, doi: 10.1039/c8gc01748d. https://doi.org/10.1039/C8GC01748D

[49] S. Khaliha et al., "Defective graphene nanosheets for drinking water purification: Adsorption mechanism, performance, and recovery," FlatChem, vol. 29, no. July, p. 100283, 2021, doi: 10.1016/j.flatc.2021.100283. https://doi.org/10.1016/j.flatc.2021.100283

[50] K. P. Loh, D. Ho, G. N. C. Chiu, D. T. Leong, G. Pastorin, and E. K. H. Chow, "Clinical Applications of Carbon Nanomaterials in Diagnostics and Therapy," Adv. Mater., vol. 30, no. 47, pp. 1-21, 2018, doi: 10.1002/adma.201802368. https://doi.org/10.1002/adma.201802368

[51] D. Passeri et al., "Biomedical applications of nanodiamonds: An overview," J. Nanosci. Nanotechnol., vol. 15, no. 2, pp. 972-988, 2015, doi: 10.1166/jnn.2015.9734. https://doi.org/10.1166/jnn.2015.9734

[52] E. Perevedentseva, Y. C. Lin, M. Jani, and C. L. Cheng, "Biomedical applications of nanodiamonds in imaging and therapy," Nanomedicine, vol. 8, no. 12, pp. 2041-2060, 2013, doi: 10.2217/nnm.13.183. https://doi.org/10.2217/nnm.13.183

[53] N. Prabhakar and J. M. Rosenholm, "Nanodiamonds for advanced optical bioimaging and beyond," Curr. Opin. Colloid Interface Sci., vol. 39, no. March, pp. 220-231, 2019, doi: 10.1016/j.cocis.2019.02.014. https://doi.org/10.1016/j.cocis.2019.02.014

[54] A. I. Shames et al., "Magnetic Resonance Study of Nanodiamonds," Synth. Prop. Appl. Ultrananocrystalline Diam., pp. 271-282, 2005, doi: 10.1007/1-4020-3322-2_21. https://doi.org/10.1007/1-4020-3322-2_21

[55] B. Pei, W. Wang, N. Dunne, and X. Li, Applications of carbon nanotubes in bone tissue regeneration and engineering: Superiority, concerns, current advancements, and prospects, vol. 9, no. 10. 2019. doi: 10.3390/nano9101501. https://doi.org/10.3390/nano9101501

[56] L. G. Delogu et al., "Ex vivo impact of functionalized carbon nanotubes on human immune cells," Nanomedicine, vol. 7, no. 2, pp. 231-243, 2012, doi: 10.2217/nnm.11.101. https://doi.org/10.2217/nnm.11.101

[57] R. L. Garnica-Gutiérrez et al., "Effect of functionalized carbon nanotubes and their citric acid polymerization on mesenchymal stem cells in vitro," J. Nanomater., vol. 2018, 2018, doi: 10.1155/2018/5206093. https://doi.org/10.1155/2018/5206093

[58] J. Huang, Y. Liu, and T. You, "Carbon nanofiber-based electrochemical biosensors: A review," Anal. Methods, vol. 2, no. 3, pp. 202-211, 2010, doi: 10.1039/b9ay00312f. https://doi.org/10.1039/b9ay00312f

[59] C. Nanotube, C. N. T. Biosensors, D. C. Ferrier, and K. C. Honeychurch, "Carbon Nanotube (CNT)-Based Biosensors," pp. 1-33, 2021.

[60] V. Vamvakaki and N. A. Chaniotakis, "Carbon nanostructures as transducers in biosensors," Sensors Actuators, B Chem., vol. 126, (2007) 193-197 https://doi.org/10.1016/j.snb.2006.11.042

[61] K. Balasubramanian and M. Burghard, "Biosensors based on carbon nanotubes," Anal. Bioanal. Chem., 385 (2006) 452-468 https://doi.org/10.1007/s00216-006-0314-8

[62] S. K. Verma, A. K. Das, S. Gantait, V. Kumar, and E. Gurel, "Applications of carbon nanomaterials in the plant system: A perspective view on the pros and cons," Sci. Total Environ., 667 (2019) 485-499 https://doi.org/10.1016/j.scitotenv.2019.02.409

[63] R. Szőllősi, Á. Molnár, S. Kondak, and Z. Kolbert, "Dual effect of nanomaterials on germination and seedling growth: Stimulation vs. phytotoxicity," Plants, (2020) 1-30 https://doi.org/10.3390/plants9121745

[64] C. Science, N. Crop, and D. Submitted, Analysis of phytotoxicity and plant growth stimulation by multi-walled carbon nanotubes; Nanomaterials,11(10) (2021) 678-690

[65] S. Park, K. S. Choi, S. Kim, Y. Gwon, and J. Kim, "Graphene oxide-assisted promotion of plant growth and stability," Nanomaterials, 10 (2020) 1-11 https://doi.org/10.3390/nano10040758

[66] F. Zhao et al., "Use of carbon nanoparticles to improve soil fertility, crop growth and nutrient uptake by corn (Zea mays l.)," Nanomaterials,11(10) (2021) 678-690 https://doi.org/10.3390/nano11102717

[67] S. Park et al., "Graphene : A new technology for agriculture Grafeno : Uma nova tecnologia para a agricultura Grafeno : Una nueva tecnología para la agricultura," Chem. Soc. Rev. 10(11), (2021) 1-11

[68] Z. Peng et al., "Advances in the application, toxicity and degradation of carbon nanomaterials in the environment: A review," Environ. Int., 134 (2019)105298 https://doi.org/10.1016/j.envint.2019.105298

[69] Y. Li et al., "A review on the effects of carbon dots in plant systems," Mater. Chem. Front., 4(2), (2020) 437-448 https://doi.org/10.1039/C9QM00614A

[70] N. E. Mahmoud and R. M. Abdelhameed, "Superiority of modified graphene oxide for enhancing the growth, yield, and antioxidant potential of pearl millet (Pennisetum glaucum L.) under salt stress," Plant Stress, 2 (2021) 100025 https://doi.org/10.1016/j.stress.2021.100025

Chapter 4

Functionalized Carbon Nanomaterials: Fabrication, Properties, and Applications

P.R. Bhilkar[1], R.S. Madankar[1], T.S. Shrirame[1,2], R.D. Utane[4], A.K. Potbhare[1], S. Yerpude[3], S.R. Thakare[5], Sarita Rai[6], A.A. Abdala[7*], and Ratiram G. Chaudhary[1*]

[1]Post Graduate Department of Chemistry, S. K. Porwal College, Kamptee-441001, India

[2]Department of Chemistry, Shri Shivaji Science and Arts, College, Chikhali, India

[3]Post Graduate Department of Microbiology, S. K. Porwal College, Kamptee-441001, India

[4]Department of Chemistry, Sant Gadge Maharaj Mahavidyalaya, Hingna, India

[5]Post Graduate Department of Chemistry, Govt. Institute of Forensic Science, Nagpur, India

[6]Department of Chemistry, Dr.H S Gour University, Sagar (MP) India

[7]Chemical Engineering Program, Texas A&M University at Qatar, Doha, Qatar

* chaudhary_rati@yahoo.com; ahmed.abdala@qatar.tamu.edu

Abstract

In recent times, carbon based nanomaterials (CNMs) like graphene, graphene oxide (GO), graphene nanoplatelet (GNP), carbon nanotubes (CNTs), carbon dots (CDs), and fullerenes are among the most promising nanomaterial due to their extraordinary physiochemical, electrical, optical, mechanical, and thermal behavior. CNMs have many potential applications, including energy storage and conversion, biomedical, catalysts, composites, and biomaterials. Nonetheless, many advanced applications require the proper functionalization of CNMs. This chapter analyses the routes for CNMs functionalization and their impact on their structure and properties. Furthermore, the current and potential application of functionalized CNMs are discussed, and the challenges and future research directions are highlighted.

Keywords

Functionalized CNMs, Carbon Nanotubes, Carbon Dots, Graphene Oxide, Potential Applications of CNMs

Contents

Functionalized Carbon Nanomaterials: Fabrication, Properties, and Applications 72

1.   Introduction..73

2.   Historical background...74

3.   Fabrication of functionalized CNMs 76

    3.1   Fabrication and functionalization of carbon nanotubes
           (CNTs) 77

    3.2   Fabrication and Functionalization of Expanded grapite
           (EG) 79

    3.3   Fabrication and functionalization of carbon dots (CDs) 79

    3.4   Fabrication and functionalization of graphene and graphene
           oxide (GO)..80

4.   Properties of functionalized CNMs...81

    4.1   Carbon nanotubes (CNTs)...82

    4.2   Expanded graphite (EG) 82

    4.3   Carbon dots (CDs) 83

    4.4   Graphene and graphene oxide (GO)...83

5.   Applications of functionalized CNMs...84

    5.1   Carbon nanotubes (CNTs)...85

    5.2   Expanded graphite (EG) 87

    5.3   Carbon dots (CDs) 89

    5.4   Graphene and graphene oxide (GO)..90

Conclusions and futuristic approach 91

References 91

1.   Introduction

Various carbon nanomaterials (CNMs) have been widely explored due to their significant properties as listed in terms of electrical conductivity, chemically stable at any pH, wider surface area for adsorption, aspect ratio, and many more. CNMs are categorised as a) zero-dimensional viz. fullerene, carbon spheres; b) one-dimensional, *viz.* carbon nanotubes (CNTs), carbon fibers (CFs); c) two-dimensional, *viz.* graphene, carbon films; d) three-dimensional, *viz.* carbon sponges, carbon foam, carbon cloth, and so on as shown in Fig. 1[1,2]. The functionalization of CNMs to enhance their properties and potential applications is a hot research area in the field of nanotechnology [3,4]. Generally, atomic or interfacial interactions between CNMs with other chemical species supremacy the properties of CNMs. CNMs have been designed and functionalized at the nanoscale level [5,7].

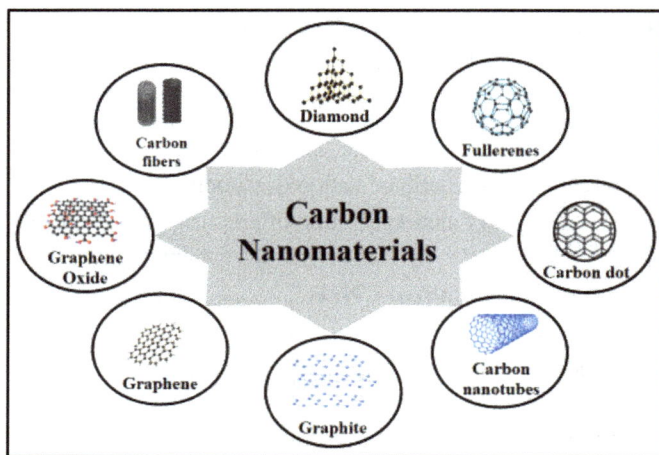

Figure 1. Different forms of CNMs

Functionalization of CNMs *via* chemical modification or structural regulation, or integration with inorganic or organic species extends further development of CMNs. Functionalization of CNMs using the methods described previously enhances their stability, surface area, and surface activity and provides molecular-level control. Functionalized CNMs (FCNMs) can be produced with a defined morphology and optimized properties to meet applications requirements, viz., producing fuels using solar energy [8], switches governed by light [9], heat-conducting systems [10], metal ion containing batteries [11], and supercapacitors [12] etc.

This chapter epitomizes the recent progress in the fabrication and physiochemical properties of functionalized CNMs. Moreover, their potential applications in the various field are also discussed.

2.   Historical background

In nanotechnology, functionalized CNMs have gotten more attention in the last few years. Earlier, only carbon is renowned for occurring in two naturally occurring allotropes, graphite, and diamond. These two allotropes differ chemically and morphologically from each other; hence, they exhibit different physical properties. Carbon was eventually discovered in other forms, including graphene, fullerenes, CNTs, carbon dots, etc. (Fig. 2).

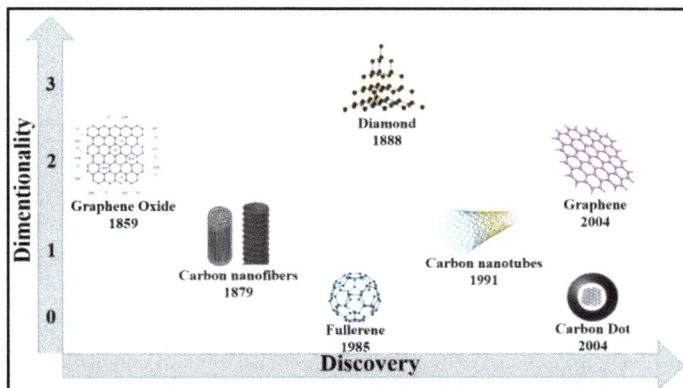

Figure 2. Dimensionality and discovery time of CNMs

All these allotropes are chemically indistinguishable but differ significantly in their physical properties [13-16]. The study of this carbon allotrope was started in 1789 by Abraham Gottlob Werner, who discovered graphite stacking flat sp^2-carbon layers. This graphite is a superb electron carrier because of the delocalization of π-electrons between sp^2-carbon [17]. After the structural identification and knowing its potential application, at the beginning of the 20th-century, graphite seemed to be the material of concentrated study. Around 1945s, graphite's physicochemical properties were studied both theoretically and experimentally. In 1896, American scientist Edward Acheson prepared the first synthetic graphite. Graphite is a flat network of sp^2 carbon stacks parallel to each other and, once it doped with metals/non-metals exhibits tremendous thermal, electrical conductivity, and chemical inertness [18,19].

Similar to graphite, diamond is also known from ancient times. Though, it is used mainly for decorative purposes. Diamond possesses the highest thermal conductivity, approximately 2200 W/mK. The atoms are held together very firmly due to presence of strong covalent bonds. Because diamond is highly thermally conductive, it is usually used in the semiconductor industry to avoid overheating silicon and other semiconducting materials [20].

The 2D-graphene having a single-layer was isolate from graphite by Geim and Novoselov in 2004. Since then, graphene has become the most researched material because of its excellent and diverse physiochemical properties with numerous potential applications [21]. Graphene is a major structural component of most CNMs (Fig. 3). Lately, the graphene/graphene-related materials have been rapidly developed to due to commercial utility globally. Primarily, graphene generates more attention in the fabrication of other

Emerging Nanomaterials and Their Impact on Society in the 21st Century   Materials Research Forum LLC
Materials Research Foundations 135 (2023) 72-99   https://doi.org/10.21741/9781644902172-4

CNMs, *viz.*, fullerenes or CNTs, making a strong base for the speedy growth of graphene-based CNMs [22].

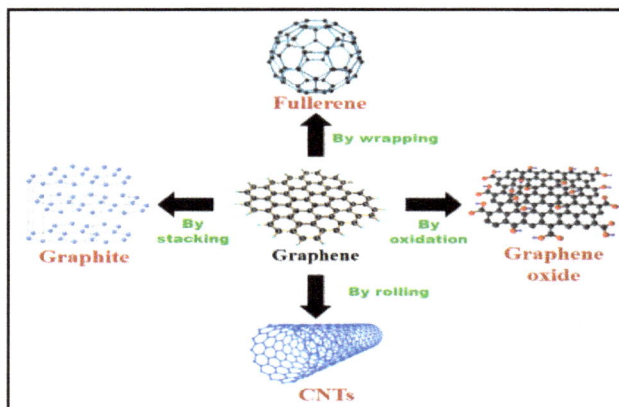

Figure 3. Graphene as the building block of other CNMs

The first report on the discovery of fullerene, various sized and shaped pieces of carbon material resembling a hollow sphere, was created in 1985 by Harry Kroto and Richard Smalley. Fullerene, also called C_{60}, has been studied widely and can be used to synthesize CNTs. This C_{60} (0D) and CNTs (1D) have unlocked a new area in the field of nanoscience material before the discovery of graphene. Both C_{60} and CNTs have an equivalent range of diameters but have variations in their properties and framework. CNTs are the strongest fiber and consist of two coaxial cylinders classified into two main categories: single-walled carbon nanotubes (SWCNTs) and multi-walled carbon nanotubes (MWCNTs). These CNTs have remarkably magnificent mechanical and electronic behavior. Because of the physicochemical properties of CNTs, it can be employed in electrical studies, adsorption studies, and biomedical applications [23, 24].

3.   Fabrication of functionalized CNMs

Despite their flexible structure, variety of dimensional ranges, and tunable surfactant chemistry, CNMs are broadly used in various applications. It is possible to chemically modify CNMs using metals like the 3d series, nitrogen, fluorine, functional groups like hydroxyls and carboxyls, or polymers. Such modifications profoundly affect their solar-energy storage, photoresponse, and electrochemical-energy storage, displaying the range of chemical-functionalization. Here in this par, we briefly describe the fabrication and modification of CNMs such as CNTs, Expanded Graphite (EG), Carbon Dots (CDs), graphene, and Graphene Oxide (GO). Functionalized CNMs such as CNTs, EG, CDs, graphene, and GO can be fabricated as follows:

3.1   Fabrication and functionalization of carbon nanotubes (CNTs)

CNTs, a helical microtubule of graphene, has been one of the most studied CNMs after its discovery in 1991. This microtubule of graphene consists of sp^2 carbon atoms with a diameter approximately equal to 1/50000th of a human hair CNTs classified as SWCNTs and MWCNTs. In contrast, MWCNTs consist of several nanotubes arranged concentrically. The unique 1D covalent tubular structure with excellent properties like mechanical, electronic, thermal conductivity, etc., makes more interest in the fabrication of CNTs. Considering the potential of CNTs and their composites, their preparation and chemical functionalization have been studied comprehensively. There are numerous routes for the fabrication of CNTs and their nanocomposites. Some of them have been mentioned in Fig. 4.

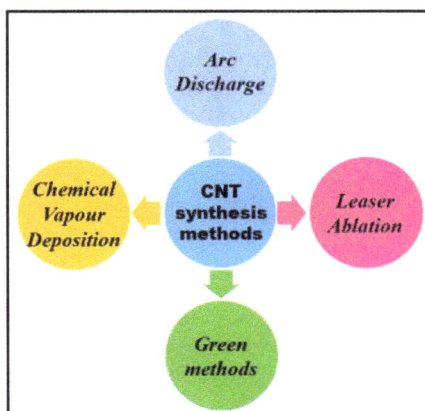

Figure 4. Different routes for the synthesis of CNTs

For the preparation of CNTs, arc discharge is a typical and earliest method. In this method, the gaseous carbon source is electrically broken down at higher temperatures to generate plasma providing growth of CNTs. This method produces fewer morphological defects in CNTs. Arc discharge chamber, two electrodes *viz.,* anode containing a catalyst, and graphite powder, whereas cathode is made up of pure graphite rod [25]. At first, two electrodes were placed under inert gaseous conditions independently. As an electric arc is applied, and a potential drop of 25V is used to produce an electric arc of 50–150 A^0 intensity [26]. The temperature between the two electrodes was set (1700-4000 °C) such that it became red hot, and plasma was formed. Due to the generation of plasma, the carbon source sublimes on the anode and condensed filament-shaped carbon product on the cathode within a very short time. The chamber was cooled in the last terminating reaction, and CNTs were collected and purified.

Usually, chemical vapor deposition (CVD) would be used for the preparation of the negligible amount of impure CNTs on the specific substrate. Concisely, the growth of CNTs can be done on solid support coated by catalyst at a higher temperature. The fabrication of CNTs can utilize various CVD techniques, including plasma enhanced CVD (PECVD), micro-wave plasma (MPECVD) or radio-frequency CVD (RF-CVD), and hot filament CVD (HFCVD), oxygen-mediated CVD, and water-mediated CVD, and floating catalyst (FCCVD). CVD provides a controlled reaction with a high yield and uniformity of the CNTs [27] .To manage the growth and framework of CNTs, the choice of catalytic material coated on a solid support is very important. Transition metal NPs *viz.*, Pt, Fe, Pd, *etc.*, are the effective catalyst used in CVD to fabricate CNTs [28]. In CVD, the sample as a carbon source, *viz.*, CH_4, CO, C_2H_2, *etc.*, was put in a quartz tube and heated at a higher temperature (>600 °C) to undergo sufficient reaction [29]. After this, an inert gas such as argon, nitrogen, etc., the reaction mixture is passed over the catalytic material coated on the solid support. Afterward, carbon precursor decays, CNTs start growing, and thin-film deposed on solid support coated with catalytic material under inert conditions. Lastly, CNTs were collected from the solid support after cooling the system [30]. By adjusting the different parameters such as carbon source, nature of catalytic materials, the flow of reaction mixture and inert gas, temperature and reaction period, framework, and yield of CNTs perchance managed [31].

Moreover, for the fabrication of SWCNTs, leaser ablation, a prominent method can be used. This method provides good yield, quality, and purity of SWCNTs [32]. In this method, Nd: YAG lasers and continuous wave-CO_2 are used. The chamber is made up of quartz placed in the middle of the furnace—the chamber contains a catalyst, a metal-graphite composite employed to synthesize CNTs, particularly SWCNTs. The catalyst is generally made up of a small concentration of metal *viz.*, Co, Ni, Cu, Pt, Co/Ni, Co/Cu, Co/Pt, Ni/Pt, Rh/Pd, and pure graphite form. [59,60] For the preparation of CNTs *via* laser ablation method, a laser is focused on the carbon source *via* a window under high temperature and inert conditions of gases. There is a pressure range of 200 to 400 Torr. Source sublimes allowed to deposit the product in the form of CNTs collected and purified [33]. Along with this method, a green approach is also utilized to fabricate CNTs as it is simple, straightforward, economical, eco-friendly, and harmless compared to other techniques. Green techniques such as sonochemical irradiation or microwave methods can be used to prepare CNTs. Moreover, in some methods, chemical reagents are replaced by natural products extracted from biogenic sources. Most of the methods mainly used a plant extract containing alkaloids, saponin, tannin, flavonoids, terpenoids, etc., to convert carbon sources to fabricate functionalized CNTs [34].

Jafer *et al.* reported fabrication of covalent functionalization of MWCNT with 5,10,15,20-(4-aminophenyl)porphyrinatonickel(II) by amide bond. The functionalized MWCNT is

Emerging Nanomaterials and Their Impact on Society in the 21st Century   Materials Research Forum LLC
Materials Research Foundations 135 (2023) 72-99   https://doi.org/10.21741/9781644902172-4

confirmed with the new peak observed in FT-IR, Raman, and XPS spectra and has enhanced the electronic properties compared to simple MWCNT [35]. Alphatova *et al.* studied a variety of natural and synthetic dispersing agents attached non-covalently to CNT surfaces by various physical adsorption ways enhanced the dispersion of SWCNTs in water. Furthermore, the nature of the dispersant recommended for noncovalent functionalization governs the toxicity of SWCNT suspensions [36].

3.2   Fabrication and Functionalization of Expanded grapite (EG)

EG is a porous and worm-like structure CNMs. EG shows superb flexibility, self-lubricity, corrosion resistivity and thermal conductivity [37, 38]. Using microwaves or high temperatures, graphite intercalation compounds (GICs) are treated chemically and physically to produce EG. For the preparation of EG chemical method is more common. Graphite flakes are treated with oxidizing agents, *viz.*, hydrogen peroxide, $KMnO_4$ or CrO_3 in a strong acid, such as HNO_3 or HCl. This chemical treatment of graphite flakes occurs at a high temperature for a specific period. After chemical treatment, the reaction mixture washed and dried. But this chemical method is non-ecofriendly and leads to harmful waste production. However, physical methods are the less harmful method as they require harsh conditions for the synthesis of EG. This method consumes more energy as it provides vacuum and high-temperature conditions for graphite flakes to convert into EG. Both chemical and physical methods required a high temperature around 1000 °C to prepare EG [39, 40].

Mondal *et al.* reported bifunctionalized graphene and EG with n-butyl lithium in one step without any separation. The two grafted functional groups played two distinct roles. However, the carboxylic groups function as dye adsorption sites, while the butyl moiety avoid the agglomeration of graphene sheets [41]. Berktas *et al.* developed EG-supported hybrid additives with cement based grouts for shallow geothermal energy systems to increase thermal conductivity. 3-aminopropyltriethoxysilane was used as a coupling agent in the functionalization process to form bridges with the EG surface by attaching amino functional groups to the silica surface [42].

3.3   Fabrication and functionalization of carbon dots (CDs)

CDs, reffered as carbon quantum dots (CQDs), are quasi-spherical nano-range particles containing carbon, oxygen, and hydrogen atoms with and diameter of 10 nm. CDs are usually categorized according to the carbon core they have: graphene quantum dots (GQDs), carbon nanodots (CDs), or polymer dots (PDs). The main approaches to preparing CDs are top-down and bottom-up, as shown in Fig. 5. Chemical, electrochemical, and physical methods can achieve these goals. A top-down synthetic approach involves disintegrating bigger carbon-structures e.g., graphite, CNTs, and nano-diamonds into CDs using the aforementioned methods, laser ablation, arc discharge, and electrochemical methods. Alternatively, bottom-up approaches involve yielding CDs from essential small precursors such as carbohydrates, citrates, amines, etc., *via* hydrothermal treatment, supported synthetics, and microwaves. Recently, green synthetic approaches have been used to fabricate CDs [43-45].

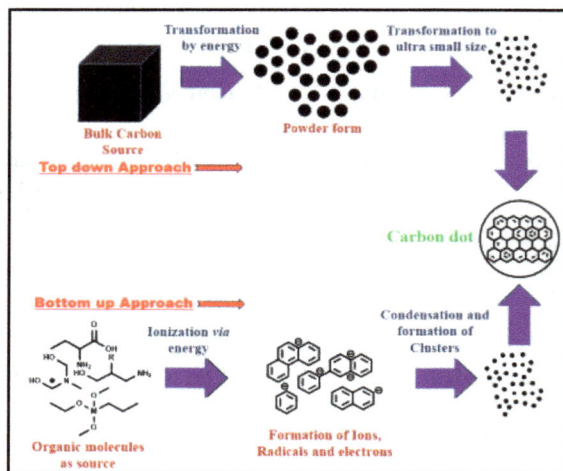

Figure. 5. Approach for the synthesis of CDs

Zhu *et al.* demonstrated a easy technique for fabrication of fluorescent CDs by heating a poly(ethylene glycol) (PEG) and saccharide solution in a microwave oven for a few minutes. These fluorescent CDs have abundant surface traps and functional groups along with bright, stable luminescence and excellent water dispersion properties [46]. Phadke *et al.* reported abiogenic synthesis of fluorescent CDs using *Azadirachta indica* (Neem) gum. Bio-linkers were attached to the surface of these CDs, making them suitable for drug attachment [47].

3.4   Fabrication and functionalization of graphene and graphene oxide (GO)

Since it was discovered using a scotch tape peeling method, graphene has gained prominence among the other materials fascinated to scientists. Graphene and its derivatives shows remarkable properties which attracted the attention of scientific community.. Several techniques have been developed to synthesize graphene from graphite as a precursor, including epitaxial growth, vacuum thermal annealing, non-catalytic synthesis, micromechanical cleavage, chemical vapor deposition, and thermal exfoliation, carbon nanotube cutting, direct ultrasound sonication, and reduction of graphene oxide. These methods either produce multilayer graphene or cost a lot of energy and resources to produce single-layer graphene. On the other hand, reduction of GO is one of the suitable methods for synthesizing graphene, generally termed reduced graphene oxide (rGO) or graphene nanosheet (GNS) [21, 48-53].

As GO was discovered much earlier than graphene, synthesized using an oxidative reaction of graphite flakes and is used to prepare other CNMs. The Hummers or modified Hummers method is typically used in GO synthesis, as shown in Fig. 6. Hummers' method involves using $NaNO_3$, $KMnO_4$, and H_2SO_4 to oxidize graphite. At the end of the reaction, remaining permanganate and manganese dioxide was removed by adding hydrogen peroxide, resulting in a bright yellow suspension. Further purification of the reaction mixture yields the GO. This method provides successful intercalation of graphene layers in the form of GO. A modified Hummers method can be implemented in one or two steps. In a one step improved Hummers method, graphite was oxidized with $KMnO_4$ with H_2SO_4 and H_3PO_4 mixture at a 9:1 weight ratio. In the two step modified Hummers method, the graphite powder is initially oxidized by H_2SO_4, $K_2S_2O_8$, and P_2O_5. Further oxidation can be done. It is a gentle approach to the oxidation of graphite [54-56].

Umekar *et al.* decorated the graphene surface with the ZnO and TiO_2 via the green route. These synthesized nanostructures enhance physical properties and radical-scavenging activity against 1,1-diphenyl-2-picrylhydrazyl free radical compared to GO [57, 58]. Abdala *et al.* reviewed the chemical functionalization of graphene and GO along with their properties and their utility in different fields [59].

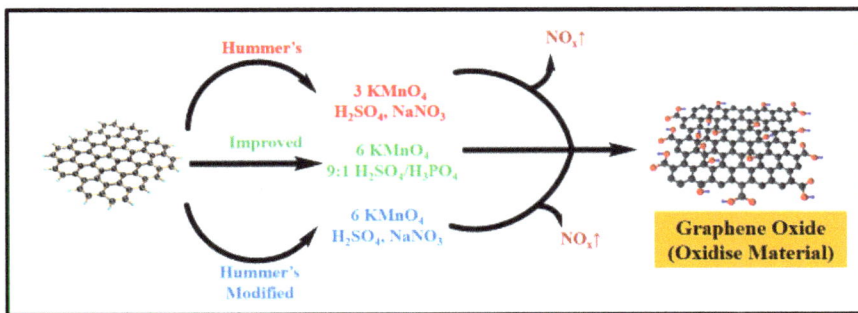

Figure 6. Typical methods for the preparation of GO

4.   Properties of functionalized CNMs

CNMs are interesting because all the materials have remarkable properties like high mechanical strength, electrochemical activity, high surface area, chemical, and thermal stability, and many more. Due to these extraordinary properties, CNMs could be employed in various fields, *viz.,* environmental application, electrochemical studies, photocatalytic activity, biomedical application, *etc.* This property can be modified based on the functionalization of CNMs with metal atoms, organic moiety, polymer, *etc.* Some of the physicochemical properties of a few CNMs are as described below:

4.1   Carbon nanotubes (CNTs)

According to a literature study, CNTs have incredibly high surface areas, large aspect ratios, and extraordinary high mechanical strength. Because of the exceptional electrical, thermal, and mechanical properties, CNTs could be used for incorporate agents. CNTs are considered among the strongest materials, particularly in the axial direction. Whereas, as the report said, it is very soft in the radial direction. Young's modulus range from 270 to 950 GPa, while the tensile strength is very high, between 11 and 63 GPa. As a result of embedded tubes deforming transversely when a load is applied to a composite structure, the radial direction elasticity of CNTs is critical. In addition, these tubular structures are flexible and do not break when bent.

The structural deformability such as defects, chirality, tubular diameter, and crystallinity of the structure dramatically affects the electrical properties of CNTs. As per the structural arrangements of sp^2-carbon, free electrons are delocalized in the molecular orbitals in a hexagonal lattice. Due to this delocalization of electron over all atoms, CNTs shows electrical properties. As a result, CNTs can be either conducting or semiconducting based on their chirality. Usually, semiconductor SWCNTs take the form of p-type semiconductors. Since electrons are transported in a ballistic fashion in SWCNTs, they can be regarded as quantum wires.

In contrast, MWCNTs are found to be relatively diffusive or quasi-ballistic. Likewise, the thermal properties of CNTs are also fascinating. CNTs can withstand high temperatures, thus making them excellent thermal conductors. The thermal conductivity of individual MWCNTs was determined to be 3,000 W/m K at room temperature, whereas that of SWCNTs was more significant than 200 W/m K [60-62].

4.2   Expanded graphite (EG)

EG, usually known as flexible graphite, is similar to graphite in terms of chemical properties but has unique mechanical properties, self-lubricity, flexibility, thermal conductivity, and corrosion resistance. Therefore, it can be employed in electrodes, storage devices, and flame retardant applications. It is a less dense ($0.002{\sim}0.005 g/cm^3$) material than bulk graphite. Hence it is lighter, malleable, and compressible as compared to graphite. This material is chemically inert to many acid and oxidizing agents and can be used in most media at any temperature. Moreover, it can resist the long temperature range (approximately -200 °C to 3000 °C), as there is no change in morphology, softening, and decomposition. EG shows remarkable thermal conductivity as it has a small thermal expansion coefficient. EG has a hexagonal plane layered structure, and hence when an external force is applied, layers slide relative to each other to produce self-lubricating property.[37, 38, 63].

4.3   Carbon dots (CDs)

CDs have mostly similar mechanical and thermal properties to CNTs and graphene. CDs are more critical for optical study as they show excellent absorbance and luminescence properties. As CDs consist of sp^2-carbon, they absorb a short wavelength region and undergo π–π^*. Typically, it shows strong absorption in the UV-visible region. The absorption phenomena differ in CDs reliant on the functionalization of the surface. CDs offer tunable photoluminescence (PL) properties due to quantum confinement effects. Because of emissive traps, CDs with bare surfaces typically have low quantum yields ($\sim$ 10%). To enhance the PL activity, surface functionalization is requried. Furthermore, CDs show chemiluminescence (CL) and electrochemiluminescence (ECL), which can be employed to design the biosensors, optical devices, and detection of metal ions. This property generates when electrons and holes are injected into the surface of a CD after direct oxidation is formed. The change or modification of surfaces of CDs with particular chemical atoms or groups enhances the activity as mentioned above[64, 65].Surface passivation enhances CQDs' fluorescence and physical properties. It was found that amine and hydroxamic acid-functionalized CDs produced green, yellow, and red emissions when introduced to different pH environments. Furthermore, this tricolor emission could be preserved when deposited in the ORMOSIL film matrix[66].

4.4   Graphene and graphene oxide (GO)

As a promptly raised star in nano-material field, graphene has fascinated many researchers globally. Graphene is is catagereios by 2-D carbon sheet considered the thinnest material in the creation, create more interest in all the potential field of science. The graphene shows highest electrical conductivity at room temperature (up to 106 s cm^{-1}) along with larger suface area 2630 m^2/g and mobility of 200,000 cm^2 V^{-1} s^{-1} at carrier density $\sim$1012 cm^{-2}. Moreover, graphene exhibit remarkable thermal conductivity ($\sim$5000 $Wm^{-1}K^{-1}$) with great mechanical properities with Young's modulus of $\sim$1 Tera Pascal and breaking strength of 42 N m^{-1}. Graphene sheet absorbs 2.3% light in wide wavelength for each layer, which hints toward the transparent nature of graphene and is highly suitable for specific opto-electronic uses. The swift growth of graphene has envisioned main potential in designing electronics, opto-electronics, and electrochemical devices and is also valuable for the biomedical field.

The benefit of using GO over graphene is its cost-effective and high yield concern. The GO is generally acquired from graphite crystals, which undergo oxidization by the action of a strong oxidizing agents as concentrated. H_2SO_4. Furthermore, using sonication, the graphite possessing oxygen atom-containing functional groups over its surface gets dispersed readily in water, increasing interlayer distances among the sheets. The GO can be supplementarily exfoliated into either single or multilayers. In this condition, the sp^2 hybridization of GO gets disrupted, resulting in the alteration of conducting property into an insulating one. To overcome the lost property of graphene can be recovered by reducing GO as reduced to rGO. The structural arrangement of rGO retrieves its hexagonal lattices

Emerging Nanomaterials and Their Impact on Society in the 21st Century   Materials Research Forum LLC
Materials Research Foundations 135 (2023) 72-99   https://doi.org/10.21741/9781644902172-4

and thus harvests graphene sheet-like structure by removing a large portion of oxygen[21, 67-70].

5.   Applications of functionalized CNMs

Since the innovation and fabrication of CNMs to ascertain novel numerous prospective applications for this material in science and technology. Furthermore, intercession hip the property of functionalized resources by the side nanoscale empowers the design of devices and materials through novel functionalities. Synthesis and solicitation of nanomaterials and their nanostructures hip the modern era of nanoscience, and nanotechnology exists interdisciplinary in nature. It is a creature by chemists, physicists, materials scientists, biologists, computer scientists, and engineers. Specific novel functionalized nanostructures such as zeolites, nanoporous materials, aerogels, nanotubes, and core-shell structures obligate en route for derived awake through their novel individualities [71]. Statistic analysis that sorts the attention-grabbing nanostructures remains that the eminence in the direction of dimensions relies on voguish nanometer assortment for the superficial outcome to quantum confinement influence. Geometric edifices such as elemental connections, electronic, optical, mechanical, magnetic, and thermal properties exist entirely posh using constituent fragment sizes in nanometer assortment. Captivating arena of nanotechnology partakes comprehensive array of solicitations such as expenditure of nanostructured constituents devours manufactured transistors by way of best-truncated haste and lasers through squat threshold current. Advantages of nanomaterials in the form of antennae take accelerated new itineraries to prospects on behalf of the discovery of analysts on object molecules [72, 73].

Carbon nanomaterials remain under the supreme expansively willful constituents because of their irreplaceable possessions. A leg on each side commences the in height explicit apparent surface, high hauler agility, better flexibility, high electrical conductivity, and optical pellucidity fostering their usage in intuiting bids. Carbon-built sensors should inform biocompatibility, good sensitivity, selectivity, and poorer recognition bounds to let slip an inclusive series of inorganic and organic molecules. Modern improvements hip the meadow of carbon-based nanomaterials intended for intuiting bids such as uni-dimensional model structure, CNTs demonstration resilient solicitation impending in scanning probe microscopy electronics (SPM), chemical and biological detecting, strengthened such as fullerenes (F), carbon onions (CO), carbon quantum dots (CQDs), nanodiamonds (NDs), carbon nanotubes (CNTs), and graphene (G). Production of these nano zeolites partakes remained argued. Their functionalization approaches in sensing a bid devour existed emphasized instead of the significant applicative meadow, and the forthcoming diagnoses and prospects ought to stand delineated.

Moreover, Carbon nanofibres (CNFs) are indispensable superbly enormous apparent capacities that sort them malleable designed for adapting and functionalization proceeding plea. An eclectic assortment of biomedical fields, bio-sensing, including cancer therapy, tissue engineering, and wound dressing by using CNFs. Some of the applications of CNMs are shown in Fig. 7.

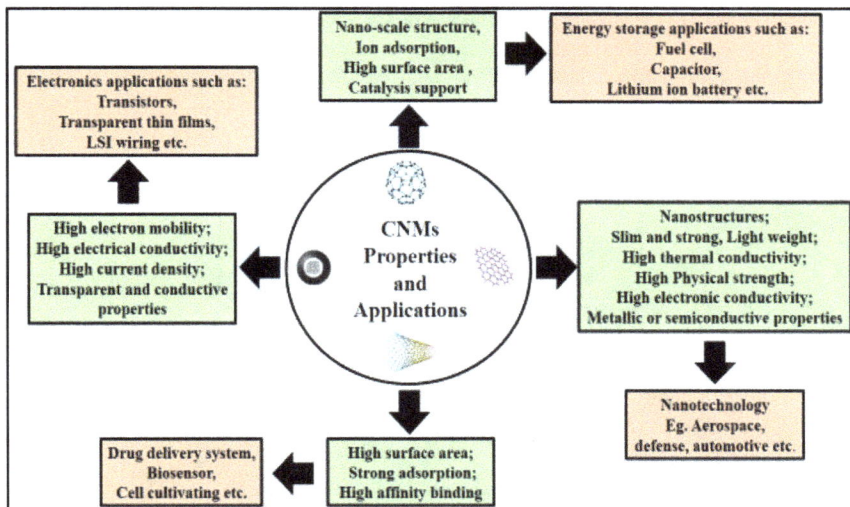

Figure 7. Applications and properties of CNMs

5.1   Carbon nanotubes (CNTs)

CNTs' exclusive commercial feasibility, such as optical, structural, thermal, mechanical, and electrical properties, led to their quick advent. It befitted an artificial intelligence (AI) juncture for diverse solicitations, including the electric range of industrial and biomedical uses[74]. Among the anticipated bids remain still a far-off dream to technical consciousness. In modern enlargements on the improvement of green approaches for the elemental functionalization of the carbon-assisted nanotubes, it provides an extra stimulus to prolong the prospect of their solicitation gamut [75, 76].

CNTs have a feeble turgid knack in numerous solvents such as alcohol, acetone, etc., and an universal solvent, water as a sign of strong intermolecular p–p interactions. Hence CNTs in industrial applications hinder the process competence. Moreover, grab, this concern by numerous methods has been settled over the decade to adapt the surface of CNTs. To advance their immovability and resolvability in water to enrich the processing of unsolvable CNTs. CNT nanofluids with remarkable properties tunable for a broad range of appliances render them helpful in synthesizing innovation [77].

Covalent functionalization such as chemical routes has been deliberate comprehensively, including the oxidation of CNTs through strong acids (e.g., nitric acid, sulfuric acid, hydroiodic acid, or a combination of them together). The novel chemical formulation of carboxylic groups onto the surface of the CNTs as the final product or for further modification by esterification or amination to set novel products. Advances in functionalization of CNTs in concrete time to bids such as bioimaging, drug delivery, gene therapy, biosensors. Three dimensional printed scaffolds for medicine, biocompatibility, thermal conductivity, energy storage, fibers and fabrics, conductive properties, air & water filtration, conductive adhesive, thermal materials, biodegradability, field emission, molecular electronics, biomedical applications, catalyst, and structural applications, etc., shown in Fig. 8 [78, 79] .

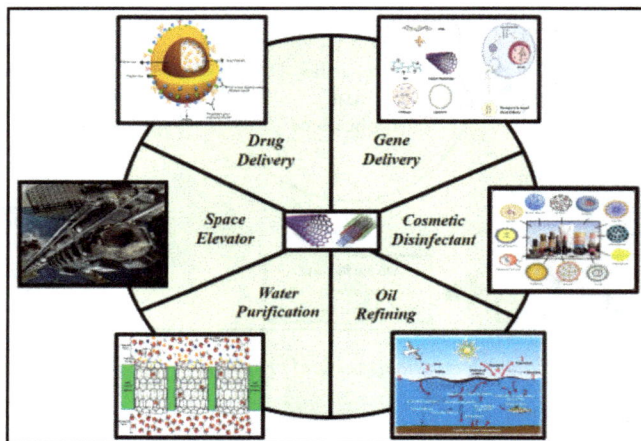

Figure 8. Some applications of Functionalized Carbon nanotubes

Free radical implanting has been a novel modus operandi in which substituted anilines, diazonium salts, and aryl or alkyl peroxides are used as the preliminary agents by covalent functionalization methods. Free radical onto the surface of CNTs which improves the solubility of CNTs as implanting of macromolecules equated to mineral acid treatments. These involve the supplement of unimportant smidgens such as nitro, hydroxyl, amino, or various onto the CNTs superficial. CNTs solubility can be enriched ominously through free-radical splicing because the hefty functional smidgens expedite the dispersal in various diluters and the despondent gradation of functionalization. In the modern era, carbon nanostructures (CNSs) obligate be situated extensively recycled in an assortment of biomedical solicitations. Moreover, a bid of CNTs on behalf of protein delivery and drug or in utensils in the vicinity distribute nucleic acids for antagonism cancer fondness. They are situated efficaciously developed voguish diagnosis in addition to non-invasively besides very much sensitive imaging maneuvers on the way to their optical properties voguish the near-infrared province.

This sort of CNTs is a precise, adaptable utensil in place of an all-encompassing assortment of solicitations as pharmaceutical biochips, novel in elevating concert daises for molecular imaging(MI), magnetic resonance (MR), neuroscience, photothermal therapy (PT), and tissue engineering (TE) [80]. Novel approach of this slog is en route for best part conversant exhausting of the functionalized carbon ingredients such for instance carbon fibers (CF), carbon nanotubes, fullerene, nanodiamonds and graphene in biomedical bids. CNTs are synthesized into covalent and noncovalent methods, covalent methods into the sidewall and endwall defects, similarly to noncovalent polymer wrapping and surface attachments of surfactants.

5.2   Expanded graphite (EG)

Industrial effluents contain a large concentration of suspended solids and high chemical oxygen demands value, toxic organic chemicals, dyes, and strong coloring agents like dyes, which pollute water bodies [81]. Various methods for treating wastewater have been reported, but they are less efficient and pose some limitations—EG attapulgite composite material is worked as an electrode for wastewater treatment. The combined effect of outstanding adsorption capacity of attapulgite and excellent electrical conductivity of EG make composite an efficient electrode in wastewater treatment. The synergistic consequence of adsorption and electrochemical oxidation is responsible for the degradation of organic pollutants from wastewater [82]. EG has been considered a potential and low-cost material for remediation of organic dyes from polluted water *via* the adsorption route owing to its exceptional performance. Hoang *et al.* investigate the adsorption capacity of Congo red and methylene blue dyes on EG fabricated using graphite flake. Adsorbent dosage of dye, concentration of pH of solution, contact time and other parameter also studied. For congo red and methylene blue, the extreme adsorption capability of EG was reported to be 201.21 mg/g and 47.62 mg/g, respectively [83].

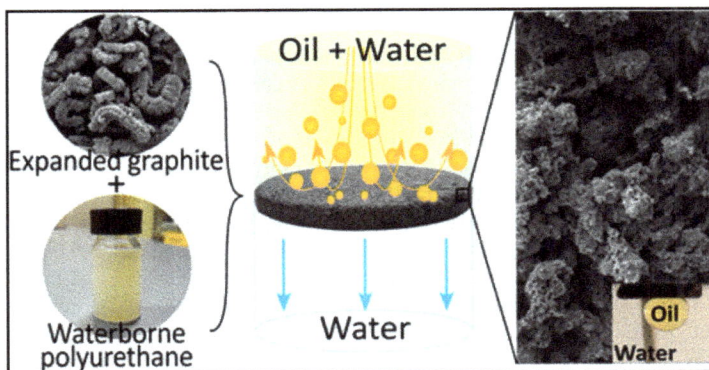

Fig. 9. EG-Polyurethane foam for water-oil filtration [84]. Reproduced with the permission of the Elsevier publishing in the Journal/Magazine via. Copyright Clearance Center (A higher resolution/colour version of this Fig. is available in the electronic copy of the article).

Recently, several approaches have been applied for efficient elimination of oil spills because of their enduring disadvantageous effect on human health and aquatic life. The EG-based polyurethane foam of different pore sizes with underwater oleophobic properties was developed for the oil-water filtration process (Fig. 9). The foam can separate surfactant-free oil-in-water mixture through the phenomenon of gravity with greater oil refusal competence and flow rates that vary depending on the foam type.

They adopted two different solvent-free synthetic routes and produced two types of foam *viz.*, PUEGr_t and PUEGr. Since PUEGr_t foam has a smaller pore size, it has higher oil rejection capability (99.99%) but lower flow rate (8547 $Lm^{-2}h^{-1}$), whereas PUEGr foam has lower oil rejection efficiency (96.85%) and very high flux (9988 $Lm^{-2}h^{-1}$). [84] Similarly, Nguyen *et al.* reported magnetic component $MnFe_2O_4$-based EG composites prepared from graphite flakes to remove oil from contaminated aqueous solution *via* the phenomenon of adsorption. They investigated the removal of heavy oils, counting crude oil and diesel oil, under the adsorption parameters like loaded oil doses, contact time, and salinity of mixing oil and water. As a result, they observed that the incorporation of $MnFe_2O_4$ causes a significant increase in magnetism of EG up to the 16 emu/g. Because of this, EG/ $MnFe_2O_4$ composite manifested a strong adsorption capacity toward crude oil (33.07 ± 0.33 g CO/g EG) and diesel oil (32.20 ± 0.46 g DO/g EG) [85].

EG, a graphite-derived material, serves as a superior Na-ion batteries anode material. EG contains a sufficiently greater interlayer distance of 4.3 A°, favoring the electrochemical intercalation of Na^+ ions. Furthermore, regulated oxidation and reduction processing may control the interlayer spacing of EG, making it attractive anode material for Na-ions batteries. The result was authenticated by cyclic voltammetry and *in situ* TEM measurements of the EG material's reversible interlayer expansion/shrinkage during the solidification and deposition processes[86].

Since EG has a higher conductivity, it acts as a good heat transfer promoter, which improves organic phase change materials' thermal conductivity. Wang *et al.* fabricated organic phase change material by blending EG and polyethylene glycol (PEG) and studied its application in energy storage[87]. PEG has been proposed as an excellent thermal energy storage material because of its coherent melting behavior, high latent heat of 187 J g^{-1}, and superior corrosion resistivity[88]. They have various applications, including electronic device protection and management, industrial heat use, active and passive building heating and cooling. Due to surface tension and capillary effect, PEG poses leakproof ability from a porous EG network. The maximum of 90 wt% of PEG dispersed in phase change material without any leakage. The EG/PEG (90 wt %) composite had a melting temperature of 61.46 °C, a larger enthalpy of 161.2 J g^{-1} and observed conductivity of 1.324 W $m^{-1}K^{-1}$. The porous structure of EG creates a thermal conductivity network, which enhances thermal conductivity [87].

EG on heating at temperatures up to 1150 °C produces vermicular-structured thermally expanded graphite (TEG). Finally, the TEG, an intumescent version of graphite, was employed to make composite materials with a variety of metal chlorides ($CuCl_2$, $FeCl_3$, and

$ZnCl_2$) and conducting polymers (poly (Styrene-co-acrylonitrile), epoxy, polyaniline, etc.) for thermal energy storage, supercapacitors, batteries, hydrogen storage, sensors, etc. [89]. Tao *et al.* fabricated an electrochemical immunosensor using poly(m-aminophenol) immobilized with anti-human immunoglobulin G (IgG) antibody (Fig 10). The amino group of poly(m-aminophenol) reacts with two aldehyde groups of glutaraldehyde which are finally coated on the surface of TEG *via* antibodies amino group. A human IgG antigen was the target antigen in this immunoassay method, and the probing antibody was goat anti-human (IgG), while poly methylene blue (PMB) served as an electron transfer mediator. The oxidation peak current of PMBH (reduced state of PMB) was proportional to IgG concentrations of 5 to 60 mg/ml in electrochemical studies, and the limit of detection LOD was found to be 0.19 µg/ml [90].

Composites of metal oxide (Ag_2O, CuO, and ZnO) distributed inside the layer and on the surface of EG were prepared. Their promising application as antimicrobial agent for successful binding as well as inactivation of pathogens was studied by Hung *et al.* Compositie were tested against Bacillus anthracis, Gram-negative and Gram-positive bacteria using minimum bactericidal concentrations (MBC), minimum inhibitory concentrations (MIC), and zones of inhibition [91].

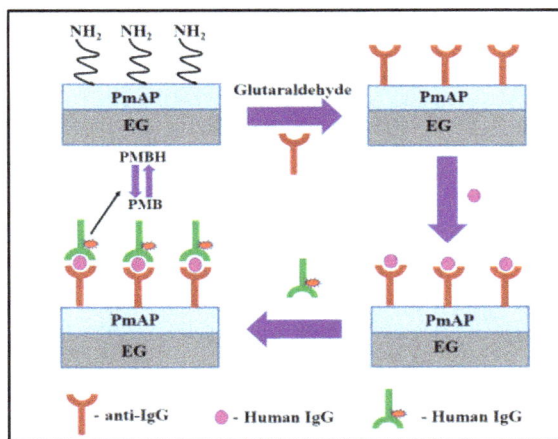

Fig. 10. Schematic illustration of the sandwich-type preparation of the electrochemical immunosensor to detect human IgG

5.3   Carbon dots (CDs)

CDs are practically sphere-shaped nanocrystals (<10 nm in size) and comprise a few molecules or atoms of nanoclusters. With their high photostability, good water solubility, and eco-friendly, simple synthesis methods that combine the optical and fluorescence properties, CDs are high-potential materials with broad applications. Further, CDs have shown no adverse toxic effect on differnt cell lines for *in-vitro* applications and have demonstrated remarkable biocompatibility for *in-vivo* biomedical applications, including biosensing, bioimaging, cancer treatment, drug delivery, and disease diagnosis. The surface of these sphere-shaped CDs contains a vast amount of -OH, -NH$_2$, -COOH, and other groups. These groups support the functionalization of CDs surfaces with many inorganic, organic, or biologically active substances to enhance their properties. These functionalized CDs are mainly used for electrochemical and optical studies.

Zuo *et al.* reviewed the fabrication and electrochemical application of functionalized fluorescent CDs. They reviewed electrochemical applications like sensing, luminescence, imaging, photocatalysis, etc. In addition to this, functionalization of the CDs with wealthy functional groups or inorganic species can enhance the electrochemical property, which shows the greater efficacy of the application described previously [92]. Similarly, Ethesabi *et al.* reviewed the elcetroluminance-based application of functionalized CDs in a very detailed manner. Also, in this review, they discussed the fabrication and mechanism of functionalized CDs. These functionalized CDs are employed in different potential applications like bioimaging, water treatment, cancer therapy, and detection of various species such as biomolecules, inorganic species, organic compounds, etc. Shen and Xia reported a one step fabrication of boronic acid functionalized CDs for use as a chemosensor. This chemosensor is accessible, selective, sensitive, and successfully employed for the blood sugar assay [93].

5.4   Graphene and graphene oxide (GO)

Graphene is a hot topic in nanomaterial research because of its excellent applications for the modernization of human society. Graphene-based NMs have caught the world's attention. The graphene-based NMs have witnessed remarkable progress in recent years exhibiting radically enhanced properties for various applications. It has been reported that the functionalization of graphene and its derivatives in NMs has improved the electrical conductivity and mechanical strength the of the composites significantly. Additionally, these NMs are generally more thermally stable, electrochemically active, and gas barrier-efficient.

Silva *et al.* presented a survey on graphene-polymer NMs for biomedical applications. A variety of graphene-polymer NMs, synthesis, and their biomedical applications have been discussed in the review article. They also reviewed the interaction of the polymer with graphene/GO, their biocompatibility, and enhancement of biomedical activity such as cellular imaging, drug delivery, tissue engineering, biosensing, wound healing, antibacterial, fungi-static, and anti-tumoral properties, etc. [94]. Moreover, in the electromembrane process, Cseri *et al.* synthesizes highly permselective anion exchange membranes made of GO and polybenzimidazolium nanocomposites. Through diazonium chemistry, 4-(tri-fluoromethylthio)phenyl groups were added to GO to enhance dispersibility in organic solvents and to provide labeling for future mapping techniques. By developing high-performance nanocomposite ion exchange membranes, electro membrane processes in a wide range of fields like desalination and water treatment will become more efficient [95]. Potbhare *et al.* fabricated graphene-based silver nanoparticles using *pseudomonas aeruginosa,* showing excellent antibacterial efficacy against humans pathogenic bacteria [96]. Chaudhary *et al.* demonstrated the ZnO-rGO NCs for the photodegradation of MB dye of about 85% in 70 minutes [97]. Umekar *et al.* synthesized zinc oxide modified photo reduced graphene oxide nanohybrid *via Clerodendrum infortunatum* [98].

Conclusions and futuristic approach

This chapter enlightens the newest literature with a clear focus on carbon-based nanomaterials. First, we analyzed the various eco-friendly methods for fabricating functionalized carbon-based nanomaterials. Then, we discuss the mechanical, electrical, thermal, morphological, luminescence, and corrosion resistance properties of FCNMs, emphasizing the impact of functionalization on their flexibility, durability, biocompatibility, mechanical permanence, surface area, and thermal stability. Moreover, encapsulated, anchored, wrapped, mixed, and layered structures of carbon-based composites and their industrial applications in semiconductors and photocatalysts are discussed. Furthermore, the advanced applications of the FCNMs as clinical agents against pathogenic bacteria. Finally, these materials show excellent electrochemical and energy storage applications in sensing, luminescence, imaging, anode material, and pseudocapacitance.

References

[1] D. Jariwala, V. K. Sangwan, L. J. Lauhon, T. J. Marks and M. C. Hersam, Carbon nanomaterials for electronics, optoelectronics, photovoltaics, and sensing. Chemical Society Reviews, 42, (2013) 2824-2860. https://doi.org/10.1039/C2CS35335K

[2] C. Buzea, I. Pacheco and K. Robbie, Nanomaterials and Nanoparticles: Sources and Toxicity. Biointerphases, 2, (2007) 17-71. https://doi.org/10.1116/1.2815690

[3] C. Pereira, C. Alves, A. Monteiro, C. Magén, A. M. Pereira, A. Ibarra, M. R. Ibarra, P. B. Tavares, J. P. Araújo, G. Blanco, J. M. Pintado, A. P. Carvalho, J. Pires, M. F. R. Pereira and C. Freire, Designing novel hybrid materials by one-pot condensation: from hydrophobic mesoporous silica nanoparticles to superamphiphobic cotton textiles. ACS Applied Materials & Interfaces, 3, (2011) 2289-2299. https://doi.org/10.1021/am200220x

[4] L. S. Ribeiro, T. Pinto, A. Monteiro, O. P. Soares, C. Pereira, C. Freire, Pereira and M. F. R. Silica nanoparticles functionalized with a thermochromic dye for textile applications. Journal of Materials Science, 48, (2013) 5085-5092. https://doi.org/10.1007/s10853-013-7296-7

[5] S. Stankovich, D. A. Dikin, G. H. Dommett, K. M. Kohlhaas, E. J. Zimney, E. A. Stach and R. S. Ruoff, Graphene-based composite materials. Nature, 442, (2006) 282-286. https://doi.org/10.1038/nature04969

[6] M. Z. Krolow, C. A. Hartwig, G. C. Link, C. W. Raubach, J. S. F. Pereira and R. S. Picoloto, Synthesis and characterization of carbon nanocomposites. NanoCarbon 2011 Springer Berlin Heidelber, 3, (2013) 33-47. https://doi.org/10.1007/978-3-642-31960-0_2

[7] L. Q. Dong, Y. Y. Feng, L. Wang and W. Feng, Azobenzene based solar thermal fuels design, properties, and applications. Chemical Society Reviews, 47, (2018) 7339-7368. https://doi.org/10.1039/C8CS00470F

[8] F. Zhang, Y. Y. Feng, M. M. Qin, T. X. Ji, F. Lv, Z. Y. Li, L. Gao, P. Long, F. L. Zhao and W. Feng, Stress-sensitive thermally conductive elastic nanocomposite based on interconnected graphite-welded carbon nanotube sponges. Carbon, 145, (2019) 378-388. https://doi.org/10.1016/j.carbon.2019.01.031

[9] L. Y. Wang, Y. Y. Feng, J. Han, F. Zhang, P. Long and W. Feng, Asymmetric selfsupporting hybrid fluorinated carbon nanotubes/carbon nanotubes sponge electrode for high-performance lithium-polysulfide battery. Chemical Engineering Journal, 349, (2018) 756-765. https://doi.org/10.1016/j.cej.2018.05.088

[10] P. Lv, P. Zhang, Y. Y. Feng, Y. Li, W. Feng and J. Electacta, High-performance electrochemical capacitors using electrodeposited MnO_2 on carbon nanotube array grown on carbon fabric. Electrochimica Acta, 78, (2012) 515-523. https://doi.org/10.1016/j.electacta.2012.06.085

[11] X. Wang, Y. Li, Y. Zhao, J. Liu, S. Tang and W. Feng, Synthesis of PANI nanostructures with various morphologies from fibers to micromats to disks doped with salicylic acid. Synthetic Metals, 160, (2010) 2008-2014. https://doi.org/10.1016/j.synthmet.2010.07.030

[12] Q-L. Yan, M. Gozin, F-Q. Zhao, A. Cohen and S-P. Pang, Highly energetic compositions based on functionalized carbon nanomaterials, Nanoscale, 8, (2016) 4799-4851. https://doi.org/10.1039/C5NR07855E

[13] A. Ferrari and J. Robertson, Interpretation of Raman spectra of disordered and amorphous carbon. Physical Review B, 61, (2000) 14095-14107. https://doi.org/10.1103/PhysRevB.61.14095

[14] L. Wei, P. K. Kuo and R. L. Thomas, Thermal conductivity of isotopically modified single crystal diamond. Physical Review Letters, 70, (1993), 3764-3767. https://doi.org/10.1103/PhysRevLett.70.3764

[15] J. Pang, A. Bachmatiuk, I. Ibrahim, L. Fu and D. Placha, CVD growth of 1D and 2D sp2 carbon nanomaterials. Journal of Materieal Science. 51, (2016) 640-667. https://doi.org/10.1007/s10853-015-9440-z

[16] S. Iijima, Helical microtubules of graphitic carbon. Nature 354, (1991) 56-58. https://doi.org/10.1038/354056a0

[17] B. Bowers, History of Electric Light and Power; Peter Peregrinus Ltd. London, UK, 55, (1982) 71-72. ISBN: 0906048680.

Emerging Nanomaterials and Their Impact on Society in the 21st Century   Materials Research Forum LLC
Materials Research Foundations 135 (2023) 72-99   https://doi.org/10.21741/9781644902172-4

[18] N.B. Brandt, S.M. Chudinov and Y.G. Ponomarev, (Eds.) Semimetals Graphite and Its Compounds; Modern Problem in Condensed Matter Sciences Series 20, Elsevier, Amsterdam, The Netherlands, 48, (1988). ISBN: 978-0444870490.

[19] K. Sugihara and H. Sato, Electrical conductivity of graphite. Journal of the Physical Society of Japan, 18, (1963) 332-341. https://doi.org/10.1143/JPSJ.18.332

[20] Y. Wu, Y. Lin, A. A. Bol, K. A. Jenkins, F. Xia, D. B. Farmer, Y. Zhu and P. Avouris, High-frequency, scaled graphene transistors on diamond-like carbon. Nature, 472, (2011) 74-78. https://doi.org/10.1038/nature09979

[21] K. S. Novoselov, A. K. Geim, S. V. Morozov and D. Jiang, Electric field effect in atomically thin carbon films. Science, 306, (2004) 666-669. https://doi.org/10.1126/science.1102896

[22] A. K. Geim and K. S. Novoselov, The rise of graphene. Nature Materials, 6, (2007) 183-191. https://doi.org/10.1038/nmat1849

[23] H. W. Kroto, J. R. Heath, S. C. O'Brien, R. F. Curl and R. E. Smalley, C60: Buckminsterfullerene. Nature, 318, (1985) 162-163. https://doi.org/10.1038/318162a0

[24] S. Iijima, Helical microtubules of graphitic carbon. Nature, 354, (1991) 56-58. https://doi.org/10.1038/354056a0

[25] D. Janas, Perfectly imperfect: a review of chemical tools for exciton engineering in single-walled carbon nanotubes, Materials Horizons, 7, (2020) 2860-2881. https://doi.org/10.1039/D0MH00845A

[26] N. Mubarak, E. Abdullah, N. Jayakumar, and J. Sahu, An overview on methods for the production of carbon nanotubes, Journal of Industrial and Engineering Chemistry, 20, (2014) 1186-1197. https://doi.org/10.1016/j.jiec.2013.09.001

[27] J. Prasek, J. Drbohlavova, J. Chomoucka, J. Hubalek, O. Jasek, V. Adam, and R. Kizek. Methods for carbon nanotubes synthesis. Journal of Materials Chemistry 21, (2011) 15872-15884. https://doi.org/10.1039/c1jm12254a

[28] S. Esconjauregui, C. M. Whelan, and K. Maex, The reasons why metals catalyze the nucleation and growth of carbon nanotubes and other carbon nanomorphologies, Carbon 47, (2009) 659-669 https://doi.org/10.1016/j.carbon.2008.10.047

[29] M. I. Ionescu, Y. Zhang, R. Li, X. Sun, H. Abou-Rachid, and L. S. Lussier, Hydrogen-free spray pyrolysis chemical vapor deposition method for the carbon nanotube growth: parametric studies, Applied Surface Science, 257, (2011) 6843-6849. https://doi.org/10.1016/j.apsusc.2011.03.011

[30] R. Bhatia and V. Prasad, Synthesis of multiwall carbon nanotubes by chemical vapor deposition of ferrocene alone, Solid State Communications, 150, (2010) 311-315. https://doi.org/10.1016/j.ssc.2009.11.023

[31] M. Golshadi, J. Maita, D. Lanza, M. Zeiger, V. Presser, and M. G. Schrlau, Effects of synthesis parameters on carbon nanotubes manufactured by template- based

chemical vapor deposition, Carbon, 80, (2014) 28-39.
https://doi.org/10.1016/j.carbon.2014.08.008

[32] N. Saifuddin, A. Raziah, and A. Junizah, Carbon nanotubes: a review on structure and their interaction with proteins, Journal of Chemistry, 2013, (2013) 18. https://doi.org/10.1155/2013/676815

[33] J. Prasek, J. Drbohlavova, J. Chomoucka, J. Hubalek, O. Jasek, V. Adam and R. Kizek, Methods for carbon nanotubes synthesis-review, Journal of Materials Chemistry, 21, (2011) 15872-15884. https://doi.org/10.1039/c1jm12254a

[34] G. Rathore, A. Gupta, and L. Singh, An eco-friendly approach for fabrication of silver nanoparticles by using ascorbic acid from various native sources, International Journal of Pharmaceutical Research, 12, (2020) 41-48. https://doi.org/10.31838/ijpr/2020.12.02.0002

[35] A. C. Jafer, V. T. Veetil, G. Prabhavathi, and R. Yamuna, Covalent functionalization and characterization of multiwalled carbon nanotubes using 5,10,15,20-tetra(4-aminophenyl)porphyrinatonickel(II), Fullerenes, Nanotubes, and Carbon Nanostructures, 26, (2018) 739-745. https://doi.org/10.1080/1536383X.2018.1492558

[36] A. L. Alpatova, W. Shan, P. Babica, B. L. Upham, A. R. Rogensues, S. J. Masten, E. Drown, A. K. Mohanty, E. C. Alocilja and V. V. Tarabara, Single-walled carbon nanotubes dispersed in aqueous media via noncovalent functionalization: effect of dispersant on the stability, cytotoxicity, and epigenetic toxicity of nanotube suspensions, Water Research, 44, (2010) 505-520. https://doi.org/10.1016/j.watres.2009.09.042

[37] J. H. Li, L. L. Feng and Z. X. Jia, Preparation of expanded graphite with 160 μm mesh of fine flake graphite, Materials Letters, 60, (2006) 746-749. https://doi.org/10.1016/j.matlet.2005.10.004

[38] S. Lee, H. min Kim, D. G. Seong and D. Lee, Synergistic improvement of flame retardant properties of expandable graphite and multiwalled carbon nanotube reinforced intumescent polyketone nanocomposites, Carbon, 143, (2019) 650-659. https://doi.org/10.1016/j.carbon.2018.11.050

[39] S. Lin, L. Dong, J. Zhang and H. Lu, Room-temperature intercalation and ~1000-fold chemical expansion for scalable preparation of high-quality graphene, Chemistry of Materials, 28, (2016) 2138-2146. https://doi.org/10.1021/acs.chemmater.5b05043

[40] H. W. Xing, L. Lang, X. Z. Jin, L. S. Jing and G. G. Quan, HClO4-graphite intercalation compound and its thermally exfoliated graphite, Materials Letters, 63, (2009) 1618-1620. https://doi.org/10.1016/j.matlet.2009.04.030

[41] T. Mondal, A. K. Bhowmick, and R. Krishnamoorti. Synthesis and characterization of bi-functionalized graphene and expanded graphite using n-butyl lithium and their use for efficient water soluble dye adsorption. Journal of Materials Chemistry A, 28, (2013) 8144-8153. https://doi.org/10.1039/c3ta11212h

[42] I. Berktas, N. Ghafar, A. Fontana, P. Caputcu, A. Menceloglu, Y. and B. S. Okan, Synergistic effect of expanded graphite-silane functionalized silica as a hybrid additive in improving the thermal conductivity of cementitious grouts with controllable water uptake. Energies, 13, (2020) 3561. https://doi.org/10.3390/en13143561

[43] Y. Wang and A. Hu, Carbon quantum dots: synthesis, properties and applications, Journal of Materials Chemistry C. 2, (2014) 6921-39. https://doi.org/10.1039/C4TC00988F

[44] J. Zhou, Tailoring multi-wall carbon nanotubes for smaller nanostructures, Carbon, 47, (2009) 829-838. https://doi.org/10.1016/j.carbon.2008.11.032

[45] J. Zhou, An electrochemical approach to fabricating honeycomb assemblies from multiwall carbon nanotubes, Carbon. 59, (2013) 130-139. https://doi.org/10.1016/j.carbon.2013.03.001

[46] H. Zhu, X. Wang, Y. Li, Z. Wang, F. Yang and X. Yang, Microwave synthesis of fluorescent carbon nanoparticles with electrochemiluminescence properties, Chemical Communications, 34, (2009) 5118-20. https://doi.org/10.1039/b907612c

[47] C. Phadke, A. Mewada, R. Dharmatti, M. Thakur, S. Pandey and M. Sharon, Biogenic Synthesis of Fluorescent Carbon Dots at Ambient Temperature Using Azadirachta indica (Neem) gum. Journal of Fluorescence, 25, (2015) 1103-1107. https://doi.org/10.1007/s10895-015-1598-x

[48] V. C. Tung, M. J. Allen, Y. Yang and R. B. Kaner, High through put solution processing of large-scale graphene, Natural Nanotechnology, 4, (2008) 25-29. https://doi.org/10.1038/nnano.2008.329

[49] H. Chen, M. B. Muller, K. J. Gilmore, G. G. Wallace and D. Li, Mechanically strong, electrically conductive, and biocompatible graphene paper, Advance Materials, 20, (2008), 3557-3561. https://doi.org/10.1002/adma.200800757

[50] Z. G. Cambaz, G. Yushin, S. Osswald, V. Mochalin and Y. Gogotsi, Noncatalytic synthesis of carbon nanotubes, graphene and graphite on SiC, Carbon, 46, (2008) 841-849. https://doi.org/10.1016/j.carbon.2008.02.013

[51] X. Li, W. Cai, J. An, S. Kim, J. Nah, D. Yang, R. Piner, A. Velamakanni, I. Jung, E. Tutuc, S. Banerjee, L. Colombo and R. Ruoff, Large-area synthesis of high-quality and uniform graphene films on copper foils, Science, 324, (2009) 1312-1314. https://doi.org/10.1126/science.1171245

[52] C. Li, X. Wang, Y. Liu, W. Wang, J. Wynn and J. Gao, Using glucosamine as a reductant to prepare reduced graphene oxide and its nanocomposites with metal nanoparticles, Journal of Nanoparticle Research, 14, (2012) 1-11. https://doi.org/10.1007/s11051-012-0875-8

[53] S. Pei, J. Zhao, J. Du, W. Ren and H.-M. Cheng, Direct reduction of graphene oxide films into highly conductive and flexible graphene films by hydrohalic acids, Carbon, 48, (2010), 4466-4474. https://doi.org/10.1016/j.carbon.2010.08.006

[54] V. Avdeev, L. Monyakina, I. Nikol'Skaya, N. Sorokina and K. Semenenko, The choice of oxidizers for graphite hydrogenosulfate chemical synthesis, Carbon, 30, (1992) 819-823. https://doi.org/10.1016/0008-6223(92)90001-D

[55] N. Sorokina, M. Khaskov, V. Avdeev and I. Nikol'Skaya, Reaction of graphite with sulfuric acid in the presence of KMnO4, Russian Journal of General Chemistry, 75, (2005) 162-168. https://doi.org/10.1007/s11176-005-0191-4

[56] J. Chen, B. Yao, C. Li and G. Shi, An improved Hummers method for eco-friendly synthesis of graphene oxide, Carbon, 64, (2013) 225-229. https://doi.org/10.1016/j.carbon.2013.07.055

[57] M. Umekar, R. Chaudhary, G. Bhusari, and A. Potbhare, Fabrication of zinc oxidedecorated phytoreduced graphene oxide nanohybrid via Clerodendrum infortunatum. Emerging Materials Research, 10, (2021) 75-84. https://doi.org/10.1680/jemmr.19.00175

[58] M. S. Umekar, R. G. Chaudhary, G. S. Bhusari, A. Mondal, A .K. Potbhare, and M. Sami, Phytoreduced graphene oxide-titanium dioxide nanocomposites using *Moringa oleifera* stick extract, Materials Today: Proceedings, 29, (2020) 709-714. https://doi.org/10.1016/j.matpr.2020.04.169

[59] S. P. Lonkar, Y. S. Deshmukh, and Ahmed A. Abdala, Recent advances in chemical modifications of graphene, Nano Research, 8, (2014) 1039-1074. https://doi.org/10.1007/s12274-014-0622-9

[60] A. Eatemadi, H. Daraee, H. Karimkhanloo, M. Kouhi, N. Zarghami, A. Akbarzadeh, M. Abasi, Y. Hanifehpour and S. W. Joo, Carbon nanotubes: properties, synthesis, purification, and medical applications, Nanoscale Research Letters, 9, (2014) 1-13. https://doi.org/10.1186/1556-276X-9-393

[61] J. Kaur, G. S. Gill, and K. Jeet, Applications of carbon nanotubes in drug delivery: a comprehensive review, in Characterization and biology of nanomaterials for drug delivery: Nanoscience and Nanotechnology in Drug Delivery, Elsevier, (2019) 113-135. https://doi.org/10.1016/B978-0-12-814031-4.00005-2

[62] L. Sidi Salah, N. Ouslimani, M. Chouai, Y. Danlée, I. Huynen, and H. Aksas, Predictive optimization of electrical conductivity of polycarbonate composites at different concentrations of carbon nanotubes: a valorization of conductive nanocomposite theoretical models, Materials, 14, (2021) 1687. https://doi.org/10.3390/ma14071687

[63] G. Prusty and S. K. Swain, Dispersion of expanded graphite as nanoplatelets in a copolymer matrix and its effect on thermal stability, electrical conductivity and permeability, New Carbon Materials, 27, (2017) 271-277. https://doi.org/10.1016/S1872-5805(12)60017-1

[64] P. Bhartiya, A. Singh, H. Kumar, T. Jaina, B. K. Singh and P. K. Dutta, Carbon dots : Chemistry, properties and applications, Journal of Indian Chemical Society, 93, (2016) 1-8.

[65] S. Y. Lim, W. Shen and Z. Gao, Carbon quantum dots and their applications, Chemical Society Reviews. 44, (2015) 362-81. https://doi.org/10.1039/C4CS00269E

[66] V. Rimal, S. Shishodia and P. K. Srivastava. Novel synthesis of high-thermal stability carbon dots and nanocomposites from oleic acid as an organic substrate. Applied Nanoscience, 10, (2020) 455-464. https://doi.org/10.1007/s13204-019-01178-z

[67] M. Liang, B. Luo; Zhi, Application of graphene and graphene-based materials in clean energy-related devices, International Journal of Energy Research. 33, (2009) 1161-1170. https://doi.org/10.1002/er.1598

[68] H. Chang and H. Wu, Graphene-based nanohybrids: preparation, functionalization, and energy and environmental application, Energy and Environmental Science, 6, (2013) 3483-3507. https://doi.org/10.1039/c3ee42518e

[69] M. Pumera, Graphene-based nanohybrids for energy storage, Energy and Environmental Science, 4, (2011) 668-674. https://doi.org/10.1039/C0EE00295J

[70] A. Rashid, Graphene-based Energy Devices, Wiley-VCH Verlag GmbH & Co. 2015, ISBN: 978-3-527-69030-5

[71] I. Lundstroem, S. Shivaraman, C. Svensson and L. Lundkvist, A hydrogen−sensitive MOS field−effect transistor, Applied Physics Letter, 26, (1975) 55. https://doi.org/10.1063/1.88053

[72] M. Zaiser, S. J. Zinkle, G. E. Lucas, R. C. Ewing and J. S. Williams, Microstructural processes in irradiated materials, Materials Research Society: Warrendale, PA, USA, (1999). ISBN: 978-1558994461.

[73] K. Balasubramanian and M. Burghard, Chemically functionalized carbon nanotubes, Carbon Nanotubes, 1, (2005) 180 -192,. https://doi.org/10.1002/smll.200400118

[74] F. Borondics, M. Bokor, P. Matus, K. Tompa, S. Pekker and E. Jakab, Reductive functionalization of carbon nanotubes. Fullerenes, Nanotubes and Carbon Nanostructures, 13, (2005) 375-382,. https://doi.org/10.1081/FST-200039375

[75] K. V. Emtsev, A. Bostwick, K. Horn, J. Jobst, G. L. Kellogg, L. Ley, J. L. McChesney, T. Ohta, S. A. Reshanov, J. Röhr; E. Rotenberg, A. K. Schmid, D. Waldmann, H. B. Weber and T. Seyller, Towards wafer-size graphene layers by atmospheric pressure graphitization of silicon carbide. Nature Materials, 8, (2009) 203-207. https://doi.org/10.1038/nmat2382

[76] G. Ciofani, V. Raffa, A. Pensabene, A. Menciassi and P. Dario, Dispersion of multi-walled carbon nanotubes in aqueous pluronic F127 solutions for biological applications. Fullerenes, Nanotubes and Carbon Nanostructures, 17, (2009) 11-25. https://doi.org/10.1080/15363830802515840

[77] M. Ramrakhiani, Nanostructures and their applications, Recent Research in Science and Technology, 4, (2012) 14-19.

[78] E.-L. Ursu, F. Doroftei, D. Peptanariu, M. Pinteala and A. Rotaru, DNA-assisted decoration of single-walled carbon nanotubes with gold nanoparticles for applications in surface-enhanced Raman scattering imaging of cells, Journal of Nanoparticle Research, 19, (2017) 181. https://doi.org/10.1007/s11051-017-3876-9

[79] W. Liu and G. Speranza, Functionalization of carbon nanomaterials for biomedical applications, Journal of Carbon Research, 5, (2019) 72. https://doi.org/10.3390/c5040072

[80] G. Speranza, Carbon nanomaterials: Synthesis, functionalization and sensing applications, Nanomaterials, 11, (2021) 967. https://doi.org/10.3390/nano11040967

[81] T. S. Shrirame, J. S. Khan, M. S. Umekar, A. K. Potbhare, P. R. Bhilkar, G. S. Bhusari, D. T. Masram, A. A. Abdala, and R. G. Chaudhary, Graphene-polymer nanocomposites for environmental remediation of organic pollutants, Metal Nanocomposites for Energy and Environmental Applications, Springer, (2022) 321-349. https://doi.org/10.1007/978-981-16-8599-6_14

[82] Y. Kong, J. Yuan, Z. Wang, S. Yao and Z. Chen, Application of expanded graphite/attapulgite composite materials as electrode for treatment of textile Wastewater, Applied Clay Science, 46 (2009) 358-362. https://doi.org/10.1016/j.clay.2009.09.006

[83] N. B. Houng, T. T. Nguyen, T. S. Nguyen, T. P. Bui, and L. G. Bach, The application of expanded graphite fabricated by microwave method to eliminate organic dyes in aqueous solution, Civil & Environmental Engineering, 6, (2019) 1584939. https://doi.org/10.1080/23311916.2019.1584939

[84] L. Vasquez, L. Campagnolo, A. Athanassious, and D. Fragouli, Expanded graphite-polyurethane foam for water- oil filtration, ACS Applied Materials & Interfaces, 11, (2019) 30207-30217. https://doi.org/10.1021/acsami.9b07907

[85] H. D. Nguyen, H. T. Nguyen, T. T. Nguyen, A. K. Thi, T. D. Nguyen, Q. T. Bui, and L. G. Bach, The preparation and characterization of $MnFe_2O_4$ -decorated expanded graphite for removal of heavy oils from water, Materials, 12, (2019) 1993. https://doi.org/10.3390/ma12121913

[86] Y. Wen, K. He, Y. Zhu, F. Han, Y. Xu, I. Matsuda, Y. Ishii, J. Cumings, and C. Wang, Expanded graphite as superior anode for sodium-ion batteries, Nature Communications, 5, (2014) 1-10. https://doi.org/10.1038/ncomms5033

[87] W. Wang, X. Yang, Y. Fang, J. Ding, and J. Yan, Preparation and thermal properties of polyethylene glycol/expanded graphite blends for energy storage, Applied Energy, 86, (2009) 1479-1483. https://doi.org/10.1016/j.apenergy.2008.12.004

[88] B. Zalba, J. M. Marin, L. F. Cabeza, and H. Mehling, Review on thermal energy storage with phase change: materials, heat transfer analysis and applications. Applied Thermal Engineering, 23, (2003) 251-83. https://doi.org/10.1016/S1359-4311(02)00192-8

[89] P. Murugan, R. D. Nagarajan, B. H. Shetty, M. Govindasamy, and A. Sundramoorthy, Recent trends in the applications of thermally expanded graphite for energy storage and sensors- a review, Nanoscale Advances, 3, (2021) 6294-6309. https://doi.org/10.1039/D1NA00109D

[90] Y. Taoa, Q. Liub, W. Li, H. Xue, Y. Qin, J. Ge, and Y. Kong, A novel electrochemical immunosensor based on poly(m-aminophenol) modified expanded graphite electrode, Synthetic Metals, 183, (2013) 50-56. https://doi.org/10.1016/j.synthmet.2013.09.010

[91] W. C. Hung, K. H. Wu, D. Y. Lyu, K. F. Cheng, W. C. Huang, Preparation and characterization of expanded graphite/metal oxides for antimicrobial application, Material Science & Engineering C, 75, (2017) 1019-1025. https://doi.org/10.1016/j.msec.2017.03.043

[92] J. Zuo, T. Jiang, X. Zhao, X. Xiong, S. Xiao and Z. Zhu, Preparation and Application of Fluorescent Carbon Dots, Journal of Nanomaterials, 2015, (2015) 13. https://doi.org/10.1155/2015/787862

[93] H. Ehtesabi, M. Amirfazli, F. Massah and Z. Bagheri, Application of functionalized carbon dots in detection, diagnostic, disease treatment, and desalination: a review, Advances in Natural Sciences: Nanoscience and Nanotechnology, 11, (2020) 025017. https://doi.org/10.1088/2043-6254/ab9191

[94] M. Silva, N. M. Alves and M. C. Paiva; Graphene-polymer nanohybrids for biomedical applications, Polymer Advance Technology, (2017) 1-14. https://doi.org/10.1002/pat.4164

[95] L. Cseri, J. Baugh, A. Alabi, A. AlHajaj, L. Zou, R. A. W. Dryfe, P. M. Budd and G. Szekely, Graphene oxide-polybenzimidazolium nanocomposite anion exchange membranes for electrodialysis, Journal of Materials Chemistry A, 6, (2018) 24728-24739. https://doi.org/10.1039/C8TA09160A

[96] A. K. Potbhare, M. S. Umekar and P. B. Chouke, M. B. Bagade, S. K. Tarik Aziz, A. A. Abdala, R. G. Chaudhary, Bioinspired graphene-based silver nanoparticles: Fabrication, characterization and antibacterial activity, Materials Today: Proceedings, 29, (2020) 720-725. https://doi.org/10.1016/j.matpr.2020.04.212

[97] R. G. Chaudhary, A. K. Potbhare, S. K. Tarik Aziz, M. S. Umekar, S. S. Bhuyar, A. Mondal, and A. Abdala, Phytochemically fabricated reduced graphene Oxide-ZnO NCs by Sesbania bispinosa for photocatalytic performances Materials Today: Proceedings, 36, (2021) 756-762. https://doi.org/10.1016/j.matpr.2020.05.821

[98] M. S. Umekar, G. S. Bhusari, A.K. Potbhare, A. Mondal, B. P. Kapgate, M. F. Desimone, and R. G. Chaudhary, Bioinspired reduced graphene oxide based nanohybrids for photocatalysis and antibacterial applications, Current Pharmaceutical Biotechnology, 22, (2021) 1759-1781. https://doi.org/10.2174/1389201022666201231115826

Emerging Nanomaterials and Their Impact on Society in the 21st Century   Materials Research Forum LLC
Materials Research Foundations 135 (2023) 100-124   https://doi.org/10.21741/9781644902172-5

Chapter 5

Smart Nanomaterials and Their Applications

Md. Kawsar Alam[1,2], Rejwana Karim[1] and A.J. Saleh Ahammad[1,*]

[1]Department of Chemistry, Jagannath University, Dhaka 1100, Bangladesh

[2]Department of Civil Engineering, Dhaka International University, Dhaka 1212, Bangladesh

*ajsahammad@chem.jnu.ac.bd

Abstract

Nanomaterials are one of the most amazing creations of human civilization because of their unprecedented potential. These materials are eminent to own excellent electrical, thermal, optical, strong mechanical strength, and catalytic properties. Gadgets or sensors made of nanoparticles are used to determine harmful substances or contaminants of the environment as well as have human biomedical health applications. They have drawn significant attention in the medical sector for possessing remarkable reactivity, extraordinary molar-extinction coefficient, specificity, long-lasting stability, and stronger absorption. In the case of certain biomedical utilization, the fabricating of nanoparticles is problematic because they lack adequate functional characteristics. Smart nanomaterials grab this attention significantly since they can be triggered by external elements like temperature, pH, light, pressure, enzyme, etc. This chapter intends to provide the classifications and applications of smart nanomaterials as tissue engineering, drug delivery, sensor, etc.

Keywords

Smart Nanomaterials, Tissue Engineering, Drug Delivery, Sensors

Contents

Smart Nanomaterials and their Applications 100

1.   Introduction...101

2.   A comparison of smart materials and common materials................102
      2.1   Advantages and disadvantages of smart nanomaterials 102
      2.2   Nanoparticles as sensor..103

3.   Types of smart nanomaterials based on stimuli 103
      3.1   Physical stimuli-responsive smart nanomaterials...........................104

3.1.1   Thermoresponsive smart nanomaterials 104

3.1.2   Piezoelectric smart nanomaterials 105

3.1.3   Electrochemical-responsive smart nanomaterials 107

3.1.4   Magneto responsive smart nanomaterials 108

3.1.5   Light responsive smart nanomaterials 109

3.2   Chemical-responsive smart nanomaterials 110

3.2.1   pH-responsive smart nanomaterials..110

3.2.2   Enzyme-responsive smart nanomaterials 115

Conclusions 116

Acknowledgments 116

References 116

1.   Introduction

From ancient periods, humanity has been learning how to survive in nature and protect themselves from the environment. Living organisms that survive in nature constantly change their characteristics and properties smartly. For example, touch and other stimuli, such as temperature and light, cause the *Mimosa pudica* plant to swiftly fold its leaves and change orientation [1]. Similarly, pinecones open and close in response to moisture [2]; with the variation of humidity, wheat awn changes its orientation [3]; in response to ambient humidity, orchid tree seedpods change into a distinct twisted structure [4]. Furthermore, the *Dionaea muscipula* can promptly close its leaves, which takes only 100 ms, and catch insects instantly [5]. The biological system's high potential ability to respond to the environment encouraged scientists to develop 'stimuli-responsive' materials.

In 1990 Toshinori Takagi first introduced the term 'intelligent material' and defined it as the material that responds to environmental changes and instantly displays changes in its function [6]. Nowadays, intelligent materials are known as 'smart material' or 'stimuli-responsive material.' Currently, the criteria for defining smart materials have been expanded. Materials that can be stimulated by particular outside sources and demonstrate different functional features are called smart materials. Temperature, electrical, magnetic field, ultrasound, pH, solvent, light, pressure, glucose, enzyme, etc. can be stimuli triggers which are illustrated in Fig. 1 [7].

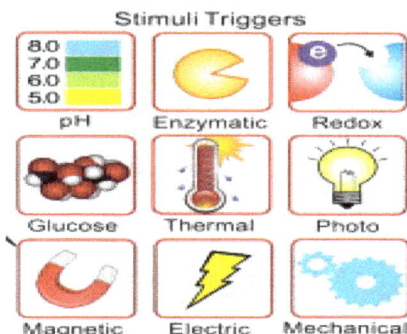

Figure 1. Various stimulating agents. Reproduced with permission [7]. Copyright 2014, Elsevier.

When exposed to appropriate stimuli, smart materials change its physical and chemical properties such as size, toughness, orientation, density, modulus, shape, surface area, permeability, and solubility. Smart materials are now enormously attractive to researchers because of their unique features and use in advanced technologies such as smart or intelligent devices.

2. A comparison of smart materials and common materials

The primary characteristics that distinguish smart materials from other materials are their unique and controllable functional features. Maximum bulk materials have fixed features that can be modified by changing functional groups; however smart nanomaterials may react to specific stimuli and occasionally display the latest functional qualities. In comparison to bigger particles, smart nanoparticles exhibit a higher surface area by mass. A particular quantity of smart nanoparticles might be substantially more responsive compared to a similar amount of substances consisting of bigger particles because growth and catalytic reactions take place on surfaces. In comparison to ordinary bulk substances, which are complex, intricate, and time-consuming, nanomaterials have a simple and rapid response.

2.1 Advantages and disadvantages of smart nanomaterials

Smart nanomaterials have many advantages. The following are the most notable advantages of smart materials:

- Can be stimulated by external triggers
- Ability to alter their own physical and chemical properties
- Immediacy (can respond quickly)

- Transiency (react to more than one environment)
- Self-actuation (own unique internal properties)
- Directness (both output and input are generated at the same site)
- Selectivity (the response to a specific environmental condition is consistent and reproducible)
- Excellent durability, reliability, energy density, bandwidth
- Smart materials embedded gadgets can promptly observe the health

Disadvantages-

- Costly, sensible, and requires appropriate storage
- Not easily accessible, long-term effects unknown
- Issues about biocompatibility and sustainable well-being [8-11]
- Appropriate skill to distinguish it among the other materials

2.2   Nanoparticles as sensor

In the research field, nanomaterial-based sensors are interesting and have potential because they have the capability to detect nanomolar to sub-picomolar level [12]. For this, scientists and engineers have a substantial interest in determining the particular contaminants that stay at an analyte, especially if their concentration exceeds the statutory limit. Due to their ease of fabrication, comparatively lower price, great sensitivity, and excellent selectivity, nanosensors have become broadly used in several fields in modern times. Smart nanomaterials have been customized into different sensors within the nanoscale on the basis of identification phenomena. The receptor and the transducer are generally the two fragments of a sensor. The receptor possesses a strong specificity characteristic that may strengthen detection sensitivity significantly. A transducer is generally responsible for taking the chemical information between receptor and analyte, which can be applied in numerous detection principles. Scientists can design sensors for the determination of a single sample or multiple samples. Multiple samples analysis is sometimes called multiplex determination [13]. Consequently, a large range of sensors for chemical, physical and biological determination has been established. Different smart nanomaterials frequently have some unique capabilities (physical, chemical, optical, electrical, catalytic, etc.), making them ideal for sensor development. As a result, an increasing concern for the maximum working field in analytical chemistry is nanomaterials-enabled sensors [14-17].

3.   Types of smart nanomaterials based on stimuli

Properties and characteristics of smart materials directly depend on the applied stimuli. When external stimuli are varied in a controlled manner, all features of smart nanomaterials are significantly altered [18-21]. Smart nanomaterials can be broadly divided into two classes based on the stimuli used. Stimuli are classified into two types: physical stimuli and chemical stimuli. The classification of smart nanomaterials is shown in Fig. 2.

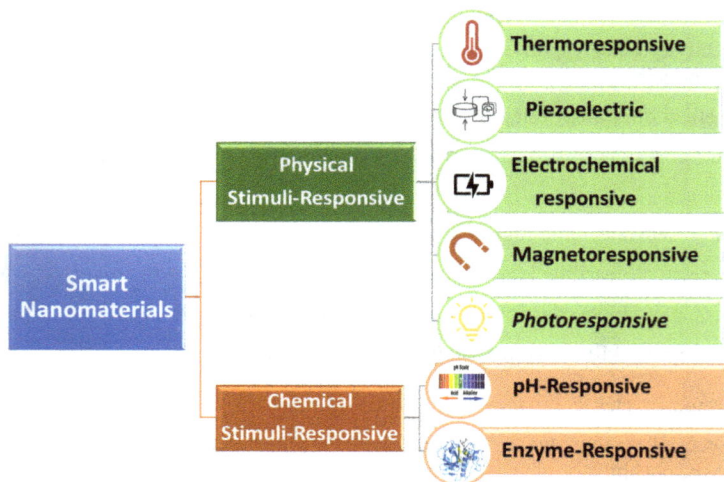

Figure 2. Types of Smart Nanomaterials based on Stimuli.

3.1   Physical stimuli-responsive smart nanomaterials

Some examples of physical stimuli-responsive smart nanomaterials are listed in Table 1.

3.1.1   Thermoresponsive smart nanomaterials

Thermoresponsive polymers (TRPs) are a kind of smart material that can respond instantly to temperature change by changing its solubility. For this unique feature, they can be used in a variety of ways. In the 1940s, for the first time, P.J. Flory and M.L. Huggins introduced the concept of polymer-solvent interaction for varying temperatures. They also explained the lower critical solution temperature (LCST) and upper critical solution temperature (UCST) [22, 23]. After that, temperature-responsive polymers received researchers' enormous interest. TRPs are mainly of two types: LCST and UCST. In general, TRPs contain some hydrophobic groups, such as methyl, ethyl, and propyl. A significant change in hydrophobic and intra/intermolecular interactions occurs when thermoresponsive materials get cooled or warmed over a critical transition point. LCST is the critical temperature below which the mixture's components are miscible, i.e., the polymer is monophasic [24]. This phenomenon happens because it is energetically favorable. On the contrary, above the LCST the polymers are biphasic, insoluble, and hydrophobic. The UCST is the critical temperature above which polymers exist in one phase only, whereas phase separation occurs below that temperature [53]. UCST is generally present in soluble polymers [54,55]. The TRP's distinct feature enables them to use for numerous purposes. For examples, tissue engineering [24], controlling radical polymerization [25], wastewater treatment [26], designing single cluster-based nanostructures [27], fluorescence sensors [28], and drug delivery [29,30]. A vast number of TRPs were successfully applied to deliver cancer drugs to the target site. The fundamentally required criteria for successful drug delivery systems are to deliver the drugs at the correct time, at the specific location, and in the proper concentrations. The problems with conventional are low solubility of drugs and not being not able to pass through the biological barriers [56]. TRPs overcome these obstacles by controlling the release of drugs with external stimuli.

In a recent study, liposomal nanoantagonist, a thermoresponsive material, was developed as photodynamic therapy for cancer treatment [57]. H. Sun et al. fabricated photothermal liposomal nanoantagonists (PLNA). To construct PLNA, they used a photosensitizer (indocyanine green) and an antagonist (L-buthionine sulfoximine). These two substances were enclosed in a temperature-responsive liposome to get the final product (PLNA). This near-infrared photothermal drug delivery nanosystem outperformed sole photodynamic therapy in terms of destroying cancer cells and preventing tumor development.

TRP-based sensors: In 2019, Zhao and his group [58] built a TRP-based electrochemical sensor. Using the ultrasonication process, a composite of carboxylated multi-walled carbon nanotubes and amino-functionalized graphene quantum dots was mixed with a thermos responsive material such as poly (styrene-b-(N-isopropylacrylamide)-b-styrene) (PS-PNIPAm-PS), and a drop of this combination was cast on the bare glassy carbon electrode to prepare the modified electrode which exhibited excellent performance to determine paracetamol. TRP became larger and blocked electron exchange at low temperatures. However, the electroactive site could be revealed due to the shrinkage of TRP at high temperature which permitted the electron exchange process of paracetamol. Therefore, on/off switching was realized by controlling the temperature. Liu et al. developed a molecularly imprinted polymer (MIP)-based resonance light scattering (RLS) thermos responsive sensor for detecting hepatitis A virus (HAV) [59].

3.1.2 Piezoelectric smart nanomaterials

The word "piezo" originates from the Greek word "piezein" which means pressure, and piezoelectricity means pressure-electricity. Smart materials which have the ability to turn mechanical pressure into electrical energy and vice-versa are known as piezoelectric materials (PEM) [60]. In 1880, Pierre and Jacques Currie first introduced the concept of the piezoelectric effect by observing the spark from quartz. PEM exhibits piezoelectricity depending on its crystal lattice structure and asymmetric nature in response to particular stimuli, for instance, mechanical pressure and electric charge. The piezoelectric effect is widely observed in nonconductive materials, such as quartz (SiO_2) and ceramics (lead zirconium titanate - PZT). According to the working principle of PEM, it exhibits a dipole moment by balancing ions in its crystal lattice when mechanical stress is applied (Fig. 3) [61].

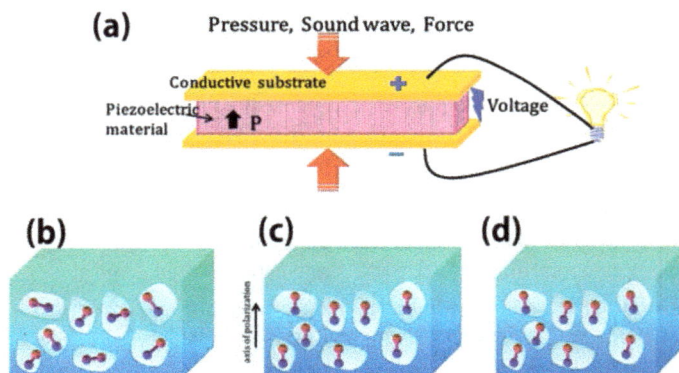

Figure 3. (a) Working principle of PEM; polarization conditions and directions (b) Without an applied force, pressure, or strain (c) During an applied force (d) After the applied force is removed. Reproduced with permission [61].

PEM uses in various electronic applications such as actuators, transducers, and sensors. Moreover, the unique controllable feature of PEM can be used industrially related to vibrational generation. Based on the piezoelectricity principle, many commercialized products such as speakers, hydrophones, and microphones were developed [62]. Over the advancement of nanotechnology, stimuli-responsive smart PEM were fabricated as thin films. Nowadays, smart PEM is mainly used in artificial targeted drug release platforms [31,39], tissue engineering [32-38], substrates for neural prostheses [40], muscle actuations [41], sensors [42,44], biosensors [40,43], and energy transductions. For example, Abidian et al. [31] built a controlled drug delivery system. In their study, the desired drug was first incorporated into poly(l-lactide) or poly(lactide-co-glycolide), a biodegradable polymer, coated by poly(3,4-ethylenedioxythiophene) nanotubes, a conducting polymer. Poly(3,4-ethylenedioxythiophene) increased the effective surface area, and activation of poly (3,4-ethylenedioxythiophene) by external electrical trigger enabled the error-free release of the desired drug.

PEM is a promising tool for tissue engineering and health monitoring. Some PEM-based wearable devices can predict the illness early by physiological monitoring. For example, Ramakrishna et al. [36] constructed aligned nanofibrous poly(L-lactic acid) (PLLA) scaffolds using the electrospinning technique. In their study, they found that the orientation of neural stem cell (NSC) elongation and neurite extension was completely identical to the orientation of PLLA fibers, and successfully employed in neural tissue engineering (NTE). Xue et al. [43] studied piezoelectric-based wearable and flexible multifunctional electronic-skin biosensors to analyze the particular metabolite's composition in perspiration. They prepared the biosensors using ZnO nanowire and specific enzymes that quickly monitor human health's physiological conditions by sensing uric acid, lactate, urea, and glucose.

3.1.3 Electrochemical-responsive smart nanomaterials

Electrochemical stimuli-responsive smart nanomaterials can usually change their physical properties by responding with an applied electric field. Therefore, these materials grab the attention of different fields such as sensors, robotics, actuators, drug delivery, etc. Electrochemical determination techniques usually determine an analyte by measuring the potential difference according to the reaction of an electrode with an analyte. Some electrochemical techniques are cyclic voltammetry, impedance spectroscopy, chronoamperometry, and chronopotentiometry, where electrodes play an important role [63]. Nanoparticles such as nano-carbons, polymers, or gold/silver, as well as antibodies, and aptamers, can be used to modify these electrodes. Electrodes are generally made of glassy carbon, gold, platinum, graphite, silver, etc. [64]. Compared to typical electrochemical approaches, the electrode's nanoscale production methodology amplifies the signal significantly and increases the signal-to-noise ratio. Electrical stimuli-responsive materials may be fabricated into a broad range of shapes as they have large extinction coefficients ($\varepsilon > 3 \times 10^{11}$ M^{-1} cm^{-1}) [65] and the simple process of adding new functions to their surface, which gives them value for manufacturing sensor. For these reasons, as a sensor, metal nanoparticles have been used extensively. Gold and silver nanoparticles in a colloidal state come in a variety of colors, depending on the colloidal nanomaterial's size. Metal nanoparticles, notably AuNPs, have been well known for their sensor application [45-47]. A silver nanoparticle is another alternative name to develop a sensor because of its anti-microbial property and dissolve ability, though it still has limitations. On the other hand, gold nanoparticles (AuNPs)have stability, biocompatibility, and their potential for sensing utilizations has been thoroughly investigated [66]. The unique physical and chemical features of AuNPs make them ideal platforms for the development of innovative sensing gadgets (Fig. 4) [66].

Some non-metallic examples of electro-responsive nanomaterials are highly conductive carbon nanotubes (CNTs) and graphenes; they also have the characteristics of vast surface regions and huge mechanical robustness [67]. The main objective of these nanomaterials is to make the electrodes, particularly glassy carbon electrodes (GCE), more sensitive to electrochemical determination [48,49]. Electrical stimuli-responsive polymers often have numerous ionizable groups which can execute the mechanical task from electrical energy. Smart nanomaterials possess excellent command over differential pulse, potential difference, and current intensity. Therefore, these smart materials are significantly used for artificially tailored medication release, energy conversion, and muscles actuation [31, 68, 69].

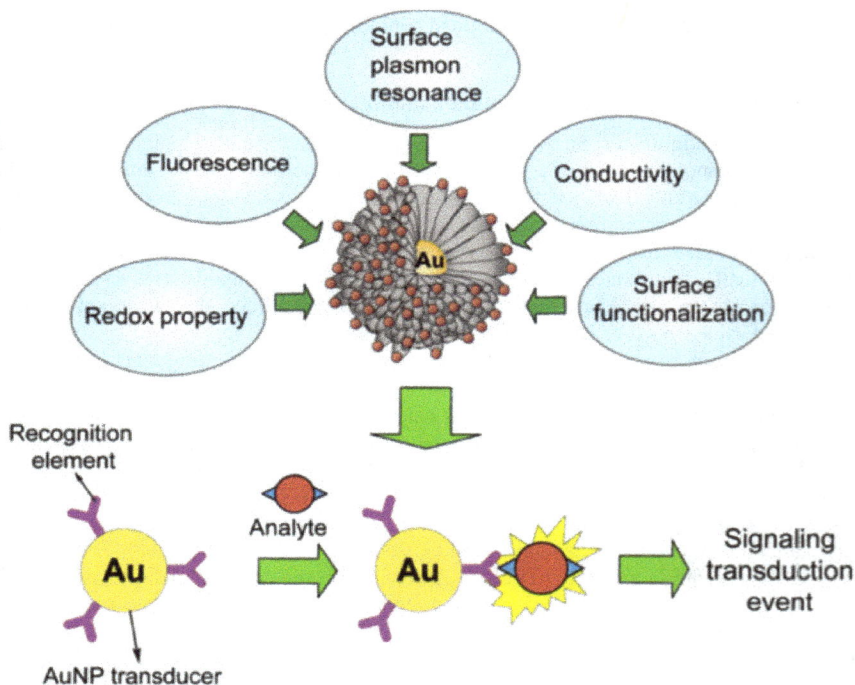

Figure 4. Physical properties of AuPs and schematic illustration of aAuNP-based detection system. Reproduced with the permission [66]. Copyright 2012, American Chemical Society.

3.1.4 Magneto responsive smart nanomaterials

The magnetic field is the stimulus agent for magneto responsive smart nanomaterials. Magnetic nanoparticles, in particular, are a soaring platform, and magneto responsive-based utilizations are receiving more attention. The advancement of these materials has enormous implications for electrical gadgets, magnetic storage, biomedical applications such as magnetic resonance imaging (MRI), magnetic hyperthermia, etc. By using different methods, including co-precipitation technique, hydrothermal method, high-temperature thermal decomposition, chemical vapor deposition, carbon arc, combustion, laser pyrolysis, microbial synthesis, electrochemical methods, etc., we can synthesize these magnetic responsive materials [70]. Magneto responsive smart nanoparticles can also potentially be used in diagnostic and therapeutic applications [70]. Magnetic responsive nanoparticles, for instance, iron, nickel, cobalt, copper, and metal oxides, offer a significant benefit in controlling cells, making cell functions and signaling easier to comprehend.

Certain magnetic responsive nanoparticles have been effectively used as an MRI contrast media, including iron, gadolinium, non-iron related smart nanoparticles, and so on [50]. Magnetic NPs are used in hyperthermia for targeting specific cancers as a developed treatment than before [71]. Gadolinium-based nanoparticles are commonly used in clinical practice as an MRI contrast media to help diagnose tumors [51]. Magnetic Resonance Imaging (MRI) is a sophisticated technology that can be used clinically to diagnose a variety of disorders since it provides real-time spatial resolution and better contrast of soft tissues with no harmful effects. However, there are still some serious disadvantages of MRI, for example, sensitivity, toxicity, and delayed imaging rate and accuracy, which should be improved by developing- enhancing, and controlling the magnetic field capable instrumentation, less cytotoxicity, stable smart nanomaterial.

3.1.5 Light responsive smart nanomaterials

Materials that can respond to external light (ultraviolet, visible, or infrared light) are called light-responsive materials. This response is also known as optical transduction, and it occurs when a light-responsive material interacts with electromagnetic radiation. Fluorescence and surface plasmon resonance-based spectroscopies are two typical optical approaches used in sensor development. Fluorescence spectroscopy uses a beam of light that comes back to its ground state after being excited to measure and identify molecules. The fluorescence signal is caused by either the nanoparticles interacting directly with the analyte or occurring a sensor conformational change. The oscillation of conduction electrons caused by smart nanomaterials' localized surface plasmon resonance stimulated by light, as shown in Fig. 5, is known as surface plasmon resonance spectroscopy, and this is a superior method for sensing biological and chemical determination [72].

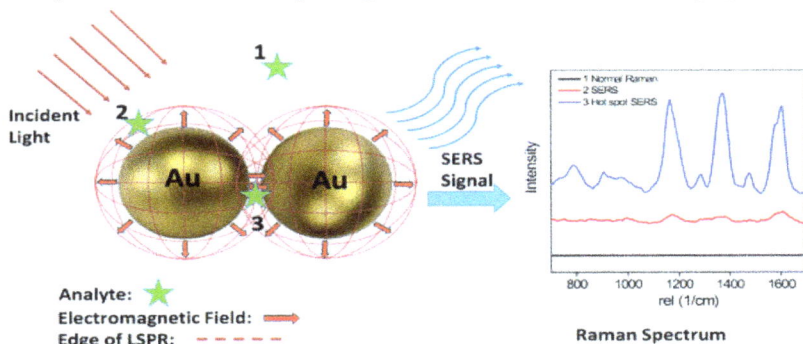

Figure 5. Schematic of surface-enhanced Raman spectroscopy phenomenon for an organic analyte on gold nanoparticles. Reproduced with permission [72].

One example of a light-responsive smart nanomaterial is graphene. Graphene derivatives have a large number of properties; light responsivity is one of them, and by transferring energy, they have the capacity to quench every fluorescence wavelength. The effectiveness of this capability is dependent on the graphene layers number and the oxidation state of those layers that means if a large number of graphene oxide is used, the fluorescence-quenching ability will be less effective. By merely tweaking the oxygen-containing moieties, it is feasible to change the maximal emitted wavelength of fluorescence from blue to near-infrared [73]. Light responsive nanomaterials have several applications due to their captivating stimulus, controllable, and tunable properties. Because regional hyperthermia destroys cancer and bacteria cells, these nanomaterials are employed in photothermal therapy. Chromophores include nitrobenzyl groups, spiropyran groups, and azobenzene groups, among others, are light-sensitive polymers [52]. Smart nanomaterials have the ability to generate active oxygen groups (hydroxyl, peroxide, singlet oxygen, and so on) and have strong water solubility that is strongly effective for photodynamic treatment. To produce harmful active oxygen groups in a particular places, photodynamic therapy typically needs a small amount of laser power. Several photodynamic processes have traditionally relied on organic photosensitizers to regulate photodynamic effects; however, smaller molar extinction coefficients, poor solubility, photobleaching, and have all shown to be limiting considerations [74]. Only the UV/vis range activates the maximum photosensitizers that could result in light poisoning and minimal tissue penetrating. Photoresponsive smart nanomaterials provide near-infrared light activation, active oxygen groups, targeting capability, and high solubility of water to overcome these hurdles, making them suitable agents to solve deep tissue concerns [75].

3.2   Chemical-responsive smart nanomaterials

Some examples of Chemical stimuli-responsive smart nanomaterials are listed in Table 2.

3.2.1 pH-responsive smart nanomaterials

pH-responsive nanomaterials change their functional properties in response to changes in pH. The life of living organisms relies on the proper level of pH. For the survival of humans, a pH level of roughly 7.4 in the serum must be closely managed. Different pH values exist in individual regions of our body, such as the pH value of saliva is between 6.5 and 7.5 (pH; stomach: (4–6.5), intestine: (5–8)) [76], at the meantime urine pH must not fall below 4.4. As an effect of increasing industrialization over 100 years, CO_2 deposition increased; as a result, the pH of the ocean water has fallen from 8.2 to 8.1. This situation affects harmfully on life in the ocean [77-78]. Because of the multifarious range of pH levels, noticeable pH-stimulus nanomaterials have been exploited. Scientists discovered that pH-stimulus silver nanoparticles cluster prepared from PEG corona and orthoester are beneficial for treating methicillin-resistant staphylococcus aureus infection in vivo wound healing tests [79]. Electronic devices (mainly switches) are commonly made using transistors (which are used to amplify current or signal).To investigate *the acidification of endocytic organelles from living cells,* researchers developed a unique

Materials Research Foundations 135 (2023) 100-124   https://doi.org/10.21741/9781644902172-5

hybrid nanoparticle arrangement that behaves like a sensor at a predetermined pH and is made up of three pH-stimulus block copolymers. The pH range of this super pH-responsive nano transistor is 6.9, 6.2, and 5.3, and it has a significant impact on biomedical applications [80]. pH-sensitive polymers classification includes those with ionizable groups and those with acid-labile bonds. Polymers that are ionizable, acidic, or alkaline and attached to an aquaphobic spine comprise the first group. All carboxylic acids and amines belong to this group, for example, polyelectrolytes [81]. The characteristics of this kind of pH-responsive material can be: exhibiting protonation/deprotonation reactions on the polymer molecules' ionizable category. Polymers (containing spine) with acid-labile covalent connections are the second type. pH is responsible (reduction) for determining the separation of polymers aggregate or dissociating polymer chains from the cleavage of the bonds. Because of the existence of covalent bonding, these polymers have a slower internal change, which makes them suitable for drug release. Within the pH scale of 0.2–0.3, a rapid transition phase of pH-responsive polymer shifts. Hyaluronic acid, alginate, poly(ethylenimine), poly (L-lysine), chitosan, etc., are ordinarily some familiar polymers that have excellent pH sensitivity [82,83]. One of the most common pH-sensitive smart nanoparticle names is chitosan, a polysaccharide containing a polymeric backbone with several main amino groups of different behavior. The creation of pH-sensitive biomaterials is one of its(chitosan) major applications [84]. Chitosan and its derivatives possess a positive surface charge in pathogenic environments (acidic states) because of the protonation of amino groups and negative/neutral oxidation state of the surface, and these types of chitosan have been used to target and provide particular areas for therapeutic effectiveness [85]. Because of the hopeful advantages of glycol chitosan, specifically in drug delivery application, the pH-sensitive glycol chitosan-coated liposome system was created by Yan et al. in 2015, particularly for tumor treatment. In comparison to the ordinary liposomes, the coated liposomes produce negative-to-positive ion reversal from pH 7.4 to pH 6.5, which reduces cellular absorption and hence results in improved antitumor efficacy by accumulating doxorubicin(DOX) [86]. Another pH-responsive polymer, poly-l-histidine (PLH), a cationic polypeptide containing imidazole groups which, is well known for the fabrication of pH-stimuli nanomaterial in biomedical treatment, especially in drug delivery systems. For example, pH-sensitive, drug-encapsulated, surface charge-shifting *poly(D, L-lactic-co-glycolic acid)-b-poly(L-histidine)-b-poly(ethylene glycol) (PLGA-PLH-PEG)* nanoparticles as a treatment of bacterial infections, were developed using PLH. Nanoparticle binding tests revealed that in acidic circumstances, (PLGA-PLH-PEG) pH-responsive nanoparticles attached to bacteria (Fig. 6) [87].

Table 1. *Applications of some physical stimuli-responsive smart nanomaterial are reported in the literature.*

Types of stimuli	Smart material	Application	Ref.
Temperature	Poly(N-isopropylacrylamide) –co-poly(glycidal methacrylate) Poly(N-vinylpyrrolidone)	Tissue engineering	[24]
Temperature	Poly(N-acryloylasparaginamide)	Controlling radical polymerization	[25]
Temperature	Poly(methacrylamide) Poly(acrylamide-coacrylonitrile)	Wastewater treatment	[26]
Temperature	Polyoxometalate	Designing single cluster-based nanostructures	[27]
Temperature	Europium Metal-Organic Framework EuL MOF	Fluorescence sensors, Detection of 2,4,6-trinitrophenol	[28]
Temperature	AuNps- PF127- HPMC	Drug delivery	[29]
Temperature	PNIPAM-PDMA-PAA	Drug delivery	[30]
Electrical	Poly(3,4-ethylenedioxythiophene)-coated poly(lactide-co-glycolide)	Drug delivery	[31]
Electrical	Polyvinylidene Fluoride	Tissue engineering	[32-35]
Electrical	Polylactic acid	Tissue engineering	[36]
Electrical	Collagen	Tissue engineering	[37, 38]
Electrical	Fe_3O_4/Polyaniline	Antimicrobial, drug delivery	[39]
Electrical	Polyaniline, poly(3,4-ethylenedioxythiophene	Neural prostheses	[40]
Stress	Poly(vinylidene fluoride)	Monitor human respiration, subtle muscle movement	[41]
Stress	Collagen films	Temperature sensor, strain gauge, wireless antenna	[42]
Stress	Enzyme/ZnO nanoarrays	Biosensing, detect lactate, glucose, uric acid, and urea in the perspiration	[43]

Stress	rGO/PVDF microdome array	Dynamic pressure sensor, temperature sensor	[44]
Electrochemical	AuNPs	Sensors	[45, 46]
Electrochemical	Silver nanoparticle	Sensors	[47]
Electrochemical	CNTs	Sensors	[48, 49]
Magnetic	Iron, nickel, cobalt, and metal oxides	Controlling cells, facilitating the understanding of cell functions and signaling	[50]
Magnetic	Gadolinium-based nanoparticles	MRI contrast agent for tumor diagnosis	[51]
Light	Poly(ethylene glycol)	Switchable fluorescent probes	[52]

Table 2. *Applications of some chemical stimuli-responsive smart nanomaterial are reported in the literature.*

Types of stimuli	Smart material	Application	Ref.
pH	Chitosan	Targeting and giving specific therapeutic viability	[85]
pH	Glycol chitosan-coated liposome	Tumor-specific drug delivery system	[86]
pH	Poly-l-histidine (PLH)	Treating bacterial infections	[87]
pH	Polyacetal	Tumor cell-specific paclitaxel delivery	[88]
pH	Polyacetal-DOX conjugates	Confirmation of tumors	[89]
Enzyme	*Nanoplatform formed from indocyanine green (ICG) and PEG-b-P(Pt co-GFLG)-b-PEGICG/Poly(Pt)*	Photo-chemotherapy	[90]
Enzyme	*Nanoplatform modified with peptide capped mesoporous silica nanoparticles* Multifunctional theranostic nanoplatform (MMTNP)	Tumor-targeted drug delivery, tumor imaging	[91]
Enzyme	*Nanoplatform of Ti substrates modified with layer-by-layer mesoporous silicananoparticles-silver nanoparticles* LBL@MSN-Ag nanoparticles	Treating bacterial infection	[92]

| Enzyme | *Adenosine triphosphate coated with silver nanoparticles ATP-Ag nanoparticles* | Antibacterial activity, anticancer activity | [93] |
| Enzyme | *Layer-by-layer assembly of poly(2-oxazoline)-based materials* | Therapeutic delivery | [94] |

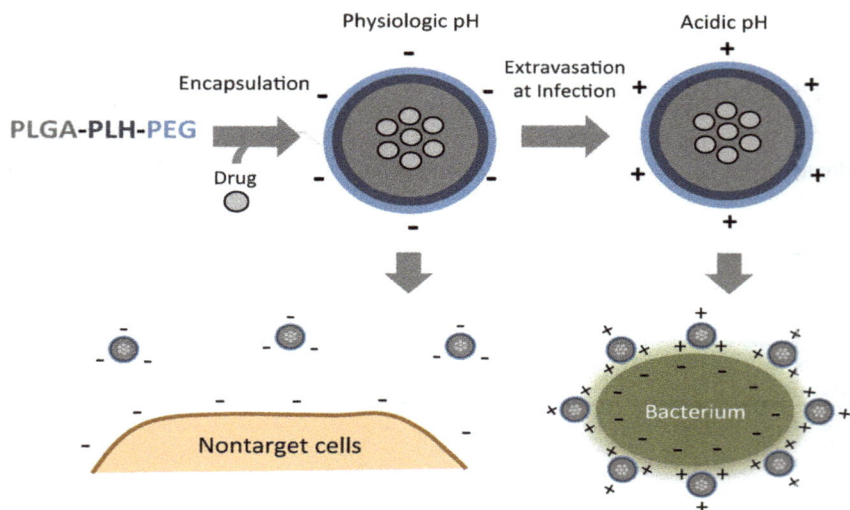

Figure 6. Schematic representation of the designed nanoparticle (NP)-mediated drug targeting to bacterial cell walls. Drugs are encapsulated into NPs using a double emulsion/solvent evaporation process. The NPs avoid uptake or binding to nontarget cells or blood components at physiologic pH 7.4 due to a slight negative charge and surface PEGylation. Inflammation at a site of infection causes increased local vascular permeability, promoting NP extravasation. The weakly acidic conditions at sites of certain infections activate the surface charge-switching mechanism, resulting in NP binding to negatively charged bacteria. Finally, controlled release of the encapsulated drug leads to antibacterial effect. Reproduced with permission [87].

Orthoester bonds imine, hydrazone, amide, ether, polyacetal and polyketide, oxime, and imine are the most commonly utilized pH-responsive linkers. Yang et al. created pH-sensitive polymeric conjugates particularly for tumor cell paclitaxel administration by covalently connecting polyethylene glycol with *4β-aminopodophyllotoxin(NPOD)* by imine linkage [88]. To create water-soluble polyacetal-DOX conjugates, Tomlinson et al.

developed amino- functionalizing linear pH-responsive amino-pendent polyacetals (APEGs). Polyacetal-DOX (pH-responsive) conjugates are suitable for drug delivery because of the small accumulation of DOX in the spleen and liver and higher formation in the tumor. As a result, these could be used as biodegradable transporters to help anticancer drugs target tumors more precisely [89].

3.2.2 Enzyme-responsive smart nanomaterials

Enzyme-responsive systems are extensively used in biomedical fields because of their enhanced selectivity, decent catalytic efficiency, perfect biological recognition, and uncomplicated decomposition. Bacteria that live in the human body secrete specific enzymes into different organs, which degrade various types of polysaccharides.

Enzyme responsive nanoplatforms have versatile applications. They have a wide range of applications in curing cancerous cells [90,91]. W. Wang et al. [90] created polymeric nanoplatforms (Fig. 7) to cure cancerous cells. The enzyme-triggered polymeric nanoplatform, ICG/Poly(Pt), was prepared by co-assembling of PEG-b-P(Pt-*co*-GFLG)-b-PEG (cisplatin) and indocyanine green (ICG). To begin, cathepsin B aids in releasing ICG and cisplatin; after that, laser enables the phototherapy, which can potentially destroy the lysosome. Finally, chemotherapy was initiated by reductive milieu.

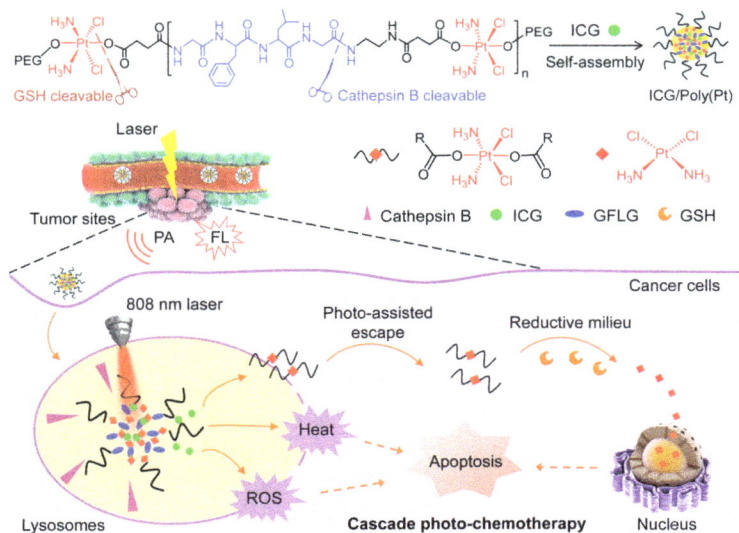

Figure 7.A schematic representation of ICG-Loaded enzyme-responsive polyprodrug nanoplatform. Reproduced with permission [90].

Nowadays, enzyme-stimulated polymer nanoplatforms are now widely used for the purpose of anti-microbial activities [92,93]. Y. Ding and his group [92] engineered an enzyme-responsive nanoplatform to treat bacterial infections caused by S. aureus. The deposition of silver nanoparticles mixed with multilayer mesoporous silica on titanium was used to create the nanoplatform. Gunawan et al. [94] introduced a carrier system based on multilayer poly(2-oxazoline) and used it to generate thrombolysis.

Conclusions

This chapter describes the growing development of smart nanomaterials as nanosensors and other applications such as drug delivery, flexible and wearable devices for health monitoring, cancer, bacteria therapeutics, etc., and focused on important recent advances in the design of nanomaterials. Nanosized smart materials belonging to carbonaceous, metallic, or inorganic and recognition agents are continuously being combined in new and creative ways for the development of analytical devices and other applications. The unique stimuli-responsive properties of smart nanomaterials have enormous utility for developing various determination techniques with high sensitivity, specificity, speed, simplicity, long-term stability, and cost-effectiveness, although it is still an urgent need for further improvement. Present smart materials can provide a versatile toolbox for controlled therapeutics by regulating the specific stimuli agents to enhance the therapeutic efficiencies though there are still some limitations. For further improvement of the disease diagnosis, dual or multi-responsive smart nanomaterials are taking thoughtful attention recently where we can trigger the multiple functions on a single smart nanomaterial. Overall, this field represents one of the future aspects of the research in sensing devices and other applications, which requires strict cooperation between different expertise such as those of physical, chemical, materials, engineering, and biological areas, boosting the multifarious vision of science and human health.

Acknowledgments

This work was partially supported by the Department of Chemistry, Jagannath University, Bangladesh, and the Department of Civil Engineering, Dhaka International University, Bangladesh.

References

[1] S. Amador-Vargas, M. Dominguez, G. León, B. Maldonado, J. Murillo, G.L. Vides, Leaf-folding response of a sensitive plant shows context-dependent behavioral plasticity, Plant Ecol. 215 (2014) 1445-1454. https://doi.org/10.1007/s11258-014-0401-4

[2] E. Reyssat, L. Mahadevan, Hygromorphs: From pine cones to biomimetic bilayers, J. R. Soc. Interface 6 (2009) 951-957. https://doi.org/10.1098/rsif.2009.0184

[3] R. Elbaum, L. Zaltzman, I. Burgert, P. Fratzl, The role of wheat awns in the seed dispersal unit, Science 316 (2007) 884-886. https://doi.org/10.1126/science.1140097

[4] R.M. Erb, J.S. Sander, R. Grisch, A.R. Studart, Self-shaping composites with programmable bioinspired microstructures. Nat. Commun. 4 (2013) 1712. https://doi.org/10.1038/ncomms2666

[5] Y. Forterre, J.M. Skotheim, J. Dumais, L. Mahadevan, How the Venus flytrap snaps, Nature 433 (2005) 421-425. https://doi.org/10.1038/nature03185

[6] T. Takagi, A concept of intelligent materials, J. Intell. Mater. Syst. Struct. 1 (1990) 149-156. https://doi.org/10.1177/1045389X9000100201

[7] Y. Lu, W. Sun, Z. Gu, Stimuli-responsive nanomaterials for therapeutic protein delivery, J. Control. Release 194 (2014) 1-19. https://doi.org/10.1016/j.jconrel.2014.08.015

[8] S. Kim, S.H. Ku, S.Y. Lim, J.H. Kim, C.B. Park, Graphene-biomineral hybrid materials, Adv. Mater. 23 (2011) 2009-2014. https://doi.org/10.1002/adma.201100010

[9] T.F. Otero, J.G. Martinez, J. Arias-Pardilla, Biomimetic electrochemistry from conducting polymers. A review: Artificial muscles, smart membranes, smart drug delivery and computer/neuron interfaces, Electrochim. Acta 84 (2012) 112-128. https://doi.org/10.1016/j.electacta.2012.03.097

[10] D. Bitounis, H. Ali-Boucetta, B.H. Hong, D.H. Min, K. Kostarelos, Prospects and challenges of graphene in biomedical applications, Adv. Mater. 25 (2013) 2258-2268. https://doi.org/10.1002/adma.201203700

[11] L. Liu, P. Li, G. Zhou, M. Wang, X. Jia, M. Liu, X. Niu, W. Song, H. Liu, Y. Fan, Increased proliferation and differentiation of pre-osteoblasts MC3T3-E1 cells on nanostructured polypyrrole membrane under combined electrical and mechanical stimulation, J. Biomed. Nanotechnol. 9 (2013) 1532-1539. https://doi.org/10.1166/jbn.2013.1650

[11] G. Aragay, F. Pino, A. Merkoçi, Nanomaterials for sensing and destroying pesticides, Chem. Rev. 112 (2012) 5317-5338. https://doi.org/10.1021/cr300020c

[12] M.R. Willner, P.J. Vikesl, Nanomaterial enabled sensors for environmental contaminants, J. Nanobiotechnol. 16 (2018) 95. https://doi.org/10.1186/s12951-018-0419-1

[13] J. Zhang, L. Wang, D. Pan, S. Song, F.Y.C. Boey, H. Zhang, C. Fan, Visual cocaine detection with gold nanoparticles and rationally engineered aptamer structures, Small 4 (2008) 1196-1200. https://doi.org/10.1002/smll.200800057

[14] S. Song, Z. Liang, J. Zhang, L. Wang, G. Li, C. Fan, Gold-nanoparticle-based multicolor nanobeacons for sequence-specific DNA analysis, Angew. Chem. Int. Ed. Engl. 48 (2009) 8670-8674. https://doi.org/10.1002/anie.200901887

[15] P. B. Chouke, K. M. Dadure, A. K. Potbhare, G. S. Bhusari, A. Mondal, K. Chaudhary, V. Singh, M. F. Desimone, R. G. Chaudhary, D. T. Masram, Biosynthesized δ-Bi2O3 nanoparticles from Crinum viviparum flower extract for photocatalytic dye degradation and molecular docking, ACS Omega, 7 (2022) 20983-20993. https://doi.org/10.1021/acsomega.2c01745

[17] P.B. Chouke, A.K. Potbhare, N.P. Meshram, M. M. Rai, K.M. Dadure, K. Chaudhary, A.R. Rai, M. Desimone, R. G. Chaudhary, D.T. Masram, Bioinspired NiO nanospheres: Exploring in-vitro toxicity using Bm-17 and L. rohita liver cells, DNA degradation, docking and proposed vacuolization mechanism. ACS Omega. 7 (2022) 6869−6884. https://doi.org/10.1021/acsomega.1c06544

[18] S.S. Das, P. Bharadwaj, M. Bilal, M. Barani, A. Rahdar, P. Taboada, S. Bungau, G.Z. Kyzas, Stimuli-responsive polymeric nanocarriers for drug delivery, imaging, and theragnosis, Polymers 12 (2020) 1397. https://doi.org/10.3390/polym12061397

[19] A K. Potbhare, R.G. Chaudhary, P.B. Chouke, A. Rai, A. Abdala, R. Mishra, M. Desimone, Graphene-based materials and their nanocomposites with metal oxides: Biosynthesis, electrochemical, photocatalytic and antimicrobial applications. Mater. Res. Forum. 83 (2020) 79-116. https://doi.org/10.21741/9781644900970-4

[20] M. Pinteala, M.J.M. Abadie, R.D. Rusu, Smart Supra- and Macro-Molecular Tools for Biomedical Applications, Materials 13 (2020) 3343. https://doi.org/10.3390/ma13153343

[21] E. Cabane, X. Zhang, K. Langowska, C.G. Palivan, W. Meier, Stimuli-responsive polymers and their applications in nanomedicine, Biointerphases 7 (2012) 9. https://doi.org/10.1007/s13758-011-0009-3

[22] P.J. Flory, Thermodynamics of high polymer solutions, J. Chem. Phys. 9 (1941) 660. https://doi.org/10.1063/1.1750971

[23] M.L. Huggins, Some properties of solutions of long-chain compounds, J. Phys. Chem. 46 (1942) 151-158. https://doi.org/10.1021/j150415a018

[24] D. Roy, W.L.A. Brooks, B. S. Sumerlin, New directions in thermoresponsive polymers, Chem. Soc. Rev.42 (2013) 7214-7243. https://doi.org/10.1039/c3cs35499g

[25] S. Glatzel, A. Laschewsky and J.F. Lutz, Well-defined uncharged polymers with a sharp UCST in water and in physiological milieu, Macromolecules 44 (2011) 413-415. https://doi.org/10.1021/ma102677k

[26] J. Seuring and S. Agarwal, First example of a universal and cost-effectiveapproach: Polymers with tunable upper critical solution temperature in water and electrolyte solution, Macromolecules 45 (2012) 3910-3918. https://doi.org/10.1021/ma300355k

[27] Q. Liu, H. Yu, Q. Zhang, D. Wang, X. Wang, Temperature-Responsive Self-Assembly of Single Polyoxometalates Clusters Driven by Hydrogen Bonds, Adv. Funct. Mater. 31 (2021) 2103561. https://doi.org/10.1002/adfm.202103561

[28] M. Zhu, X.Z. Song, S.Y. Song, S.N. Zhao, X. Meng, L.L Wu, C. Wang, H.J. Zhang, A Temperature-Responsive Smart Europium Metal-Organic Framework Switch for Reversible Capture and Release of Intrinsic Eu3+ Ions, Adv. Sci. 2 (2015) 1500012. https://doi.org/10.1002/advs.201500012

[29] M.G. Arafa, R.F. El-Kased, M.M. Elmazar, Thermoresponsive gels containing gold nanoparticles as smart antibacterial and wound healing agents, Sci. Rep. 8 (2018) 13674. https://doi.org/10.1038/s41598-018-31895-4

[30] Y. Chen, Y. Gao, L.P. da Silva, R.P. Pirraco, M. Ma, L. Yang, R.L. Reis, J. Chen, A thermo-/pH-responsive hydrogel (PNIPAM-PDMA-PAA) with diverse nanostructures and gel behaviors as a general drug carrier for drug release, Polym. Chem. 9 (2018) 4063-4072. https://doi.org/10.1039/C8PY00838H

[31] M.R. Abidian, D.H. Kim, D.C. Martin, Conducting-Polymer Nanotubes for Controlled Drug Release, Adv. Mater. 18 (2006) 405-409. https://doi.org/10.1002/adma.200501726

[32] N. Royo-Gascon, M. Wininger, J.I. Scheinbeim, B.L. Firestein, W. Craelius, Piezoelectric substrates promote neurite growth in rat spinal cord neurons, Ann. Biomed. Eng. 41 (2013) 112-122. https://doi.org/10.1007/s10439-012-0628-y

[33] R.F. Valentini, T.G. Vargo, J.A. Gardella Jr., P. Aebischer, Electrically charged polymeric substrates enhance nerve fibre outgrowth In vitro, Biomaterials 13 (1992) 183-190. https://doi.org/10.1016/0142-9612(92)90069-Z

[34] P. Aebischer, R.F. Valentini, P. Dario, C. Domenici, P.M. Galletti, Piezoelectric guidance channels enhance regeneration in the mouse sciatic nerve after axotomy, Brain Res. 436 (1987) 165-168. https://doi.org/10.1016/0006-8993(87)91570-8

[35] H. Delaviz, A. Faghihi, A.A. Delshad, M. hadiBahadori, J. Mohamadi, A. Roozbehi, Repair of peripheral nerve defects using a polyvinylidene fluoride channel containing nerve growth factor and collagen gel in adult rats, Cell J. 13 (2011) 137-142.

[36] F. Yang, R. Murugan, S. Wang, S. Ramakrishna, Electrospinning of nano/micro scale poly(L-lactic acid) aligned fibers and their potential in neural tissue engineering, Biomaterials 26 (2005) 2603-2610. https://doi.org/10.1016/j.biomaterials.2004.06.051

[37] R.C. de Guzman, J.A. Loeb, P.J. VandeVord, Electrospinning of matrigel to deposit a basal lamina-like nanofiber surface, J. Biomater. Sci. Polym. Ed. 21 (2010) 1081-1101. https://doi.org/10.1163/092050609X12457428936116

[38] T.J. O'Shaughnessy, H.J. Lin, W. Ma, Functional synapse formation among rat cortical neurons grown on three-dimensional collagen gels, Neurosci. Lett. 340 (2003) 169-172. https://doi.org/10.1016/S0304-3940(03)00083-1

[39] M.K. Hossain, H. Minami, S.M. Hoque, M.M. Rahman, M.K. Sharafat, M.F. Begum, M.E. Islam, H. Ahmad, Mesoporous electromagnetic composite particles: Electric current responsive release of biologically active molecules and antibacterial

properties, Colloids Surf. B: Biointerfaces 181 (2019) 85-93.
https://doi.org/10.1016/j.colsurfb.2019.05.040

[40] K.J. Otto, C.E. Schmidt, Neuron-targeted electrical modulation, Science 367 (2020) 1303-1304. https://doi.org/10.1126/science.abb0216

[41] Z. Liu, S. Zhang, Y.M. Jin, H. Ouyang, Y. Zou, X.X. Wang, L.X. Xie, Z.J. Li, Flexible piezoelectric nanogenerator in wearable self-powered active sensor for respiration and healthcare monitoring, Semicond. Sci. Technol. 32 (2017) 064004. https://doi.org/10.1088/1361-6641/aa68d1

[42] S. Moreno, M. Baniasadi, S. Mohammed, I. Mejia, Y. Chen, M.A. Quevedo-Lopez, N. Kumar,S. Dimitrijevich, M. Minary-Jolandan, Biocompatible collagen films as substrates for flexible implantable electronics, Adv. Electron. Mater. 1 (2015) 1500154. https://doi.org/10.1002/aelm.201500154

[43] W. Han, H. He, L. Zhang, C. Dong, H. Zeng, Y. Dai, L. Xing, Y. Zhang, X. Xue, A Self-Powered Wearable Non-invasive Electronic-Skin for Perspiration Analysis Basing on Piezo-Biosensing Unit Matrix of Enzyme/ZnO Nanoarrays, ACS Appl. Mater. Interfaces 9 (2017) 29526-29537. https://doi.org/10.1021/acsami.7b07990

[44] J. Park, M. Kim, Y. Lee, H.S. Lee, H. Ko, Fingertip skin-inspired microstructured ferroelectric skins discriminate static/dynamic pressure and temperature stimuli, Sci. Adv. 1 (2015) e1500661. https://doi.org/10.1126/sciadv.1500661

[45] Y.Y. Chen, H.T. Chang, Y.C. Shiang, Y.L. Hung, C.K. Chiang, C.C. Huang, Colorimetric assay for lead ions based on the leaching of gold nanoparticles, Anal. Chem. 81 (2009) 9433-9439. https://doi.org/10.1021/ac9018268

[46] Y.F. Lee, F.H. Nan, M.J. Chen, H.Y. Wu, C.W. Ho, Y.Y. Chen, C.C. Huang,Detection and removal of mercury and lead ions by using gold nanoparticle-based gel membrane, Anal Methods. 4 (2012) 1709-1717. https://doi.org/10.1039/c2ay05872c

[47] C.N. Lok, C.M. Ho, R. Chen, Q.Y. He, W.Y. Yu, H. Sun, P.K.H. Tam, J.F. Chiu, C.M. Che, Silver nanoparticles: partial oxidation and antibacterial activities, J. Biol. Inorg. Chem. 12 (2007) 527-534. https://doi.org/10.1007/s00775-007-0208-z

[48] H. Fan, Y. Li, D. Wu, H. Ma, K. Mao, D. Fan, B. Du, H. Li, Q. Wei,Electrochemical bisphenol A sensor based on N-doped graphene sheets, Anal. Chim. Acta. 711 (2012) 24-28. https://doi.org/10.1016/j.aca.2011.10.051

[49] J. Dong, X. Fan, F. Qiao, S. Ai, H. Xin, A novel protocol for ultra-trace detection of pesticides: Combined electrochemical reduction of Ellman's reagent with acetylcholinesterase inhibition, Anal. Chim. Acta 761 (2013) 78-83. https://doi.org/10.1016/j.aca.2012.11.042

[50] O. Perlman, H. Azhari, MRI and ultrasound imaging of nanoparticles for medical diagnosis, in: C. Kumar CSSR (edS) Nanotechnology characterization tools for

biosensing and medical diagnosis, Springer, Berlin, Heidelberg, 2018, pp. 333-365. https://doi.org/10.1007/978-3-662-56333-5_8

[51] Y. Shen, F.L. Goerner, C. Snyder, J.N. Morelli, D. Hao, D. Hu, X. Li, V.M. Runge, T1 relaxivities of gadolinium-based magnetic resonance contrast agents in human whole blood at 1.5, 3, and 7 T, Investig. Radiol. 50 (2015) 330-338. https://doi.org/10.1097/RLI.0000000000000132

[52] R. Klajn, Spiropyran-based dynamic materials, Chem. Soc. Rev. 43 (2014) 148-184. https://doi.org/10.1039/C3CS60181A

[53] Q. Zhang, C. Weber, U.S. Schubert, R. Hoogenboom, Thermoresponsive polymers with lower critical solution temperature: from fundamental aspects and measuring techniques to recommended turbidimetry conditions, Mater. Horizons 4 (2017) 109-116. https://doi.org/10.1039/C7MH00016B

[54] G. Marcelo, L.R. Areias, M.T. Viciosa, J.M.G. Martinho, J.P.S. Farinha, PNIPAm-based microgels with a UCST response, Polymer 116 (2017) 261-267. https://doi.org/10.1016/j.polymer.2017.03.071

[55] F.F. Sahle, M. Gulfam, T.L. Lowe, Design strategies for physical-stimuli-responsive programmable nanotherapeutics, Drug Discov. Today 23 (2018) 992-1006. https://doi.org/10.1016/j.drudis.2018.04.003

[56] L.J. Jeanneret, The targeted delivery of cancer drugs across the blood-brain barrier: chemical modifications of drugs or drug-nanoparticles?, Drug Discov. Today 13 (2008) 1099-1106. https://doi.org/10.1016/j.drudis.2008.09.005

[57] H. Sun, M. Feng, S. Chen, R. Wang, Y. Luo, B. Yin, J. Li, X. Wang, Near-infrared photothermal liposomal nanoantagonists for amplified cancer photodynamic therapy, J. Mater. Chem. B 8 (2020) 7149-7159. https://doi.org/10.1039/D0TB01437K

[58] P. Zhao, M. Ni, C. Chen, Z. Zhou, X. Li, C. Li, Y.Xie, J. Fei, Stimuli-enabled switch-like paracetamol electrochemical sensor based on thermosensitive polymer and MWCNTs-GQDs composite nanomaterial, Nanoscale11 (2019) 7394-7403. https://doi.org/10.1039/C8NR09434A

[59] Y. Liu, T. Shen, L. Hu, H. Gong, C. Chen, X. Chen, Development of athermosensitive molecularly imprinted polymer resonance light scattering sensor for rapid and highly selective detection of hepatitis A virus in vitro, Sens.Actuators B Chem. 253 (2017) 1188-93. https://doi.org/10.1016/j.snb.2017.07.166

[60] W. P. Mason, Piezoelectricity, its history and applications, J. Acoust. Soc. Am. 70 (1981) 1561-1566. https://doi.org/10.1121/1.387221

[61] S. Chandrasekaran, C. Bowen, J. Roscow, Y. Zhang, D.K. Dang, E.J. Kim, R.D.K. Misra, L. Deng, J.S. Chung, S.H. Hur, Micro-scale to nano-scale generators for energy harvesting: Self powered piezoelectric, triboelectric and hybrid devices, Phys. Rep. 792 (2018) 1-33. https://doi.org/10.1016/j.physrep.2018.11.001

[62] V.Y. Topolov, C.R. Bowen, P. Bisegna, New aspect-ratio effect in three-component composites for piezoelectric sensor, hydrophone and energy-harvesting applications, Sens. Actuators A Phys. 229 (2015) 94-103. https://doi.org/10.1016/j.sna.2015.03.025

[63] D. Grieshaber, R. MacKenzie, J. Vörös, E. Reimhult, Electrochemical biosensors-sensor principles and architectures, Sensors 8 (2008) 1400-58. https://doi.org/10.3390/s8031400

[64] O.A. Sadik, A.O. Aluoch, A. Zhou, Status of biomolecular recognition using electrochemical techniques, Biosens. Bioelectron. 24 (2009) 2749-2765. https://doi.org/10.1016/j.bios.2008.10.003

[65] S. Link, M.A. El-Sayed, Spectral properties and relaxation dynamics of surface plasmon electronic oscillations in gold and silver nanodots and nanorods, J. Phys. Chem. B 103 (1999) 8410-8426. https://doi.org/10.1021/jp9917648

[66] K. Saha, S.S. Agasti, C. Kim, X. Li, V.M. Rotello Gold nanoparticles in chemical and biological sensing, Chem. Rev. 112 (2012) 2739-2779. https://doi.org/10.1021/cr2001178

[67] W. Yang, K.R. Ratinac, S.P. Ringer, P. Thordarson, J.J Gooding, F. Braet, Carbon nanomaterials in biosensors: should you use nanotubes or graphene?, Angew. Chem. Int. Ed. Engl. 49 (2010) 2114-2138. https://doi.org/10.1002/anie.200903463

[68] J.H. Ha, H.H. Shin, H.W. Choi, J.H. Lim, S.J. Mo, C.D. Ahrberg, J.M. Lee, B.G. Chung, Electro-responsive hydrogel-based microfluidic actuator platform for photothermal therapy, Lab Chip 20 (2020) 3354-3364. https://doi.org/10.1039/D0LC00458H

[69] N.H. Nassab, D. Samanta, Y. Abdolazimi, J.P. Annes, R.N. Zare, Electrically controlled release of insulin using polypyrrole nanoparticles, Nanoscale 9 (2017) 143-149. https://doi.org/10.1039/C6NR08288B

[70] V.F. Cardoso, A. Francesko, C. Ribeiro, M. Bañobre-López, P. Martins, S. Lanceros-Mendez, Advances in magnetic nanoparticles for biomedical applications, Adv. Healthc. Mater. 7 (2018) 1700845. https://doi.org/10.1002/adhm.201700845

[71] K. Mahmoudi, A. Bouras, D. Bozec, R. Ivkov, C. Hadjipanayis, Magnetic hyperthermia therapy for the treatment of glioblastoma: a review of the therapy's history, efficacy and application in humans, Int. J. Hyperth. 34 (2018) 1316-1328. https://doi.org/10.1080/02656736.2018.1430867

[72] H. Wei, S.M.H. Abtahi, P.J. Vikesland, Plasmonic colorimetric and SERS sensors for environmental analysis, Environ. Sci.: Nano 2 (2015) 120-135. https://doi.org/10.1039/C4EN00211C

[73] E. Morales-Narváez, L. Baptista-Pires, A. Zamora-Gálvez, A. Merkoçi, Graphene-Based Biosensors: Going Simple, Adv. Mater. 29 (2017) 1604905. https://doi.org/10.1002/adma.201604905

[74] M. Lan, S. Zhao, W. Liu, C.S. Lee, W. Zhang, P. Wang, Photosensitizers for photodynamic therapy, Adv. Healthc. Mater. 8 (2019) 1900132. https://doi.org/10.1002/adhm.201900132

[75] R. Vankayala, K.C. Hwang, Near-infrared-light-activatable nanomaterial-mediated phototheranostic nanomedicines: an emerging paradigm for cancer treatment, Adv. Mater. 30 (2018) 1706320. https://doi.org/10.1002/adma.201706320

[76] A.A. Date, J. Hanes, L.M. Ensign, Nanoparticles for oral delivery: Design, evaluation and state-of-the-art, J. Control. Release 240 (2016) 504-526. https://doi.org/10.1016/j.jconrel.2016.06.016

[77] A. Waugh, A. Grant, Anatomy and Physiology in Health and Illness, tenth ed., Philadelphia, Pa, USA: Churchill Livingstone, Elsevier, 2007.

[78] O. Hoegh-Guldberg, P.J. Mumby, A.J. Hooten, R.S. Steneck, P. Greenfield, E. Gomez, C.D. Harvell, P.F. Sale, A.J. Edwards, K. Caldeira, N. Knowlton, Coral reefs under rapid climate change and ocean acidification, Science 318 (2007) 1737-1742. https://doi.org/10.1126/science.1152509

[79] X. Xie, T.C. Sun, J. Xue, Z. Miao, X. Yan, W. Fang, Q. Li, R. Tang, Y. Lu, L. Tang, Z. Zha, Ag nanoparticles cluster with pH-triggered reassembly in targeting anti-microbial applications, Adv. Funct. Mater. 30 (2020) 2000511. https://doi.org/10.1002/adfm.202000511

[80] Y. Wang, C. Wang, Y. Li, G. Huang, T. Zhao, X. Ma, Z. Wang, B.D. Sumer, M.A. White, J. Gao, Digitization of Endocytic pH by Hybrid Ultra-pH-Sensitive Nanoprobes at Single-Organelle Resolution, Adv. Mater. 29 (2017) 1603794. https://doi.org/10.1002/adma.201603794

[81] P.A. Lund, D.D. Biase, O. Liran, O. Scheler, N.P. Mira, Z. Cetecioglu, E.N. Fernandez, S. Bover-Cid, R. Hall, M. Sauer, C. O'Byrne, Understanding how microorganisms respond to acid pH is central to their control and successful exploitation, Front. Microbiol. 11 (2020) 556140. https://doi.org/10.3389/fmicb.2020.556140

[82] M.R. Aguilar, J.S. Román, 1 - Introduction to smart polymers and their applications, in: M.R. Aguilar, J.S. Román (Eds.), Smart Polymers and Their Applications, Woodhead Publishing, Cambridge, UK, 2014, pp. 1-11 https://doi.org/10.1533/9780857097026.1

[83] F. Reyes-Ortega, 3 - pH-responsive polymers: properties, synthesis and applications, in: M.R. Aguilar, J.S. Román (Eds.), Smart Polymers and Their Applications, Woodhead Publishing, Cambridge, UK, 2014, pp. 45-92. https://doi.org/10.1533/9780857097026.1.45

[84] B. Sultankulov, D. Berillo, K. Sultankulova, T. Tokay, A. Saparov, Progress in the Development of Chitosan-Based Biomaterials for Tissue Engineering and Regenerative Medicine, Biomol. 9 (2019) 470. https://doi.org/10.3390/biom9090470

[85] L. Yan , S.H. Crayton , J.P. Thawani , A. Amirshaghaghi ,A.Tsourkas , Z. Cheng, A pH-Responsive Drug-Delivery Platform Based on Glycol Chitosan-CoatedLiposomes, Small 11 (2015) 4870-4874. https://doi.org/10.1002/smll.201501412

[86] L. Yan, S.H. Crayton, J.P. Thawani, A. Amirshaghaghi, A. Tsourkas, Z. Cheng, A pH-Responsive Drug-Delivery Platform Based on Glycol Chitosan-Coated Liposomes, Small 11 (2015) 4870-4874. https://doi.org/10.1002/smll.201501412

[87] A.F. Radovic-Moreno, T.K. Lu, V.A. Puscasu, C.J. Yoon, R. Langer, O.C. Farokhzad, Surface charge-switching polymeric nanoparticles for bacterial cell wall-targeted delivery of antibiotics, ACS Nano 6 (2012) 4279-4287. https://doi.org/10.1021/nn3008383

[88] Y. Kang, W. Ha, Y.Q. Liu, Y. Ma, M.M. Fan, L.S. Ding, S. Zhang, B.J. Li, pH-responsive polymer-drug conjugates as multifunctional micelles for cancer-drug delivery, Nanotechnology 25 (2014) 335101. https://doi.org/10.1088/0957-4484/25/33/335101

[89] R. Tomlinson, J. Heller, S. Brocchini, R. Duncan, Polyacetal− doxorubicin conjugates designed for pH-dependent degradation, Bioconjug. Chem. 2003, 14, 1096-1106. https://doi.org/10.1021/bc030028a

[90] W. Wang, G. Liang, W. Zhang, D. Xing, X. Hu, Cascade-Promoted Photo-Chemotherapy against Resistant Cancers by Enzyme-Responsive Polyprodrug Nanoplatforms, Chem. Mater. 30 (2018) 3486-3498. https://doi.org/10.1021/acs.chemmater.8b01149

[91] J.J. Hu, L.H. Liu, Z.Y. Li, R.X. Zhuo, X.Z. Zhang, MMP-responsive theranostic nanoplatform based on mesoporous silica nanoparticles for tumor imaging and targeted drug delivery. J. Mater. Chem. B 4 (2016) 1932-1940. https://doi.org/10.1039/C5TB02490K

[92] Y. Ding, Y. Hao, Z. Yuan, B. Tao, M. Chen, C. Lin, P. Liu, K. Caib, A dual-functional implant with an enzyme-responsive effect for bacterial infection therapy and tissue regeneration, Biomater. Sci. 8 (2020) 1840-1854. https://doi.org/10.1039/C9BM01924C

[93] L.P. Datta, A. Chatterjee, K. Acharya, P. De, M. Das, Enzyme responsive nucleotide functionalized silver nanoparticles with effective anti-microbial and anticancer activity, New J. Chem. 41 (2017) 1538-1548. https://doi.org/10.1039/C6NJ02955H

[94] S.T. Gunawan, K. Kempe, T. Bonnard, J. Cui, K. Alt, L.S. Law, X. Wang, E. Westein, G.K. Such, K. Peter, C.E. Hagemeyer, F. Caruso, Multifunctional Thrombin-Activatable Polymer Capsules for Specific Targeting to Activated Platelets, Adv. Mater. 27 (2015) 5153-5157. https://doi.org/10.1002/adma.201502243

Emerging Nanomaterials and Their Impact on Society in the 21st Century   Materials Research Forum LLC
Materials Research Foundations 135 (2023) 125-151   https://doi.org/10.21741/9781644902172-6

Chapter 6

Emerging Nanomaterials in Drug Delivery and Therapy

Mahmuda Nargis[1,2*], Raju Ahmed[3], Abu Bin Ihsan[1]

[1]Department of Pharmaceutical Engineering, Faculty of Engineering, Toyama Prefectural University, 5180 Kurokawa, Imizu, Toyama 939-0398, Japan

[2]Department of Pharmacy, Dhaka International University, Dhaka, Bangladesh

[3]College of Public Health, University of Iowa, Iowa City, IA 52242, United States

*mnmou926@gmail.com

Abstract

Medical science has been influenced in a number of ways by the development of new tools for manipulating materials at the nanoscale. Physicochemical advantages like the greater surface area to mass ratio, ultra-small size, and high level of reactivity have set the nanomaterials apart from bulk materials with the same composition. Due to these unique characteristics, nanomaterials are massively used to get over the drawbacks of traditional medicinal and diagnostic drugs and offer outstanding panoramas. In this chapter, we will focus on the developments of nanomaterials in the fields of drug delivery and disease therapy, with their enormous potential and future prospects.

Keywords

Nanoparticles, Liposomes, Micelles, Hydrogel, Lipoprotein, Disease Control, Tumor, Medical Science, Diagnostic

Contents

Emerging Nanomaterials in Drug Delivery and Therapy............................125

1. Introduction..126

2. Nanomaterials in drug delivery and therapy.....................................127

    2.1   Liposomes...127

    2.2   Micelles..129

    2.3   Lipoprotein-based nanomaterials 130

    2.4   Hydrogel 130

2.5   Dendrimers 132

2.6   Carbon nanotubes (CNTs)..133

3.   **Barriers and challenges**..**134**

Conclusions and future perspectives..**136**

References **137**

1.   Introduction

The demand for nanomaterials is increasing very rapidly throughout the world. In 2015, the market for nanomaterials was estimated to be worth \$2.6 trillion. Potentials for the development of new evolutionary applications in medical science have been created by the emergence of new manufacturing techniques that allow the manipulation of materials at nanoscale levels. As a result, novel solutions for existing health issues such as antibiotic-resistant infections, development of vaccine, and cancer therapy are possible [1]. Materials having a diameter of less than 100 nanometers are known as nanomaterials. In addition to tissue engineering, it has the potential to manipulate the frontiers of nanomedicine including microfluidics, biosensors, drug delivery and microarray testing [2-5].

The nanomaterials could be obtained either by engineering as a byproduct of mechanical or industrial operations, or by natural processes [6]. Currently, there are thousands of nanomaterials and they are classified according to their origin, shape or intent of use. It is important to mention that nanomaterials differ from their counterparts in terms of chemical, physical, biological, and therapeutic properties. When they are in the form of nanoparticles, they are used a lot in medical science to diagnose and treat diseases.

Nanoparticles are usually small nanospheres because they are formed by materials designed at the atomic or molecular level [7]. As a result, they can mobile more easily within the human body than larger materials. Since atomic or molecularly tailored materials make up nanoparticles, they are often tiny nanospheres. As a consequence, it is easier for them to move within the human body compared to larger objects. They must have the ability to be compatible with the biological system without expressing any side effects [8]. These unique features of nanoparticles enable them to include both diagnostic and therapeutic substances into a single nanoparticle. Due to these characteristics, nanoparticles provide an assessment of their therapeutic efficiency by monitoring the distribution, accumulation, and quantifying the drugs at the target site [9,10] because of their small sizes, nanoparticles can traverse the BBB, and other physiological barriers and elude the detection process of the reticuloendothelial system, preventing their destruction. New methods for making nanovaccines and nanoadjuvants are provided by these properties of nanoparticles. New horizons for medical science are opened with their use in cancer treatment and gene therapy [11,12]. Conjugation, encapsulation, or adsorption may be used to load nanoparticles with specific chemicals, depending on the application. A variety of chemicals, such as pharmaceuticals or chemotherapeutic agents or imaging compounds, or biological molecules such as antigens and antibodies may be delivered using these devices. Light and heat may also be sent to a specific location by using these devices. Nanoparticles may be classified as organic, inorganic, polymeric, or metallic based on their chemical composition. They are further classified into solid lipid nanoparticles (SLNs), carbon nanotubes, dendrimers, micelles, and liposomes, which are often used as vehicles for therapeutic and drug delivery [13-15].

In this chapter we will focus on the advances of different nanomaterials, their implementations in the field of drug delivery and therapy with stressing on the leverage of biomaterials and bio-medical engineering inventions to overcome biological blockades and heterogeneity of the patient. We also highlight the biological obstacles that have hampered the broad success of their applications, as well as a critical examination of rational nanoparticles design that has attempted to overcome them.

2.   Nanomaterials in drug delivery and therapy

The emergence of diverse nanoscale materials has given medical research options for evolving new applications. This has provided a road to new approaches for long-standing health issues including cancer treatment, vaccine development, and drug-resistant organisms. Numerous nanoscale materials have been invented with several clinical advantages in the medical field. A total of 51 nanoparticles and 77 different compounds have been given FDA approval for clinical testing [16]. Polymeric and liposomal components make up a significant fraction of the total. Drug delivery systems may also make use of more complicated materials including micelles, metals, and protein-based nanoparticles. Prior to the employment of an appropriate method of drug administration, it must be capable of guaranteeing increased bioavailability and improved therapeutic effectiveness at the target location [17,18]. For this reason, a broad range of substances in various structural forms has been combined with pharmaceuticals to create efficient nano-drug delivery systems. Preclinical and clinical research shows that they might be used to treat a variety of disorders [19-21]. There are a growing variety of nanomaterials being explored as a means of administering medication, and these materials have shown considerable promise as diagnostic and therapeutic tools [22,23].

2.1   Liposomes

Liposomes were revealed by Alec Bangham in 1960. Liposomes are closed spherical vesicles composed of phospholipids and steroids enclosing an aqueous core, usually 50–450 nm in size [24-26]. In the pharmaceutical and cosmetic industries, they are the first colloidal drug carriers [27] that are employed for targeted delivery, gene therapy [28-30] and transportation of a variety of compounds. Drugs may be easily incorporated into their membrane structure, which is comparable to that of cell membranes, making them a better delivery vehicle than other methods. Carriers are the most popular because of their numerous features, including biocompatibility, biodegradability, low toxicity, ability to encapsulate several components, and approval for multiple clinical studies. There are four forms of liposomes: ordinary, PEGylated, ligand-targeted, and theranostic liposomes [31]. Thin layer hydration, mechanical agitation, solvent evaporation, solvent injection, and the surfactant solubilization techniques are the most common methods of synthesis [32]. Liposomes may be loaded with drugs either actively (drug encapsulated after liposome formation) or passively (drug encapsulated during liposome manufacture) [33]. Through a process known as diffusion, the medications are transported to plasma proteins [34]. Using active and passive targeting methodologies, liposomes may be able to target certain cells. In vitro and in vivo, PEGylated liposomes outperform ordinary liposomes at passively targeting cancer cells. To increase medicine delivery to tumors and prevent adverse effects, liposomes have been shown in clinical trials [35]. Metastatic breast cancer and AIDS-related Kaposi's sarcoma may be treated with PEGylated liposomes conjugated with anthracyclines, doxorubicin and daunorubicin [36-38]. PEGylation liposomes have a detrimental impact on the target location known as the PEG dilemma, which makes it difficult to use medicines or other macromolecules in real life [39,40]. Targeting ligands are attached to the PEG tails to form active targeting liposomes. For the liposome-drug combination liposomes to reach their target location, they must be targeted to certain cells. They may be constructed to release their contents in the enzyme-rich, low pH environment by using pH-triggered techniques [41,42]. Left ventricular dysfunction and symptoms of congestive heart failure are reduced when HER2-targeted liposomal doxorubicin is used [43]. The anti-PD-1 antibody has changed cancer treatment, although its efficacy is presently lacking. Incorporating siRNA targeting the vascular endothelial growth factor (VEGF) receptor, cRGD-MEND and GALA-MEND anti-PD-1 therapies have been successful in modifying the tumor microenvironment. For the treatment of Alzheimer's disease, PEGylated liposomes functionalized with OX26MAb (anti-transferring receptor monoclonal antibody) and 19B8MAb (anti-amyloid beta peptide antibody) are utilized. Immune-based liposomes, according to Nakanishi et al., are more effective when OX26MAb is conjugated to a streptavidin-biotin complex [44]. Receptor-mediated endocytosis is facilitated by active targeting liposomes [45]. Liposome cellular absorption is poor due to receptor and liposome saturation and cannot give delivered cargo with appropriate therapeutic benefits. In light of this, researchers have looked into the use of PEGylated liposomes modified with two different ligands in order to improve liposome internalization at the target region and increase cellular absorption of macromolecules or chemotherapeutics [46,47].

To combat the metabolic syndrome, Harashima et al. are developing novel nanomedicines that induce death in adipose and liver endothelial cells or alter their function. PTNPs (ligand-targeted liposomal nanocarriers) produced by this team precisely target adipose endothelial cells (AECs) by binding to prohibitin receptors [48]. In a dose-dependent way, PTNP treatment with the apoptosis-initiating protein cytochrome C (CytC) effectively caused apoptosis in AECs, with an anti-obesity effect [49]. AEC-targeted delivery vehicles were further created by finding several unique aptamer-based ligands against AECs by this research group [50,51]. Liposomal nanocarriers for hepatic delivery of short interfering

RNA (siRNA) and a library of pH-sensitive cationic lipids have been developed successfully in the effective drug delivery system [52-54].

2.2 Micelles

Amphiphilic compounds, which have hydrophobic and hydrophilic moieties, produce micelles in solution. When a particular concentration of amphiphilic molecules is reached, they self-assemble to create these structures. To form micelles, it is necessary to maintain a certain concentration (CMC) of the amphiphilic compounds. Temperature and the solvent system are both important when it comes to micelle formation since they influence parameters like amphiphile concentration, amphiphilic molecule hydrophobic/hydrophilic domain size, and micelle formation rate [55]. Typically, the micelles have a hydrophilic shell and a hydrophobic core, allowing for the incorporation of both hydrophilic and hydrophobic medicines. Polymeric micelles have piqued attention as a drug delivery method due to their high stability, low cytotoxicity, and capacity to distribute drugs in a regulated and sustained way while also increasing drug accumulation at the target location [56]. The monomer ratio in block copolymers may be adjusted through hydrophobic interactions, electrostatic interactions, metal complexation and hydrogen bonding of component block copolymers to produce suitable micelles [57]. Because of their comparable mesoscopic size ranges to viruses and lipoproteins, polymeric micelles are less likely to be cleared from the body, resulting in greater medication bioavailability [58]. Peptides, sugars and other compounds are functionalized for receptor-mediated drug delivery. To evaluate the toxicity and tolerance of Genexol-PM (PEG-poly(D,L-lactide)-paclitaxel) to conventional paclitaxel, a phase I trial found that Genexol-PM had a better toxicity and tolerance profile in patients with advanced refractory malignancies [59]. Micellar systems coupled with polyoxygen/polyallylglycidyl ether have been shown to have extended circulation and effective doxorubicin release at tumor sites [60]. A camptothecin-loaded micellar formulation using polyethylene glycol-poly(aspartate ester) block copolymers demonstrated prolonged circulation in the bloodstream for the treatment of tumors [61]. Using Doxorubicin-loaded poly(ethylene glycol) micelles, 1,2-dioleoyl-3-trimethylammonium propane showed anticancer effects against bladder cancer [62]. There are intriguing applications in the treatment of solid tumors using cholesterol-modified microparticles containing curcumin (mPEG–PLA-Ch-CUR), which have a better encapsulation efficiency. Lipid and polyion complex-based micelles, which target tumor necrosis factor receptors, have shown promising therapeutic effectiveness by exhibiting high agonist activity [63]. In order to reach the posterior ocular tissues, polymeric micelles are useful for ocular medication administration [64]. PEG-b-PC micelle-encapsulated dasatinib treats proliferative vitreoretinopathy (PVR) without harming ARPE-19 cells in any way [65]. Plasmid-based micelles of two chemotherapeutic medicines (Dox and Tax) combined with polyethylene glycol and poly (D,L-lactic acid-co-glycolide) showed synergistic cytotoxicity and increased therapeutic effectiveness against breast cancer with triple-negative status [66]. Recently, polyglycoside-based materials and alternating peptides in the field of self-assembled nanocarriers have attracted the focus of researchers [67-74].

2.3   Lipoprotein-based nanomaterials

These naturally occurring nanoparticles, which are composed of hydrophobic cores of non-polar lipids, cholesterol esters and triglycerides, are known as lipoproteins. Membranes of free cholesterol and apolipoproteins cover the hydrophobic center of the cell. It is the body's job to generate and digest lipoproteins, which help to keep levels of lipids and cholesterol in check. Lipoprotein research was first described in 1930. Five primary categories of plasma lipoproteins are chylomicrons (CM), very low-density lipoproteins (VLDL), intermediate density lipoproteins (IDL), low density lipoproteins (LDL), and high-density lipoproteins (HDL) [75-79]. It has recently emerged that lipoproteins may be used to treat a wide range of disorders since they are non-immunogenic, biocompatible, biodegradable, and harmless [80,81]. LDL and HDL bind to receptors on distinct cell types that play a role in the metabolism of lipids. The medication delivery researchers have paid attention to both of these issues. Synthetic lipoprotein nanoparticles with transport characteristics and biocompatibility similar to natural lipoproteins may be created by combining protein subunits with lipids [82-84]. Ultracentrifugation, size-exclusion chromatography, tangential flow filtration, density gradient tests, and immune-based assays are all methods used in labs to isolate lipoproteins. By using ethanol injection and cholate dialysis, six distinct recombinant human apolipoproteins are combined to make synthetic HDL. Lipoproteins are drug delivery vehicles because of their great drug loading capacity. r11-DOX-loaded LDL is a delivery vehicle for radionucleotides used in the treatment of cancers and other malignancies.

2.4   Hydrogel

Swelling in water or biological fluids does not compromise the structure of hydrophilic three-dimensional polymeric networks, known as "hydrogels." [85]. Crosslinked polymers through covalent bonds physically or chemically produce them [86]. To keep polymers from dissolving, hydrophilic functional groups in the polymers help them attract water. Introducing non-reversible covalent connections to a polymer matrix is a common way to make hydrogels, whether via reversible interactions, non-reversible chemical processes, or UV/photopolymerization. The gelation process is dependent on time and concentration and is triggered by an external stimulus (pH, temperature or light) [87]. They are biocompatible, harmless, and biodegradable. Therefore, they are extensively employed in pharmaceutical and medical areas for a variety of purposes, including drug delivery, tissue engineering, wound healing, biosensors, contact lens material, artificial skin and the lining of artificial hearts, etc. They may also be employed in 3D cell culture [88,89]. For oral, rectal, ophthalmic, epidermal, and subcutaneous drug delivery applications, hydrogels are employed. Biomacromolecules (proteins and DNA, as well as hydrophilic or hydrophobic medicines) may be encapsulated in hydrogels, which makes them ideal candidates for controlled drug release [90]. Specific drug delivery applications may be achieved by using hydrogels that are sensitive to pH and temperature [91].

Photo-responsive supramolecular hydrogel (hyaluronic acid grafted with cyclodextrins and azobenzene) has recently been explored [92]. Hyaluronic acid is a typical modulator of wound healing. When exposed to UV light, the gel produces a 2–3-fold increase in the amount of EGF released. This may be done by alternating between visible and UV light. These hydrogels showed excellent biocompatibility, viability (~90%), and healing efficiency (96 percent healing). Another HA-based hydrogel showed better cell viability and wound healing efficiency. Metal-ligand coordination connections formed by hyaluronic acid and bisphosphonate (BP) groups and silver ions (Ag+) in this work were used to target bacteria in the wound site [93].

A PEG-PAEU copolymer hydrogel was designed for the regulated release of therapeutic proteins and has proven effective in extending the short half-life of proteins. Clinical use of therapeutic proteins is limited by their short half-life, which makes them appealing as potential treatments for human disorders. As a therapeutic protein, urate oxidase is being tested in this trial to treat hyperuricemia [94]. Hexamethylene diisocyanate, Pluronic F127, and hyaluronic acid have been utilized to generate an injectable thermoresponsive hydrogel system that transitions from sol to gel at 37°C. It was shown that this hydrogel system, which was biocompatible and degraded, lasted for more than 28 days in cancer therapy [95].

One of the most promising vehicles for anticancer therapy against A549 and HepG2 cells is a hydrogel that contains salecan and poly (methacrylic acid) (PMAA) [96]. The pH of the gel affects the amount of drug released from it. In comparison to control groups, paclitaxel-loaded polysaccharide hydrogel showed tumor inhibition and the best anticancer effectiveness [97]. Gold nanocrystal-modified cyclodextrin hydrogel was shown to have pH-dependent sustained release of DOX through host–guest interaction. The authors suggested that the creation of these hybrid hydrogels would bring novel and therapeutically helpful techniques for medical applications [98].

Carboxymethyl cellulose/graphene oxide (CMC/GO) nanohybrid hydrogel beads were physically crosslinked with FeC_3H_2O and employed for the controlled release of an anticancer medication (DOX) [99]. In terms of loading capacity and DOX release, this gel is superior. Gold nanocomposite supramolecular hybrid hydrogels for drug delivery were made using a brand-new approach (a simple one-step ligand-exchange process). Anticancer medicines DOX were released slowly from synthetic oligopeptide-based thermoreversible, pH-sensitive hydrogels at physiological pH, offering a possible therapeutic vehicle for the future [100].

Hydrogel-based soft contact lenses for controlled medication release have become popular in recent years [101]. Additionally, they are used in the treatment of eye infections in order to extend the duration of drug release while also increasing the bioavailability of the medication at the site of action [102]. Since its structural resemblance to the tissue's extracellular matrix (ECM), hydrogels have attracted considerable attention in cartilage and bone tissue engineering [103]. As a less invasive alternative to implantation surgery, injectable hydrogels have piqued the interest of researchers working on cartilage and bone tissue treatments. They may be molded into any desired form to accommodate uneven flaws. [104-106]. It has been discovered that PVA-chitosan hydrogel containing mesenchymal stem cells may help regenerate cartilage [107]. An early-stage osteoarthritis therapeutic option has been found in hydrogel made from polymerized hydrophilic zwitterionic monomers such as 2-methacryloyloxyethyl phosphorylcholine and ethylene glycol dimethacrylate [108]. Besides these, some newly invented hydrogels are now in the focus of the researchers due to their additional intriguing properties e.g. tough, reproducible and self-healing. [109-116].

2.5   Dendrimers

Organometallic macromolecules, dendrimers have a three-dimensional structure. Both the surface and interior characteristics of dendrimers may be altered by manipulating their structure. Despite the fact that, the first dendrimer polypropylenimine (PPI) was discovered in 1978, the dendrimers with well-defined structures were developed in 1980 [117]. Depending on their composition, dendrimers may be divided into three types: core, interior, and shell. Dendrimer form is determined by the core, host–guest characteristics are determined by the interior, and the surface may be polymerized or changed with functional groups. Dendrimer morphology is determined by the core and the quantity and kind of branching units inside [118]. Dendrimers of lower generations are pliable and open, whereas those of higher generations create more dense, three-dimensional forms. Also, the structure's stiffness is influenced by the machine's iteration number [119]. A dendrimer's form depends on the ionic strength, pH, and charged groups on its surface in aquatic settings.

Dendrimers may be produced in a convergent or divergent manner Newkome et al. [120] and Vögtle et al. [121] took the diverging path. The polyfunctional cores react with monomer units that have a single reactive site and numerous protected or unreactive groups in this method of polymerization. Unreactive or protected groups may be activated for subsequent reactions with the cores by adding monomer units to the reaction mixture. Fréchet et al. were the first to use this convergent method [122]. In this approach, the dendrimer forms a dendron from the surface inward and interacts with an appropriate core to complete the synthesis. The convergent technique is more favorable than the divergent approach because only a small number of active sites are present per reaction, which reduces structural flaws in the result. A second technique, the Fréchet et al. two stage convergent method, was initially described in 1990 [123] by these authors. Using macromonomer units synthesized convergently, this approach allows for the creation of monodisperse, higher generation dendrimers in a more rapid manner than previously thought. Dendrimers have recently been synthesized using lego chemistry, click chemistry, and the double exponential growth approach. Dendrimers may be divided into polyamidoamine dendrimers, poly(propylene imine) dendrimers, glycodendrimers, liquid crystalline dendrimers, and peptide dendrimers depending on the kinds of core and peripheral groups. There are a number of advantages to using dendrimers in biomedicine, including their high biological compatibility and great structural homogeneity, as well as their hyperbranching and well-delineated globular shapes [124]. This class of water-soluble polymers, known as Polyamidoamine (PAMAM) dendrimers, has an unusual tree-like branching architecture and a compact spherical form in solution. Oral medication administration using them has been widely studied. Many arms and amine groups on the surface of PAMAM make it a powerful drug carrier [125-128]. There has been research on dendrimers as an anti-HIV-1 agent. When it comes to anionic dendrimers, VivaGel is one of the most successful. It's now being tested in phase I/II clinical studies [129]. Glycerol and succinic acid-based dendrimers were employed to deliver 10-hydroxy-camptothecin in human breast adenocarcinoma (MCF-7), colorectal adenocarcinoma HT-29, non-small cell lung carcinoma (NCI-H460), and glioblastoma cells [130,131]. By releasing highly reactive singlet oxygen upon excitation, dendrimers may trigger apoptosis and necrosis in tumor cells [132, 133]. An excellent transdermal delivery mechanism for medicines with limited solubility and extremely hydrophobic moieties was shown using dendrimers.

2.6   Carbon nanotubes (CNTs)

There are nanoparticles in the form of tubes called carbon nanotubes (CNTs) that are hexagonal in structure and made up of hybridized carbon atoms. Allotrope [134] of carbon, fullerenes, are made up of Carbon Nanotubes (CNTs). Discovery of CNTs by Iijima in 1991 and publication of his findings [135]. A single sheet of graphene (single-walled carbon nanotubes) or numerous sheets of graphene (multiwalled carbon nanotubes) 76 may be rolled up to make them. Covalent and non-covalent interactions may be used to load drugs into CNTs, either on the surface or inside the core. A breakthrough in materials science and engineering has been made possible because of the extraordinary mechanical strength of carbon nanotubes (CNTs) [136]. It is possible to use CNTs to encapsulate medicinal substances in nanotubes, such as medicines and proteins and DNA. Healthy tissues benefit from this property since it reduces the cytotoxicity of the drug. With their nanoneedle-like shape, CNTs enter the cytoplasm of target cells quickly and effectively without harming them. Because CNTs have a low disintegration rate and are hydrophobic in nature, their applications are restricted [137]. CNTs may be modified in a variety of ways, including adsorption, electrostatic interaction, and covalent interaction, in order to minimize or eliminate these side effects. There are hydrophilic biocompatible polymers like polyethylene oxide (PEG) that may be utilized to boost the solubility of CNTs in the body, as well as their systemic retention and time in circulation [138,139]. Ibuprofen was loaded into PEGylated multi wall carbon nanotubes using a drug delivery technique [140]. When it comes to carrying nucleic acids [141,142], proteins [143,144] pharmaceutical compounds [145,146] antibodies and low molecular weight targeted agents, CNTs have shown to be an effective carrier. Using near-infrared photothermal ablation, which raises tumor temperatures in response to light intensity, is a viable treatment option for these tumors. The molecular dynamics approach makes CNTs easier to use by facilitating small molecule transport, which is a crucial tool in gene therapy. SWCNTs modified with carboxylated AS1411 aptamers and coupled with PEI and PEG have been utilized to deliver shRNA to tumor cells. These demonstrated good selectivity for the target location and reduced the development of gastric cancer cells that were high in nucleolin [147].

3.   Barriers and challenges

Drug transport and nanoparticle accumulation to target areas are hampered by biological and physical obstacles, reducing their practical use. Even under normal physiological parameters, this phenomenon happens [148]. It is difficult to distinguish and quantify these obstacles since they vary patient-to-patient and occur at the systemic, microenvironment, and cellular levels. Various attempts have been made to include numerous functionalities and moieties into the overall nanoparticle design but many of these solutions have failed to fulfill these demands. Nonspecific distribution and inadequate therapeutic accumulation are only two of the many challenges facing drug developers today. Traditional nanoparticles must be redesigned in order to overcome these obstacles to medicine delivery. It is impossible to achieve site-specific therapeutic delivery unless nanocarrier design takes into consideration most, if not all, of the biological hurdles that a particle encounters during intravenous injection. It's possible to introduce new design components by addressing each of these obstacles one at a time.

In terms of medicine delivery, nanomaterials have the capacity to encapsulate hydrophilic and hydrophobic molecules, as well as improve pharmaceutical stability and personalized dosage. They've previously shown their worth in the treatment of cancer patients. It's difficult to deliver the maximum dosage to where it's needed, though, because of things like making sure it doesn't spread throughout the body. Nanomaterials' interactions with cells and tissues may be influenced by a variety of their physical features. They've been shown to affect cytotoxicity, drug release, targeting, and imaging contrast effectiveness based on factors including size, shape, and stiffness, as well as surface chemistry. As previously stated, nanoparticles encounter challenges based on the method of administration and the type and progression of the patient's disease [149]. It is possible to bypass some of the difficulties of systemic distribution by using local delivery methods, although even more invasive and sophisticated strategies have their own problems. Furthermore, systemic administration is more prevalent in NP applications since local delivery may be useful in disorders with pathology restricted to identifiable, accessible locations, such as certain solid tumors or traumatic traumas. Due to their surface properties, many nanoparticles are coated with PEG in order to slow down their excretion. PEGylation enhances circulation duration by altering the size and solubility of nanoparticles and insulating the nanoparticle's surface from enzymes and antibodies that might induce breakdown, secretion, and clearance. Although this physical barrier is there, macrophages and other immune cells may still be recognized. Additionally, when significant quantities of PEGylated nanoparticles are present, anti-PEG antibodies may be produced, resulting in their rapid clearance [150].

Nanoparticles are subjected to varying flow rates in the circulation, which generates shear stress, which may destroy the platforms or their payload and impede extravasation. To prevent NPs from attaching to vessel walls and spreading to target tissues, these fluid forces may remove nanoparticle's surface coatings [151,152].

Nanoparticle's physical and chemical properties may impact their clearance from the circulation, although interactions with the mononuclear phagocytic system (MPS) or the reticuloendothelial system (RES) are more prevalent. These systems include macrophages, monocytes, and dendritic cells, which suck up nanoparticles and deposit them in the spleen and liver. Because of this quicker clearing, clearance of cationic nanoparticles is the quickest, followed by anionic nanoparticles, whereas neutral and slightly negative nanoparticles have the longest half-lives in circulation. In order for nanoparticles to reach their intended place, they must first be expelled from the body [153]. It was also influenced by pathogenic settings, such as the vascular of a tumor. Non-specific distribution caused by extravasation creates a translational obstacle for applications that need precise localization.

Administration routes may be optimized to increase biodistribution. Nanoparticles, in particular, may be affected by the manner through which they are injected, and numerous studies have looked at how these routes alter their destiny [154,155]. When PLGA nanoparticles are injected intravenously, they tend to accumulate in the liver and the spleen, but when they are injected subcutaneously or intranodally, they tend to accumulate in the lymph nodes in the area where they were injected [156]. Some immunotherapeutic applications may benefit from nanoparticles reaching the lymphatic system before entering the circulation through these alternate delivery pathways [156,157]. Another method of preventing extravasation via the use of nanoparticles is pulmonary administration, especially through the use of nanoparticle inhalation. Because it avoids hepatic first-pass metabolism and increases distribution of dendrimer-based nanoparticles to lung and lymph nodes as compared to intravenous delivery, this technique is preferable.

Tight connections between BBB endothelium and epithelial cells (in intravenous treatment) and the digestive tract (in oral delivery) operate as physical barriers to nanoparticle dispersion. To get to the CNS, nanoparticles must be taken up by BBB endothelial cells through receptor-mediated endocytosis and exocytosed to the opposite side [153,158]. Drugs may be efficiently transported into tumor tissue and distributed throughout the body through receptor-mediated transcytosis [159,160]. Due to differences in plasma membrane transporters on BBB endothelial cells, it is difficult to bridge this barrier. Nanoparticle transport may be aided by transporters like glucose transporters, which are overexpressed on the BBB, and by molecules like vascular cell adhesion molecule 1.

Conditions in microenvironments are very different from those found in nature, and this results in considerable changes to the physical features and stability of nanoparticles. Zoning changes in pH and acidity may be seen throughout the digestive system. The gastrointestinal system is an unstable habitat for many nanoparticles because of the enzymes that induce degradation and these conditions [161]. There are a number of ways in which sick states may alter the gastrointestinal microenvironment, which can lead to variations in the way biomaterials behave in the body. Dendrimer/dextran biomaterials were revealed to be disease-dependently compatible with microenvironments in colon cancer and colitis [162].

The restricted absorption and penetration of nanoparticles in the tumor microenvironment are due in part to factors such as vascular density, interstitial fluid pressure, and extracellular matrix (ECM) [163-165]. So why are nanoparticles able to accumulate so effectively in cancerous tissue? The answer has been widely debated, with just a few well-established patterns relating nanoparticle design to tumor delivery. For example, tumor development may be aided by nanoparticle properties such as hydrodynamic diameters greater than 100 nm, rod-shaped structures, near-neutral charges, or inorganic material compositions.

When nanoparticles come into contact with their target cells, they face a number of obstacles to absorption and intracellular trafficking, reducing their effectiveness. There are several challenges that nanoparticles must overcome in order to enter cells and correct intracellular trafficking, as well as how cellular heterogeneity influences nanoparticle interaction. To increase the nanoparticles' drug delivery capabilities, scientists are exploring other methods. Nanoparticles, nanotubes, nanodisks, nanoshells, and nanowires are just some of the geometries and structures that may be created. Non-spherical geometries with high aspect ratios (length/diameter) are superior at delivering medicines.

Conclusions and future perspectives

In recent years, the use of nanomaterials in illness diagnosis and treatment, vaccine development, and disease prevention has expanded significantly. With the use of nanomaterials, medication administration may have several benefits, including greater water solubility, longer resistance time, decreased side effects by directing the drug to the targeted location, reduced dose-dependent toxicity, and protection against early release. The development of nanomaterial-based drug delivery is motivated by many factors, including the need for better existing therapies, the ability to deliver drugs to particular locations, and increased patient compliance. Preclinical and clinical studies have shown promising results for a large number of these therapies. Nanomedicines offer several benefits over conventional therapies, including lower toxicity and better effectiveness owing to controlled drug release and superior pharmacokinetics and pharmacodynamics. When it comes to long-term, site-specific medication delivery, a suitable drug carrier is required. The FDA has authorized a variety of nanomaterial-based formulations, including nanocrystal formulations, polymeric nanoparticles (PLGA), polymeric micelles, and lipid-based nanoparticles including liposomes [166]. Other nanoparticles, including inorganic, protein-based, and metallic nanoparticles, are awaiting clearance. Nanomedicines include Abraxane® (paclitaxel), Caelyx®, Myocet® (doxorubicin), and Mepact® (mifamurtide), as well as Emend® (aprepitant), Rapamune®, and Myocet®, all of which are now available on the market (sirolimus). One-size-fits-all nanomedicines have a significant problem when it comes to actual use. Toxicology profile, dose standardization, and assessment of adverse effects in the liver, kidneys, spleen and lungs are all critical in clinical studies.

With a better understanding of the biological processes governing these barriers and how they evolve in different disease states, as well as advances in materials science, nanoparticles capable of sequentially negotiating these barriers will continue to be developed. This could lead to not only the effective translation of novel treatments, but also the advancement of nanoparticle-based drug delivery from a promising sector to a feasible and widely used technique for the treatment of a variety of diseases. This chapter examined a wide range of materials that are already being utilized or have the potential to be employed as delivery vehicles for medications. Because of their unique qualities, using them alone or in conjunction with other medicines has increased therapeutic efficacy. Currently, there is no conclusive evidence on the short- and long-term impacts of nanomedicines on humans. Although some of these materials have failed in clinical trials, new and innovative materials those are now under research show significant potential, indicating that new therapeutic alternatives may be available in the near future.

References

[1] A. Manuja, B. Kumar, R.K. Singh, Nanotechnology developments: opportunities for animal health and production, Nanotechnol. Dev. 2,1 (2012) e4. https://doi.org/10.4081/nd.2012.e4

[2] M.S. Arayne, N. Sultana, F. Qureshi, Review: nanoparticles in delivery of cardiovascular drugs, Pak. J. Pharm. Sci. 20 (2007) 340-348.

[3] J.K. Patra, K-H. Baek, Green nanobiotechnology: factors affecting synthesis and characterization techniques, J. Nanomater. 2014 (2014) 417305. https://doi.org/10.1155/2014/417305

[4] R.R. Joseph, S.S. Venkatraman, Drug delivery to the eye: what benefits do nanocarriers offer? Nanomedicine 12 (2017) 683-702. https://doi.org/10.2217/nnm-2016-0379

[5] A.Z. Mirza, F.A. Siddiqui, Nanomedicine and drug delivery: a mini review, Int. Nano. Lett. 4 (2014) 94. https://doi.org/10.1007/s40089-014-0094-7

[6] J. Jeevanandam, A. Barhoum, Y. S. Chan, A. Dufresne, M. K. Danquah, Review on nanoparticles and nanostructured materials: history, sources, toxicity and regulations, Beilstein J. Nanotechnol. 9 (2018) 1050-1074. https://doi.org/10.3762/bjnano.9.98

[7] G.R. Rudramurthy, M.K. Swamy, U.R. Sinniah, A. Ghasemzadeh, Nanoparticles: alternatives against drug resistant pathogenic microbes, Molecules 21 (2016) 836. https://doi.org/10.3390/molecules21070836

[8] A. Jurj, C. Braicu, L-A. Pop, C. Tomuleasa, C.D. Gherman, I. Berindan- Neagoe, The new era of nanotechnology, an alternative to change cancer treatment, Drug Design Dev. Ther. 11 (2017) 2871-2890. https://doi.org/10.2147/DDDT.S142337

[9] S.C. Baetke, T. Lammers, F. Kiessling, Applications of nanoparticles for diagnosis and therapy of cancer, Br. J. Radiol. 88, 1054 (2015) 20150207. https://doi.org/10.1259/bjr.20150207

[10] A.I. Irimie, C. Braicu, R. Cojocneanu-Petric, I. Berindan-Neagoe, R. S. Campian. Novel technologies for oral squamous carcinoma biomarkers in diagnostics and prognostics, Acta. Odontol. Scand. 73, 3 (2015)161-168. https://doi.org/10.3109/00016357.2014.986754

[11] V. Ceña, P. Játiva, Nanoparticle crossing of blood-brain barrier: a road to new therapeutic approaches to central nervous system diseases, Nanomedicine (Lond). 13 (2018) 1513-1516. https://doi.org/10.2217/nnm-2018-0139

[12] Y. Zhou, Z. Peng, E.S. Seven, R.M. Leblanc, Crossing the blood-brain barrier with nanoparticles, J. Control. Release 270 (2018) 290-303. https://doi.org/10.1016/j.jconrel.2017.12.015

[13] N.K. Singh, S. K. Singh, D. Dash, P. Gonugunta, M. Misra, P. Maiti, CNT Induced β-phase in polylactide: unique crystallization, biodegradation, and biocompatibility, J. Phys. Chem. C. 117 (2013) 10163-10174. https://doi.org/10.1021/jp4009042

[14] D. Lombardo, M.A. Kiselev, M.T. Caccamo, Smart nanoparticles for drug delivery application: development of versatile nanocarrier platforms in biotechnology and nanomedicine, J. Nanomater. 2019 (2019) 3702518. https://doi.org/10.1155/2019/3702518

[15] A. Mishra, S. K. Singh, D. Dash, V.K. Aswal, B. Maiti, M. Misra, P. Maiti, Self-assembled aliphatic chain extended polyurethane nanobiohybrids: Emerging hemocompatible biomaterials for sustained drug delivery, Acta. Biomater. 10 (2014) 2133-2146. https://doi.org/10.1016/j.actbio.2013.12.035

[16] D. Bobo, K.J. Robinson, J. Islam, K.J. Thurecht, S.R. Corrie, Nanoparticle-based medicines: a review of FDA-approved materials and clinical trials to date, Pharm. Res. 33, 10 (2016) 2373-2387. https://doi.org/10.1007/s11095-016-1958-5

[17] S.M. Moghimi, A.C. Hunter, J.C. Murray, Nanomedicine: current status and future prospects, FASEB J. 19 (2005) 311-330. https://doi.org/10.1096/fj.04-2747rev

[18] J.L. Markman, A. Rekechenetskiy, E. Holler, J.Y. Ljubimova, Nanomedicine therapeutic approaches to overcome cancer drug resistance, Adv. Drug Deliv. Rev. 65 (2013)1866-1879. https://doi.org/10.1016/j.addr.2013.09.019

[19] G. Kapusetti, N. Misra, V. Singh, S. Srivastava, P. Roy, K. Dana, P. Maiti, Bone cement based nanohybrid as a super biomaterial for bone healing, J. Mater. Chem. B 2 (2014) 3984-3997. https://doi.org/10.1039/C4TB00501E

[20] A. Sharma, U.S. Sharma, Liposomes in drug delivery: progress and limitations, Int. J. Pharm. 154 (1997) 123-140. https://doi.org/10.1016/S0378-5173(97)00135-X

[21] K.S. Soppimath, T.M. Aminabhavi, A.R. Kulkarni, W.E. Rudzinski, Biodegradable polymeric nanoparticles as drug delivery devices, J. Control. Release 70 (2001) 1-20. https://doi.org/10.1016/S0168-3659(00)00339-4

[22] J.K. Patra, G. Das, L.F. Faceto, E.V.R.C.M. del P. Rodriguez-Torres, L.S. Acosta-Torres, L.A. Diaz-Torres, R. Grilo, M.K. Swamy, S. Sharma, S. Habtemariam, H-S.

Shin, Nano based drug delivery systems: recent developments and future prospects, J. Nanobiotechnology 16 (2018) 71. https://doi.org/10.1186/s12951-018-0392-8

[23] A. Biswas, A. Shukla, P. Maiti, Biomaterials for interfacing cell imaging and drug delivery: An overview, Langmuir 35, 38, (2019) 12285-12305. https://doi.org/10.1021/acs.langmuir.9b00419

[24] N.V. Beloglazova, O.A. Goryacheva, E.S. Speranskaya, T. Aubert, P.S. Shmelin, V.R. Kurbangaleev, I.Yu. Goryacheva, S.De. Saeger, Silica-coated liposomes loaded with quantum dots as labels for multiplex fluorescent immunoassay, Talanta 134 (2015) 120-125. https://doi.org/10.1016/j.talanta.2014.10.044

[25] A. Bunker, Poly(ethylene glycol) in drug delivery, why does it work, and can we do better? All atom molecular dynamics simulation provides some answers, Phys. Procedia. 34 (2012) 24-33. https://doi.org/10.1016/j.phpro.2012.05.004

[26] F. Danhier, O. Feron, V. Préat, To exploit the tumor microenvironment: passive and active tumor targeting of nanocarriers for anti-cancer drug delivery, J. Control Release 148, 2 (2010) 135-146. https://doi.org/10.1016/j.jconrel.2010.08.027

[27] L. Tao, A. Faig, K.E. Uhrich, Liposomal stabilization using a sugar- based, PEGylated amphiphilic macromolecule, J. Colloid Interface Sci. 431 (2014) 112-116. https://doi.org/10.1016/j.jcis.2014.06.004

[28] I. Sugiyama, Y. Sadzuka, Enhanced antitumor activity of different double arms polyethyleneglycol-modified liposomal doxorubicin, Int. J. Pharm. 441 (2013) 279-284. https://doi.org/10.1016/j.ijpharm.2012.11.032

[29] G. Bozzuto, A. Molinari, Liposomes as nanomedical devices, Int. J. Nanomed. 10 (2015) 975-999. https://doi.org/10.2147/IJN.S68861

[30] G.H. Shin, J.T. Kim, H.J. Park. Recent developments in nanoformulations of lipophilic functional foods, Trends Food Sci. Technol. 46, 1 (2015)144-157. https://doi.org/10.1016/j.tifs.2015.07.005

[31] L. Sercombe, T. Veerati, F. Moheimani, S.Y. Wu, A.K. Sood, S. Hua, Advances and challenges of liposome assisted drug delivery, Front. Pharm. 6 (2015) 286. https://doi.org/10.3389/fphar.2015.00286

[32] N.G. Kotla, B. Chandrasekar, P. Rooney, G. Sivaraman, A. Larrañaga, K.V. Krishna, A. Pandit, Y. Rochev, Biomimetic lipid-based nanosystems for enhanced dermal delivery of drugs and bioactive agents, ACS Biomater. Sci. Eng. 3 (2017) 1262-1272. https://doi.org/10.1021/acsbiomaterials.6b00681

[33] A. Akbarzadeh, R. Rezaei-Sadabady, S. Davaran, S.W. Joo, N. Zarghami, Y. Hanifehpour, M. Samiei, M. Kouhi, K. Nejati-Koshki, Liposome: classification, preparation, and applications, Nanoscale Res. Lett. 8 (2013) 102. https://doi.org/10.1186/1556-276X-8-102

[34] B.K. Lee, Y.H. Yun, K. Park, Smart nanoparticles for drug delivery: boundaries and opportunities, Chem. Eng. Sci. 125 (2015) 158-164. https://doi.org/10.1016/j.ces.2014.06.042

[35] J.S. Refuerzo, J.F. Alexander, F. Leonard, M. Leon, M. Longo, B. Godin, Liposomes: a nanoscale drug carrying system to prevent indomethacin passage to the fetus in a pregnant mouse model, Am. J. Obstet. Gynecol. 212, 4 (2015) 508.e1-e7. https://doi.org/10.1016/j.ajog.2015.02.006

[36] M. Rawat, D. Singh, S. Saraf, S. Saraf. Nanocarriers: promising vehicle for bioactive drugs, Biol. Pharm. Bull. 29, 9 (2006) 1790-1798. https://doi.org/10.1248/bpb.29.1790

[37] M.L. Adams, A. Lavasanifar, G.S. Kwon, Amphiphilic block copolymers for drug delivery, J. Pharm. Sci. 92, 7 (2003) 1343-55. https://doi.org/10.1002/jps.10397

[38] E.V. Batrakova, T.Y. Dorodnych, E.Y. Klinskii, E.N. Kliushnenkova, O.B. Shemchukova, O.N. Goncharova, S.A. Arjakov, V.Y. Alakhov, A.V. Kabanov, Anthracycline antibiotics non-covalently incorporated into the block copolymer micelles: in vivo evaluation of anti-cancer activity, Br. J. Cancer. 74, 10 (1996) 1545-1552. https://doi.org/10.1038/bjc.1996.587

[39] H. Hatakeyama, H. Akita, K. Kogure, M. Oishi, Y. Nagasaki, Y. Kihira, M. Ueno, H. Kobayashi, H. Kikuchi, H. Harashima, Development of a novel systemic gene delivery system for cancer therapy with a tumor-specific cleavable PEG-lipid, Gene Ther. 14, 1 (2007) 68-77. https://doi.org/10.1038/sj.gt.3302843

[40] H. Hatakeyama, H. Akita, H. Harashima, A multifunctional envelope type nano device (MEND) for gene delivery to tumours based on the EPR effect: a strategy for overcoming the PEG dilemma, Adv. Drug. Deliv. Rev. 63, 3 (2011) 152-160. https://doi.org/10.1016/j.addr.2010.09.001

[41] T. Jiang, Z. Zhang, Y. Zhang, H. Lv, J. Zhou, C. Li, L. Hou, Q. Zhang, Dual-functional liposomes based on pH-responsive cell-penetrating peptide and hyaluronic acid for tumor-targeted anticancer drug delivery, Biomaterials 33 (2012) 9246-9258. https://doi.org/10.1016/j.biomaterials.2012.09.027

[42] X. Guo, F.C. Szoka, Steric stabilization of fusogenic liposomes by a low-pH sensitive PEG−diortho ester−lipid conjugate, Bioconjug. Chem. 12 (2001) 291-300. https://doi.org/10.1021/bc000110v

[43] J.G. Reynolds, E. Geretti, B.S. Hendriks, H. Lee, S.C. Leonard, S.G. Klinz, C.O. Noble, P.B. Lücker, P.W. Zandstra, D.C. Drummond, K.J. Olivier Jr, U.B. Nielsen, C. Niyikiza, S.V. Agresta, T.J. Wickham, HER2-targeted liposomal doxorubicin displays enhanced anti-tumorigenic effects without associated cardiotoxicity, Toxicol. Appl. Pharmacol. 262, 1 (2012)1-10. https://doi.org/10.1016/j.taap.2012.04.008

[44] T. Nakanishi, S. Fukushima, K. Okamoto, M. Suzuki, Y. Matsumura, M. Yokoyama, T. Okano, Y. Sakurai, K. Kataoka, Development of the polymer micelle carrier system for doxorubicin, J. Control Release 74, 1 (2001) 295-302. https://doi.org/10.1016/S0168-3659(01)00341-8

[45] I.A. Khalil, K. Kogure, H. Akita, H. Harashima, Uptake pathways and subsequent intracellular trafficking in nonviral gene delivery, Pharmacol. Rev. 58, 1 (2006) 32-45. https://doi.org/10.1124/pr.58.1.8

[46] G. Kibria, H. Hatakeyama, H. Harashima, Cancer multidrug resistance: mechanisms involved and strategies for circumvention using a drug delivery system, Arch. Pharm. Res. 37 (2013) 4-15. https://doi.org/10.1007/s12272-013-0276-2

[47] Y. Sakurai, T. Hada, H. Harashima, Preparation of a cyclic RGD: Modified liposomal SiRNA formulation for use in active targeting to tumor and tumor endothelial cells, Methods Mol. Biol. 1364 (2016) 63-69. https://doi.org/10.1007/978-1-4939-3112-5_6

[48] M.N. Hossen, K. Kajimoto, H. Akita, M. Hyodo, H. Harashima, Vascular-targeted nanotherapy for obesity: Unexpected passive targeting mechanism to obese fat for the enhancement of active drug delivery, J. Control Release 163 (2012) 101-110. https://doi.org/10.1016/j.jconrel.2012.09.002

[49] M.N. Hossen, K. Kajimoto, H. Akita, M. Hyodo, T. Ishitsuka, H. Harashima, Therapeutic assessment of cytochrome C for the prevention of obesity through endothelial cell-targeted nanoparticulate system, Molecular Therapy 21, 3 (2013) 533-541. https://doi.org/10.1038/mt.2012.256

[50] M. Nargis, Development of an innovative drug delivery system targeted to adipose vessel utilizing novel nucleic acid aptamer for control of obesity, Hokkaido University Collection of Scholarly and Academic Papers. (2014) DOI: 10.14943/doctoral.k11558.

[51] K. Kajimoto, M.N. Hossen, H. Harashima, Antiangiogenic nanotherapy for the control of obesity, Nanomedicine (Lond), 8, 5 (2013) 671-673. https://doi.org/10.2217/nnm.13.27

[52] A. Akhter, Y. Hayashi, Y. Sakurai, N. Ohga, K. Hida, H. Harashima, A liposomal delivery system that targets liver endothelial cells based on a new peptide motif present in the ApoB-100 sequence, Int. J. Pharm. 456, 1 (2013) 195 -201. https://doi.org/10.1016/j.ijpharm.2013.07.068

[53] A. Akhter, Y. Hayashi, Y. Sakurai, N. Ohga, K. Hida, H. Harashima, Ligand density at the surface of a nanoparticle and different uptake mechanism: two important factors for successful siRNA delivery to liver endothelial cells, Int. J. Pharm. 475 (2014) 227-237. https://doi.org/10.1016/j.ijpharm.2014.08.048

[54] K. Hashiba, Y. Sato, H. Harashima, pH-labile PEGylation of siRNA-loaded lipid nanoparticle improves active targeting and gene silencing activity in hepatocytes, J. Control Release 262 (2017) 239-246. https://doi.org/10.1016/j.jconrel.2017.07.046

[55] G. Gaucher, M-H Dufresne, V. P. Sant, N. Kang, D. Maysinger, J-C Leroux, Block copolymer micelles: preparation, characterization and application in drug delivery, J. Control. Release 109 (2005) 169-188. https://doi.org/10.1016/j.jconrel.2005.09.034

[56] Z. Ahmad, A. Shah, M. Siddiq, H.-B. Kraatz, Polymeric micelles as drug delivery vehicles, RSC Adv. 4 (2014) 17028-17038. https://doi.org/10.1039/C3RA47370H

[57] K. Kataoka, A. Harada, Y. Nagasaki, Block copolymer micelles for drug delivery: design, characterization and biological significance, Adv. Drug Deliv. Rev. 64 (2012) 37-48. https://doi.org/10.1016/j.addr.2012.09.013

[58] Y. Matsumura, H. Maeda, A new concept for macromolecular therapeutics in cancer chemotherapy: mechanism of tumoritropic accumulation of proteins and the antitumor agent smancs, Cancer Res. 46 (1986) 6387-6392.

[59] T.-Y. Kim, D-W Kim, J-Y. Chung, S. G. Shin, S-C. Kim, D. S. Heo, N. K. Kim, Y-J. Bang, Phase I and pharmacokinetic study of Genexol-PM, a cremophor- free, polymeric micelle-formulated paclitaxel, in patients with advanced malignancies, Clin. Cancer Res. 10 (2004) 3708-3716. https://doi.org/10.1158/1078-0432.CCR-03-0655

[60] D. Vetvicka, M. Hruby, O. Hovorka, T. Etrych, M. Vetrik, L. Kovar, M. Kovar, K. Ulbrich, B. Rihova, Biological evaluation of polymeric micelles with covalently bound doxorubicin, Bioconjugate Chem. 20 (2009) 2090-2097. https://doi.org/10.1021/bc900212k

[61] M. Watanabe, K. Kawano, M. Yokoyama, P. Opanasopit, T. Okano, Y. Maitani, Preparation of camptothecin-loaded polymeric micelles and evaluation of their incorporation and circulation stability, Int. J. Pharm. 308 (2006) 183-189. https://doi.org/10.1016/j.ijpharm.2005.10.030

[62] P. Kumari, O.S. Muddineti, S.V. Rompicharla, P. Ghanta, B.B.N.A. Karthik, B. Ghosh, S. Biswas, Cholesterol-conjugated poly(D, L-lactide)-based micelles as a nanocarrier system for effective delivery of curcumin in cancer therapy, Drug Deliv. 24 (2017) 209-223. https://doi.org/10.1080/10717544.2016.1245365

[63] R. N. Gilbreth, S. Novarra, L. Wetzel, S. Florinas, H. Cabral, K. Kataoka, J. Rios-Doria, R. J. Christie, M. Baca, Lipid- and polyion complex-based micelles as agonist platforms for TNFR superfamily receptors, J. Control Release 234 (2016) 104-114. https://doi.org/10.1016/j.jconrel.2016.05.041

[64] A. Mandal, R. Bisht, I.D. Rupenthal, A.K. Mitra, Polymeric micelles for ocular drug delivery: from structural frameworks to recent preclinical studies, J. Control Release 248 (2017) 96-116. https://doi.org/10.1016/j.jconrel.2017.01.012

[65] Q. Li, K.L. Lai, P.S. Chan, S.C. Leung, H.Y. Li, Y. Fang, K.K. To, C.H.J. Choi, Q.Y. Gao, T.W. Lee, Micellar delivery of dasatinib for the inhibition of pathologic cellular processes of the retinal pigment epithelium, Coll. Surf. B 140 (2016) 278-286. https://doi.org/10.1016/j.colsurfb.2015.12.053

[66] S. Su, Y. Ding, Y. Li, Y. Wu, G. Nie, Integration of photothermal therapy and synergistic chemotherapy by a porphyrin self-assembled micelle confers che-mosensitivity in triple-negative breast cancer, Biomaterials 80 (2016) 169-178. https://doi.org/10.1016/j.biomaterials.2015.11.058

[67] R. Miyazaki, M. Nargis, A.B. Ihsan, N. Nakajima, M. Hamada, Y. Koyama, Effects of glycon and temperature on self-assembly behaviors of α-galactosyl ceramide in water, Langmuir 37 (2021) 7936-7944. https://doi.org/10.1021/acs.langmuir.1c00545

[68] M. Nargis, A.B. Ihsan, Y. Koyama, Thermo-responsive structure and dye-encapsulation of micelles comprising bolaamphiphilic quercetin polyglycoside, Langmuir 36 (2020) 10764-10771. https://doi.org/10.1021/acs.langmuir.0c01564

[69] M. Nargis, A.B. Ihsan, Y. Koyama, Impact of sugar chain length on micellization of quercetin-3-o -glycosides, Chem. Lett. 49, 8 (2020) 896-899. https://doi.org/10.1246/cl.200270

[70] M. Nargis, A.B. Ihsan, Y. Koyama, Bolaamphiphilic properties and pH-dependent micellization of quercetin polyglycoside, RSC Adv. 9 (2019) 33674-33677. https://doi.org/10.1039/C9RA05711K

[71] A.B. Ihsan, M. Nargis, Y. Koyama, Effects of the hydrophilic-lipophilic balance of alternating peptides on self-assembly and thermo-responsive behaviors, Int. J. Mol. Sci. 20 (2019) 4604-4614. https://doi.org/10.3390/ijms20184604

[72] A.B. Ihsan, Y. Koyama, Impact of polypeptide sequence on thermal properties for diblock, random, and alternating copolymers containing a stoichiometric mixture of glycine and valine, Polymer 161 (2019) 197-204. https://doi.org/10.1016/j.polymer.2018.12.021

[73] A.B. Ihsan, Y. Koyama, T, Taira, T. Imura, Thermo-responsive structure and surface activity of kinetically stabilized micelle composed of fluorinated alternating peptides in organic solvent, ChemistrySelect 3 (2018) 4173-4178. https://doi.org/10.1002/slct.201800590

[74] Y. Koyama, A. B. Ihsan, T. Taira, T. Imura, Fluorinated polymer surfactants bearing alternating peptide skeleton prepared by three-component polycondensation, RSC Adv. 8 (2018) 7509-7513. https://doi.org/10.1039/C8RA00581H

[75] T.-S. Nguyen, P.M.M. Weers, V. Raussens, Z. Wang, Amphotericin B induces interdigitation of apolipoprotein stabilized nanodisk bilayers, Biochim. Biophys. Acta. 1778 (2008) 303-312. https://doi.org/10.1016/j.bbamem.2007.10.005

[76] H. Huang, W. Cruz, J. Chen, G. Zheng, Learning from biology: synthetic lipoproteins for drug delivery, Wiley Interdiscip. Rev. Nanomed. Nanobiotechnol. 7(3) (2015) 298-314. https://doi.org/10.1002/wnan.1308

[77] M. Su, W. Chang, K. Shi, D. Wang, M. Wang, T. Xu, W. Yan, Preparation and activity analysis of recombinant human high-density lipoprotein, Assay Drug Dev. Technol. 10(5) (2012) 485-91. https://doi.org/10.1089/adt.2012.467

[78] B.A. Kingwell, M.J. Chapman, A. Kontush, N.E. Miller, HDL-targeted therapies: progress, failures and future, Nat. Rev. Drug Discov. 13 (2014) 445-464. https://doi.org/10.1038/nrd4279

[79] T. Gordon, W.P. Castelli, M.C. Hjortland, W.B. Kannel, T.R. Dawber, High density lipoprotein as a protective factor against coronary heart disease, Am. J. Med. 62 (1977) 707-714. https://doi.org/10.1016/0002-9343(77)90874-9

[80] G.J. Miller, N.E. Miller, Plasma-high-density-lipoprotein concentration and development of ischæmic heart-disease, Lancet 305 (1975) 16-19. https://doi.org/10.1016/S0140-6736(75)92376-4

[81] J. Jia, Y. Xiao, J. Liu, W. Zhang, H. He, L. Chen, M. Zhang, Preparation, characterizations, and in vitro metabolic processes of paclitaxel-loaded discoidal recombinant high-density lipoproteins, J. Pharm. Sci. 101,8 (2012) 2900-2908. https://doi.org/10.1002/jps.23210

[82] R.G. Parmar, M. Busuek, E.S. Walsh, K.R. Leander, B.J. Howell, L. Sepp-Lorenzino, E. Kemp, L.S. Crocker, A. Leone, C.J. Kochansky, B.A. Carr, R.M. Garbaccio, S.L. Colletti, W. Wang, Endosomolytic bioreducible poly(amido amine disulfide) polymer conjugates for the in vivo systemic delivery of siRNA therapeutics. Bioconjug. Chem. 24, 4 (2013) 640-647. https://doi.org/10.1021/bc300600a

[83] T. Murakami, W. Wijagkanalan, M. Hashida, K. Tsuchida, Intracellular drug delivery by genetically engineered high-density lipoprotein nanoparticles. Nanomedicine, 5 (2010) 867-879. https://doi.org/10.2217/nnm.10.66

[84] K. Suda, T. Murakami, N. Gotoh, R. Fukuda, Y. Hashida, M. Hashida, A. Tsujikawa, N. Yoshimura, High-density lipoprotein mutant eye drops for the treatment of posterior eye diseases, J. Control. Release 266 (2017) 301-309. https://doi.org/10.1016/j.jconrel.2017.09.036

[85] N. Peppas, P. Bures, W. Leobandung, H. Ichikawa, Hydrogels in pharmaceutical formulations, Eur. J. Pharm. Biopharm. 50 (2000) 27-46. https://doi.org/10.1016/S0939-6411(00)00090-4

[86] X. Su, M. J. Tan, Z. Li, M. Wong, L. Rajamani, G. Lingam, X. J. Loh, Recent progress in using biomaterials as vitreous substitutes, Biomacromolecules 16 (2015) 3093-3102. https://doi.org/10.1021/acs.biomac.5b01091

[87] S.R.V. Tomme, G. Storm, W.E. Hennink, In situ gelling hydrogels for pharmaceutical and biomedical applications, Int. J. Pharm. 355 (2008) 1-18. https://doi.org/10.1016/j.ijpharm.2008.01.057

[88] N. Peppas, R. Langer, New challenges in biomaterials, Science 263 (1994) 1715-1720. https://doi.org/10.1126/science.8134835

[89] A.S. Hoffman, B.D. Ratner, Synthetic hydrogels for biomedical applications. In Hydrogels for Medical and Related Applications (ed. ACS Symposium Series) 1-36 (American Chemical Society, Washington, DC, USA, 1976). https://doi.org/10.1021/bk-1976-0031.ch001

[90] C.-C. Lin, A.T. Metters, Hydrogels in controlled release formulations: network design and mathematical modeling, Adv. Drug Deliv. Rev. 58 (2006) 1379-1408. https://doi.org/10.1016/j.addr.2006.09.004

[91] A.S. Hoffman, Hydrogels for biomedical applications, Adv. Drug Deliv. Rev. 64 (2012), 18-23. https://doi.org/10.1016/j.addr.2012.09.010

[92] W. Zhao, Y. Li, X. Zhang, R. Zhang, Y. Hu, C. Boyer, F. J. Xu, Photo-responsive supramolecular hyaluronic acid hydrogels for accelerated wound healing, J. Controlled Release 323 (2020) 24-35. https://doi.org/10.1016/j.jconrel.2020.04.014

[93] L. Shi, Y. Zhao, Q. Xie, C. Fan, J. Hilborn, J. Dai, D.A. Ossipov, Self-healing polymeric hydrogel formed by metal-ligand coordination assembly: Design, fabrication, and biomedical applications, Adv. Healthc. Mater. 7 (2018) 1-9. https://doi.org/10.1002/marc.201800837

[94] J. Cho, S.H. Kim, B. Yang, J. M. Jung, I. Kwon, D.S. Lee, Albumin affibody-outfitted injectable gel enabling extended release of urate oxidase-albumin conjugates for hyperuricemia treatment, J. Controlled Release 324 (2020) 532-544. https://doi.org/10.1016/j.jconrel.2020.05.037

[95] Y.-Y. Chen, H.-C. Wu, J.-S. Sun, G.-C. Dong, T.-W. Wang, Injectable and thermoresponsive self-assembled nanocomposite hydrogel for long-term anticancer drug delivery, Langmuir 29 (2013) 3721-3729. https://doi.org/10.1021/la400268p

[96] X. Qi, W. Wei, J. Li, Y. Liu, X. Hu, J. Zhang, L. Bi, W. Dong, Fabrication and characterization of a novel anticancer drug delivery system: salecan/poly(methacrylic acid) semi-interpenetrating polymer network hydrogel, ACS Biomater. Sci. Eng. 1 (2015) 1287-1299. https://doi.org/10.1021/acsbiomaterials.5b00346

[97] H. Jang, K. Zhi, J. Wang, H. Zhao, B. Li, X. Yang, Enhanced therapeutic effect of paclitaxel with a natural polysaccharide carrier for local injection in breast cancer, Int. J. Biol. Macromol. 148 (2020) 163-172. https://doi.org/10.1016/j.ijbiomac.2020.01.094

[98] J. Yu, W. Ha, J.-N. Sun, Y.-P. Shi, Supramolecular hybrid hydrogel based on host-guest interaction and its application in drug delivery, ACS Appl. Mater. Interfaces 6 (2014) 19544-19551. https://doi.org/10.1021/am505649q

[99] M. Rasoulzadeh, H. Namazi, Carboxymethyl cellulose/graphene oxide bio-nanocomposite hydrogel beads as anticancer drug carrier agent, Carbohydr. Polym. 168 (2017) 320-326. https://doi.org/10.1016/j.carbpol.2017.03.014

[100] J. Naskar, G. Palui, A. Banerjee, Tetrapeptide-based hydrogels: for encapsulation and slow release of an anticancer drug at physiological pH, J. Phys. Chem. B 113 (2009) 11787-11792. https://doi.org/10.1021/jp904251j

[101] F. Alvarez-Rivera, A. Concheiro, C. Alvarez-Lorenzo, Eur. J. Pharm. Biopharm. 122 (2018) 126-136. https://doi.org/10.1016/j.ejpb.2017.10.016

[102] S. Deepthi, J. Jose, Novel hydrogel-based ocular drug delivery system for the treatment of conjunctivitis, Int. Ophthalmol. 39, 6 (2019) 1355-1366. https://doi.org/10.1007/s10792-018-0955-6

[103] B.V. Slaughter, S. S. Khurshid, O. Z. Fisher, A. Khademhosseini, N. A. Peppas, Hydrogels in regenerative medicine. Adv Mater. 21 (2009) 3307-3329. https://doi.org/10.1002/adma.200802106

[104] K.J. Walker, S.V. Madihally, Anisotropic temperature sensitive chitosan-based injectable hydrogels mimicking cartilage matrix, J. Biomed. Mater. Res. B. Appl. Biomater. 103 (2015) 1149-1160. https://doi.org/10.1002/jbm.b.33293

[105] R. Jin, L.S.M. Teixeira, P.J. Dijkstra, M. Karperien, C.A. van Blitterswijk, Z.Y. Zhong, J. Feijen, Injectable chitosan-based hydrogels for cartilage tissue engineering, Biomaterials 30 (2009) 2544-2551. https://doi.org/10.1016/j.biomaterials.2009.01.020

[106] Y. Wei, Y. Hu, W. Hao, Y. Han, G. Meng, D. Zhang, Z. Wu, H. Wang, A novel injectable scaffold for cartilage tissue engineering using adipose-derived adult stem cells, J. Orthop. Res. 26 (2008) 27-33. https://doi.org/10.1002/jor.20468

[107] H. Dashtdar, M.R. Murali, A.A. Abbas, A.M. Suhaeb, L. Selvaratnam, L.X. Tay, PVA-chitosan composite hydrogel versus alginate beads as a potential mesenchymal stem cell carrier for the treatment of focal cartilage defects, Knee Surg. Sports Traumatol. Arthrosc. 23 (2015) 1368-1377. https://doi.org/10.1007/s00167-013-2723-5

[108] B.G. Cooper, R.C. Stewart, D. Burstein, B.D. Snyder, M.W. Grinstaff, A tissue-penetrating double network restores the mechanical properties of degenerated articular cartilage, Angew. Chem. 55 (2016) 4226-4230. https://doi.org/10.1002/anie.201511767

[109] A.B. Ihsan, T.L. Sun, S. Kuroda, M.A. Haque, T. Kurokawa, T. Nakajima, J.P. Gong, A phase diagram of neutral polyampholyte - From solution to tough hydrogel, J. Mat. Chem. B 1 (2013) 4555-4562. https://doi.org/10.1039/c3tb20790k

[110] T.L. Sun, T. Kurokawa, S. Kuroda, A.B. Ihsan, T. Akasaki, K. Sato, M.A. Haque, T. Nakajima, J.P. Gong Physical Hydrogels Composed of Polyampholytes Demonstrate High Toughness and Viscoelasticity, Nat. Mater. 12 (2013) 932-937. https://doi.org/10.1038/nmat3713

[111] A.B. Ihsan, T.L. Sun, T. Kurokawa, S.N. Karobi, T. Nakajima, T. Nonoyama, C.K. Roy, F. Luo, J.P. Gong, Self-healing behaviors of tough polyampholyte hydrogels, Macromolecules 49 (2016) 4245-4252. https://doi.org/10.1021/acs.macromol.6b00437

[112] F. Luo, T.L Sun, T. Nakajima, T. Kurokawa, Y. Zhao, K. Sato, A.B. Ihsan, X. Li, H. Guo, J.P Gong, Oppositely charged polyelectrolytes form tough, self-healing and rebuildable hydrogels, Adv. Mater 27 (2015) 2722-2727. https://doi.org/10.1002/adma.201500140

[113] F. Luo, T.L. Sun, T. Nakajima, T. Kurokawa, A.B. Ihsan, X. Li, H. Guo, J.P. Gong, Free reprocessability of tough and self-healing hydrogels based on polyion complex, ACS Macro. Lett. 4 (2015) 961-964. https://doi.org/10.1021/acsmacrolett.5b00501

[114] F. Luo, T.L. Sun, T. Nakajima, D.R. King, T. Kurokawa, Y. Zhao, A.B. Ihsan, X. Li, H. Guo, J.P. Gong, Strong and tough polyion-complex hydrogels from oppositely charged polyelectrolytes: A comparative study with polyampholyte hydrogels, Macromolecules 49 (2016) 2750-2760. https://doi.org/10.1021/acs.macromol.6b00235

[115] F. Luo, T.L. Sun, T. Nakajima, T. Kurokawa, Y. Zhao, A.B. Ihsan, H. Guo, X. Li, J. P. Gong Crack blunting and advancing behaviors of tough and self-healing polyampholyte hydrogel, Macromolecules 47 (2014) 6037-6046. https://doi.org/10.1021/ma5009447

[116] C.K. Roy, H. Guo, T.L. Sun, A.B. Ihsan, T. Kurokawa, M. Takahata, T. Nonoyama, T. Nakajima, J.P. Gong, Self-adjustable adhesion of polyampholyte hydrogels, Adv. Mater. 27 (2015) 7344-7348. https://doi.org/10.1002/adma.201504059

[117] D.A. Tomalia, Birth of a new macromolecular architecture: dendrimers as quantized building blocks for nanoscale synthetic polymer chemistry, Prog. Polym. Sci. 30 (2005) 294-324. https://doi.org/10.1016/j.progpolymsci.2005.01.007

[118] P.G. de Gennes, H. Hervet, Statistics of « starburst » polymers, J. Phys. Lett. (1983) 44, 351-360. https://doi.org/10.1051/jphyslet:01983004409035100

[119] A.W. Bosman, H.M. Janssen, E.W. Meijer, About Dendrimers: Structure, physical properties, and applications, Chem. Rev. 99 (1999) 1665-1688. https://doi.org/10.1021/cr970069y

[120] G.R. Newkome, Z. Yao, G.R. Baker and V.K. Gupta, Cascade molecules: a new approach to micelles, A Arborol. J. Org. Chem. 50 (1985) 2003-2004. https://doi.org/10.1021/jo00211a052

[121] E. Buhleier, W. Wehner, F. Vogtle, Cascade and Nonskid-Chain-Like Syntheses of Molecular Cavity Topologies, Synthesis 2 (1978) 155-158. https://doi.org/10.1055/s-1978-24702

[122] C.J. Hawker and J.M.J. Fréchet, Preparation of polymers with controlled molecular architecture. A new convergent approach to dendritic macromolecules, J. Am. Chem. Soc. 112 (1990) 7638-7647 https://doi.org/10.1021/ja00177a027

[123] K.L. Wooley, C.J. Hawker, J.M.J. Fréchet, Hyperbranched macromolecules via a novel double-stage convergent growth approach, J. Am. Chem. Soc. 113 (1991) 4252-4261. https://doi.org/10.1021/ja00011a031

[124] J.M.J. Fréchet, Functional polymers and dendrimers: reactivity, molecular architecture, and interfacial energy, Science 263, 5154 (1994) 1710-1715. https://doi.org/10.1126/science.8134834

[125] D.A. Tomalia, Starburst/cascade dendrimers: fundamental building blocks for a new nanoscopic chemistry set, Advanced Materials 6, 7-8 (1994) 529-539. https://doi.org/10.1002/adma.19940060703

[126] M. El-Sayed, M. Ginski, C. Rhodes, H. Ghandehari, Transepithelial transport of poly(amidoamine) dendrimers across Caco-2 cell monolayers, J. Control Release 81, 3 (2002) 355-365. https://doi.org/10.1016/S0168-3659(02)00087-1

[127] N. Vijayalakshmi, A. Ray, A. Malugin, H. Ghandehari, Carboxyl-terminated PAMAM-SN38 conjugates: synthesis, characterization, and in vitro evaluation, Bioconjugate Chem. 21, 10 (2010) 1804-1810. https://doi.org/10.1021/bc100094z

[128] D.S. Wilbur, P.M. Pathare, D.K. Hamlin, K.R. Buhler, R.L. Vessella, Biotin reagents for antibody pretargeting. 3. Synthesis, radioiodination, and evaluation of biotinylated starburst dendrimers, Bioconjugate Chem. 9, 6 (1998) 813-825. https://doi.org/10.1021/bc980055e

[129] N.A. Peppas, Star polymers and dendrimers: prospects of their use in drug delivery and pharmaceutical applications, Controlled Release Society Newsletter, 12 (1995) 12-13.

[130] A. Pérez-Anes, G. Spataro, Y. Coppel, C. Moog, M. Blanzat, C.-O. Turrin, A.-M. Caminade, I. Rico-Lattes, J.-P. Majoral, Phosphonate terminated PPH dendrimers: influence of pendant alkyl chains on the in vitro anti-HIV-1 properties, Org. Biomol. Chem. 7 (2009) 3491-3498. https://doi.org/10.1039/b908352a

[131] M.T. Morgan, M.A. Carnahan, C.E. Immoos, A.A. Ribeiro, S. Finkelstein, S.J. Lee, M.W. Grinstaff, Dendritic molecular capsules for hydrophobic compounds, J. Am. Chem. Soc. 125 (2003) 15485-15489. https://doi.org/10.1021/ja0347383

[132] N. Nishiyama, H.R. Stapert, G-D. Zhang, D. Takasu, D-L. Jiang, T. Nagano, T. Aida, K. Kataoka, Light-harvesting ionic dendrimer porphyrins as new photosensitizers for photodynamic therapy, Bioconjugate Chem. 14 (2003) 58-66. https://doi.org/10.1021/bc025597h

[133] C. Yiyun, N. Man, T. Xu, R. Fu, X. Wang, X. Wang, L. Wen, Transdermal delivery of nonsteroidal anti-inflammatory drugs mediated by polyamidoamine (PAMAM) dendrimers, J. Pharm. Sci. 96 (2007) 595-602. https://doi.org/10.1002/jps.20745

[134] W. Zhang, Z. Zhang, Y. Zhang, The application of carbon nanotubes in target drug delivery systems for cancer therapies, Nanoscale Res. Lett. 6 (2011) 555-576. https://doi.org/10.1186/1556-276X-6-555

[135] S. Iijima, Helical microtubules of graphitic carbon, Nature (London) 354 (1991) 56-58. https://doi.org/10.1038/354056a0

[136] S.K.S Kushwaha, S. Ghoshal, A.K. Rai, S. Singh, Carbon nanotubes as a novel drug delivery system for anticancer therapy: a review, Braz. J. Pharm. Sci. 49, 4 (2013) 629-643. https://doi.org/10.1590/S1984-82502013000400002

[137] C. Gherman, M.C. Tudor, B. Constantin B, T. Flaviu, R. Stefan, B. Maria, S. Chira, C. Braicu, L. Pop, R. C. Petric, I. Berindan-Neagoeet, Pharmacokinetics evalua- tion of carbon nanotubes using FTIR analysis and histological analysis, J. Nanosci. Nanotechnol. 15, 4 (2015) 2865-2869. https://doi.org/10.1166/jnn.2015.9845

[138] B.V. Farahani, G.R. Behbahani, N. Javadi, Functionalized multi walled carbon nanotubes as a carrier for doxorubicin: drug adsorption study and statistical optimization of drug loading by factorial design methodology, J. Braz. Chem. Soc. 27, 4 (2016) 694-705. https://doi.org/10.5935/0103-5053.20150318

[139] C.D.M. Fletcher, J.A. Bridge, P. Hogendoorn, F. Mertens, 2013. WHO classification of tumours of soft tissue and bone. (4th ed.) Lyon: IARC Press.

[140] J.M. Worle-Knirsh, K. Pulskamp, H.F. Krug, Oops they did it again! Carbon nanotubes hoax scientists in viability assays, Nano Lett. 6, 6 (2006) 1261-8. https://doi.org/10.1021/nl060177c

[141] N.W.S. Kam, T.C. Jessop, P.A. Wender, H. Dai Nanotube molecular transporters: internalization of carbon nanotube-protein conjugates into mammalian cells, J. Am. Chem. Soc. 126 (2004) 6850-6851. https://doi.org/10.1021/ja0486059

[142] N.W. Kam, H. Dai, Carbon nanotubes as intracellular protein transporters: generality and bio- logical functionality, J. Am. Chem. Soc. 127 (2005) 6021-6026. https://doi.org/10.1021/ja050062v

[143] J.Y. Chen, S. Chen, X. Zhao, L.V. Kuznetsova, S.S. Wong, I. Ojima, Functionalized single-walled carbon nanotubes as rationally designed vehicles for tumor targeted drug delivery, J. Am. Chem. Soc. 130 (2008) 16778-85. https://doi.org/10.1021/ja805570f

[144] Z. Liu, X.M. Sun, N. Nakayama-Ratchford, H.J. Dai, Supramolecular chemistry on water soluble carbon nanotubes for drug loading and delivery, ACS Nano. 1 (2007) 50-56. https://doi.org/10.1021/nn700040t

[145] Q.X. Mu, D.L. Broughton, B. Yan, Endosomal leakage and nuclear translocation of multiwalled carbon nanotubes: developing a model for cell uptake, Nano Lett. 9 (2009) 4370-4375. https://doi.org/10.1021/nl902647x

[146] N.W.S. Kam, H. Dai, Carbon nanotubes as intracellular protein transporters: generality and bio- logical functionality, J. Am. Chem. Soc. 127, 16 (2005) 6021-6026. https://doi.org/10.1021/ja050062v

[147] S. Deb, H.K. Patra, P. Lahiri, A.Kr. Dasgupta, K. Chakrabarti, U. Chaudhuri, Multistability in platelets and their response to gold nanoparticles, Nanomedicine 7 (2011) 376-384. https://doi.org/10.1016/j.nano.2011.01.007

[148] S. Wilhelm, A.J. Tavares, Q. Dai, S. Ohta, J. Audet, H.F. Dvorak, W.C.W. Chan, Analysis of nanoparticle delivery to tumours, Nat. Rev. Mater. 1 (2016) 16014. https://doi.org/10.1038/natrevmats.2016.14

[149] E. Blanco, H. Shen, M. Ferrari, Principles of nanoparticle design for overcoming biological barriers to drug delivery, Nat. Biotechnol. 33 (2015) 941-951. https://doi.org/10.1038/nbt.3330

[150] M.D. McSweeney, T. Wessler, L.S.L. Price, E. C. Ciociola, L.B. Herity, J.A. Piscitelli, W.C. Zamboni, M. G. Forest, Y. Cao, S.K. Lai. A minimal physiologically based pharmacokinetic model that predicts anti-PEG IgG-mediated clearance of PEGylated drugs in human and mouse, J. Control. Rel. 284 (2018) 171-178. https://doi.org/10.1016/j.jconrel.2018.06.002

[151] L. Hosta-Rigau, B. Städler, Shear stress and its effect on the interaction of myoblast cells with nanosized drug delivery vehicles, Mol. Pharm. 10 (2013) 2707-2712. https://doi.org/10.1021/mp4001298

[152] M. Cooley, A. Sarode, M. Hoore, D.A. Fedosov, S. Mitragotri, A.S. Gupta, Influence of particle size and shape on their margination and wall-adhesion: implications in drug delivery vehicle design across nano-to-micro scale, Nanoscale 10 (2018) 15350-15364. https://doi.org/10.1039/C8NR04042G

[153] C. von Roemeling, W. Jiang, C.K. Chan, I.L. Weissman, B.Y.S. Kim, Breaking down the barriers to precision cancer nanomedicine, Trends Biotechnol. 35 (2017) 159-171. https://doi.org/10.1016/j.tibtech.2016.07.006

[154] Q. Zhong, O.M. Merkel, J.J. Reineke, S.R.P. da Rocha, Effect of the route of administration and PEGylation of poly(amidoamine) dendrimers on their systemic and lung cellular biodistribution, Mol. Pharm. 13 (2016) 1866-1878. https://doi.org/10.1021/acs.molpharmaceut.6b00036

[155] L. Battaglia, P.P. Panciani, E. Muntoni, M.T. Capucchio, E. Biasibetti, P. De Bonis, S. Mioletti, M. Fontanella, S. Swaminathan, Lipid nanoparticles for intranasal administration: application to nose-to-brain delivery, Expert Opin. Drug Deliv. 15 (2018) 369-378. https://doi.org/10.1080/17425247.2018.1429401

[156] Y. Dölen, M. Valente, O. Tagit, E. Jäger, E.A.W. Van Dinther, N.K. van Riessen, M. Hruby, U. Gileadi, V. Cerundolo, C.G. Figdor. Nanovaccine administration route is critical to obtain pertinent iNKt cell help for robust anti-tumor T and B cell responses, Oncoimmunology 9 (2020) 1738813. https://doi.org/10.1080/2162402X.2020.1738813

[157] D.N. McLennan, C.J.H. Porter, S.A. Charman, Subcutaneous drug delivery and the role of the lymphatics, Drug Discov. Today Technol. 2 (2005) 89-96. https://doi.org/10.1016/j.ddtec.2005.05.006

[158] C. Saraiva, C. Praça, R. Ferreira, T. Santos, L. Ferreira, L. Bernardino, Nanoparticle-mediated brain drug delivery: overcoming blood-brain barrier to treat neurodegenerative diseases, J. Control. Release 235 (2016) 34-47. https://doi.org/10.1016/j.jconrel.2016.05.044

[159] Q. Zhou, S. Shao, J. Wang, C. Xu, J. Xiang, Y. Piao, Z. Zhou, Q. Yu, J. Tang, X. Liu, Z. Gan, R. Mo, Z. Gu, Y. Shen, Enzyme-activatable polymer-drug conjugate

augments tumour penetration and treatment efficacy. Nat. Nanotechnol. 14 (2019) 799-809. https://doi.org/10.1038/s41565-019-0485-z

[160] S. Sindhwani, A.M. Syed, J. Ngai, B.R. Kingston, L. Maiorino, J. Rothschild, P. MacMillan, Y. Zhang, N.U. Rajesh, T. Hoang, J.L.Y. Wu, S. Wilhelm, A. Zilman, S. Gadde, A. Sulaiman, B. Ouyang, Z. Lin, L. Wang, M. Egeblad, W.C.W. Chan, The entry of nanoparticles into solid tumours, Nat. Mater. 19 (2020) 566-575. https://doi.org/10.1038/s41563-019-0566-2

[161] L.M. Ensign, R. Cone, J. Hanes, Oral drug delivery with polymeric nanoparticles: the gastrointestinal mucus barriers, Adv. Drug Deliv. Rev. 64 (2012) 557-570. https://doi.org/10.1016/j.addr.2011.12.009

[162] N. Oliva, M. Carcole, M. Beckerman, S. Seliktar, A. Hayward, J. Stanley, N.M. Parry, E.R. Edelman, N. Artzi, Regulation of dendrimer/dextran material performance by altered tissue microenvironment in inflammation and neoplasia, Sci. Transl. Med. 7 (2015) 272ra11. https://doi.org/10.1126/scitranslmed.aaa1616

[163] Y. Zan, Z. Dai, L. Liang, Y. Deng, L. Dong, Co-delivery of plantamajoside and sorafenib by a multi-functional nanoparticle to combat the drug resistance of hepatocellular carcinoma through reprograming the tumor hypoxic microenvironment, Drug Deliv. 26 (2019) 1080-1091. https://doi.org/10.1080/10717544.2019.1654040

[164] Q. Dai, S. Wilhelm, D. Ding, A.M. Syed, S. Sindhwani, Y. Zhang, Y.Y. Chen, P. MacMillan, W.C.W. Chan, Quantifying the ligand-coated nanoparticle delivery to cancer cells in solid tumors, ACS Nano 12 (2018) 8423-8435. https://doi.org/10.1021/acsnano.8b03900

[165] J. Witten, K. Ribbeck, The particle in the spider's web: transport through biological hydrogels, Nanoscale 9 (2017) 8080-8095. https://doi.org/10.1039/C6NR09736G

[166] Z. Tang, X. Zhang, Y. Shu, M. Guo, H. Zhang, W. Tao, Insights from nanotechnology in COVID-19 treatment, Nano Today 36 (2021) 101019. https://doi.org/10.1016/j.nantod.2020.101019

Emerging Nanomaterials and Their Impact on Society in the 21st Century   Materials Research Forum LLC
Materials Research Foundations 135 (2023) 152-177   https://doi.org/10.21741/9781644902172-7

Chapter 7

Hybrid Nanomaterials: Historical Developments, Classification and Biomedical Applications

Jarin Ibedita Edi[1], Abdul Muhaymin[2,3], Md. Abu Bin Hasan Susan[4*]

[1]Scholars' School and College, Dhanmondi, Dhaka, Bangladesh

[2]CAS Key Laboratory for Biomedical Effects of Nanomaterials and Nanosafety & CAS Center for Excellence in Nanoscience, National Center for Nanoscience and Technology of China, Beijing 100190, PR China

[3]University of Chinese Academy of Sciences, Beijing 100049, PR China

[4]Department of Chemistry, Dhaka University, Dhaka 1000, Bangladesh

* susan@du.ac.bd

Abstract

Hybrid nanomaterials (HNs) have exceptional physical and chemical properties and combine superior qualities of both inorganic and organic materials to exploit them to desirable chemistry through imposing multifunctionality in a single material. Strategies of synthesis of HNs involve the fabrication of components, either organic or inorganic, through the insertion of molecules or nano-objects or polymerization of precursors. The interconnected porous network of HNs also enables a range of applications. In this chapter, we have discussed historical development, strategies of synthesis, and, classification of HNs with emphasis on advancement for biomedical applications.

Keywords

Hybrid Nanomaterials, Classification, Historical Development, Strategies for Synthesis, Biomedical Applications

Contents

Hybrid Nanomaterials: Historical Developments, Classification and Biomedical Applications..152

1. **Introduction**..153

2. **History of hybrid nanomaterials**..154

3. **Classification of hybrid nanomaterials**...155

   3.1   First-class hybrid nanomaterials 156

   3.2   Second-class hybrid nanomaterials 157

4.   **Strategies for synthesis of hybrid nanomaterials 157**

   4.1   In situ formation 159

   4.2   Sol-gel process...160

   4.3   Electrocrystallization 160

   4.4   Hydrothermal method 160

   4.5   Wet chemistry approach 160

   4.6   Polymerization of organic monomers with preformed inorganic components...160

   4.7   Simultaneous incorporation of components 161

5.   **Applications of hybrid nanomaterials 161**

   5.1   Mesoporous silica based hybrid nanoparticles...............................161

   5.2   Quantum dot based hybrid nanomaterials 165

   5.3   `Nanoscale metal-organic frameworks based hybrid nanomaterials..168

   5.4   Iron oxide nanoparticle based hybrid system 170

Conclusions...172

Acknowledgement..172

Competing financial interests 172

References...173

1. Introduction

Hybrid nanomaterials (HNs) have been one of the most fascinating domains of research in the recent decade. Materials of this kind have expanded the scope of research by showcasing exceptional physical and chemical properties within the nanoscale [1]. Their advanced characteristics, task specificity, and miniaturization quality have made them distinct from other materials bringing about a rapid rise in their growth and demand. The multidisciplinary of this scientific trend, as well as the integration of the advantageous and superior properties of both organic and inorganic materials, is pivotal to developing multifunctional materials based on building blocks [2-17].

In general, HNs comprise two or more components that are diverse. The components are typically organic, such as polymers, pharmaceutical ingredients, organic substances with specific functional groups, biomolecules, ligands, etc., and inorganics such as metals as particles, clusters, ions, sulfides, oxides, and salts and non-metals and their derivatives. The incorporation or integration of these components to develop HNs is carried out by following a number of important interactions [11]. Specific interactions start from the binding of nanoscale to the building of molecules through π-complexation, covalent bonds, etc. and also involve agreeable interactions in different modes, leading to microstructuring. Other interactions are supramolecular in nature, like self-assembly resulting from electrostatic and dispersion interactions, hydrogen bonding, etc. [12].

A variety of interactions between the components provide HNs with outstanding new properties and multifunctional nature. The most interesting fact is that the properties of HNs can be tuned or changed by changing their basic structural compositions, chemical bonding between the components, structural morphology, etc. Consequently, more enhanced, developed and innovative HNs can be introduced with improved characteristics such as controlled wetting features, mechanical integrity, superior thermal stability, electron conductivity, light emission, gas permeability, etc. The HNs have thus been finding new promising applications in various sectors such as photovoltaics, energy storage, fuel cells, remediation of pollutants, catalysis, microelectronics, diagnostics, smart coatings, optics, and sensing [6, 9, 13].

In this chapter, we reviewed the history of different types of HNs developed so far along with the protocols for synthesis. Eyes have been pointed on the properties of HNs and their tunability for a range of promising applications, with emphasis on the use of selected advanced hybrid material types such as mesoporous silica nanoparticle (MSN)-, quantum dot (QD)-, nanoscale metal-organic framework (NMOF)- and iron oxide nanoparticle (IONP)- based HNs as biomedicines. Critical analysis has been conducted to identify and accordingly address the issues associated with the use of HNs for biomedical applications and the future perspective of HNs has been highlighted.

2.   History of hybrid nanomaterials

HNs have a long history of productivity. The concept of HNs has been materialized in the 8th century. The pigment, Maya blue used in Mexico is often considered the prototype of a hybrid material. It was fabricated by incorporating natural blue indigo encapsulated within the channels of a clay mineral, palygorskite [14]. The hybrid material involves stabilization of the indigo derivative into magnesium aluminium phyllosilicate [15].

Hybrid materials have been commercialized since the 1950s. The major uses were as constituents of the paint industry, where inorganic nano-pigments are suspended within the organic solvents. The sol-gel process has been the most basic and important chemical process for organic and inorganic components in hybrid materials and underpinned the foundation of organic-inorganic HNs at the nanoscale.

Different formulations and strategies have so far been used for the fabrication of HNs [5]. In general, the strategy entails constructing innovative materials and technologies by integrating basic functional organic/inorganic, molecular, or extended blocks to develop materials with significant features. The hybrid approach for material synthesis is called a Lego©-like approach. Furthermore, the modern approach is to integrate hybrid systems into miniaturization devices [14, 16]. The rapid growth and development of HNs have by now made a nice platform for the administration of drugs and smart materials responsive to a stimulus and the development of sensors. This has gradually expanded to a range of new applications to cater to the needs of the civilization including medicine, smart and intelligent materials, and energy.

The development of a fundamental knowledge base in one discipline of science almost always leads to breakthroughs in other fields. The historical development of HNs is a clear reflection of the notion. For instance, cold fusion, despite intensive research by physicists, did not get much success to achieve its original goal. But a dramatic development was noted through the introduction of metal hydrides. The findings of the incredible performance of palladium hydrides in reduction processes have immediately been relevant in the disciplines of catalysis and electrochemistry [17].

There have been attempts to combine nanotechnology with a potent technique for synthesis. The combination is in principle based on the exploitation of supramolecular chemistry and the principle of self-assembly to develop higher-order functional materials for nanodevices. The success inspired further advancement and interdisciplinary activities at the interface of nanotechnology, life sciences, organic chemistry, and biotechnology have opened the route for nanoarchitectonics. Enzymes have been confined to nano-sized hybrid materials to ensure stabilization and the next feasible degree of complexity has been envisaged. Bioconjugate materials developed through crossbreeding of responsive polymers and biomacromolecules have proved to be useful in creating a novel synthetic tool to commercialize HNs [17].

3. Classification of hybrid nanomaterials

HNs are chemical conjugates of organic and/or inorganic components. They are formed by an intimate blending of inorganic and organic materials or both [18]. The organic components are usually naturally occurring materials, which are mostly carbon in nature [19]. They play a very important role in the advancement of hybrid nanomaterials. There are a lot of great advantages of organic materials. They offer basic adaptability, structural flexibility, convenience in processing, tunable electrical characteristics, photoconductivity, and efficient luminescence [20]. Inorganic components, on the other hand, are generally chemical compounds excluding carbon [19]. They have enhanced and high charge mobility, thermal stability, and dielectric constant and exhibit excellent features of magnetic and electrically conductive materials with high mechanical stability. At the nano-dimension, the incorporation of the inorganic components with organic materials produces exceptional HNs for manifold applications [20].

Inorganic materials can be modified by organic moieties following various processes. They are modified or altered through the introduction of charge at the surface of ligatures for functionalization. Some exceptional colloidal polymers can be stabilized by organic moieties. These include nanoparticles, nanostars, nanoflowers, nanotubes, etc. The prime requirement is electrostatic stabilization to ensure a homogenous and consistent distribution avoiding the formation of clusters. Similarly, inorganic constituents are utilized to alter or modify organic matrices or lattices. The inorganic constituents used are small particles of varying morphologies. These are generally made of metals, semiconductors, carbon-based, minerals, clays, and ceramics and may be two-dimensional. Inorganic components are subjected to coordination or integration into organic moieties or frameworks or matrices of biological/chemical species. Both polymeric and monomeric species, layers-by-layers materials, and synthetic molecules can serve as chemical matrices. These also include brushes and hydrogels and block copolymers, both within the frame of coatings or vehicles. The biological matrices are of many different varieties including microorganisms, bacteria, molecules, carbohydrates, polysaccharides, nucleic acids, proteins, and lipids [21]. Fig. 1 summarizes different types of HNs depending on the organic and inorganic components used.

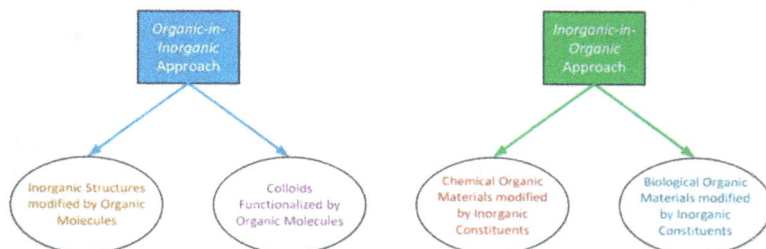

Figure 1. Types of hybrid materials based on the adding of inorganic and organic components and vice versa [21]. (Reprinted from [21]).

Based on the intensity of the interactions between the building blocks, both inorganic and organic, Kickelbick categorized hybrid materials into Class I which involves weaker interactions and Class II concerning stronger interactions [18]. In consideration of the chemical bonding or links built up between the constituent materials, or the chemical root of the interface, HNs are now classified as first-class and second-class HNs.

3.1   First-class hybrid nanomaterials

First-class HNs involve weak chemical interactions. They are based mainly on hydrogen bonds, Coulomb forces, dipole-dipole forces and London dispersion forces [21]. They are made by combining inorganic compounds with organic polymeric species (e.g., zeolite/low-density polyethylene) or combining organic with an inorganic network (for example, ring-opening metathesis of a polymer with a network of silica). A first-class HN was synthesized using a twofold mix of poly(2,6-dimethyl-1,4-phenylene ether)/poly(styrene-co-acrylonitrile) (PPE/SAN). This mix culminated in the combined advantageous properties of two blended materials and formed a hybrid composite with a set of desired properties. However, a first-class hybrid material has a weak interfacial interaction between the components. This could result in issues such as incompatibility and decreased mechanical qualities, such as inadequate fracture toughness [18].

3.2   Second-class hybrid nanomaterials

The second-class of hybrid nanomaterials is composed mainly of solid chemical bonds which are strong instead of chemically weak interactions. They are based on phase synergism achieved through ionic-covalent bonds, Lewis acid-base interaction and covalent bonds. Depending on the relative sustainability of the synergistic effects between the components, this class of materials determines the types of organic functionalization or complicated organic binding with transition metal ions that are necessary for the preparation or anchoring of organic moiety to the inorganic material [21].

Formation of second-class HNs involves covalent bonding of inorganic clusters such as metal oxide clusters to organic polymers (e.g., poly(methyl methacrylate (PMMA)) or double network of organic and inorganic components through covalent bonds (e.g., chitosan/silica scaffolds). The interactions of organic components (PPE, SAN, and SBM i.e. polystyrene-block-polybutadiene-block-poly(methyl methacrylate)) on atomic or molecular level represent interactions of this kind of HNs. ABC triblock terpolymers, specifically SBM serves the purpose of compatibilizing for improvement of interaction between PPE and SAN at the interface. Polystyrene (PS) and PMMA block thermodynamically interact and upon mixing polybutadiene (PB) irregularly disperses at the interface to form a "raspberry" morphology [18, 22-23].

4.   Strategies for synthesis of hybrid nanomaterials

The preparation of organic-inorganic HNs has been the focus for more than three decades. The strategies for synthesis include the polymerization of suitable precursors or insertion of molecular, or nano-objects like nanoslabs, polymers, functional nanoparticles, alkoxides or metal salts, etc. The reactions are carried out mostly at 20-300 °C at ambient or autogenous pressurization conditions in aqueous or natural solvents. Usually, such moderate and mild conditions are preferred for many reactions involving supramolecular, polymer, organic and organometallic, and intercalation chemistry. Organic or biological components are introduced to prepare organic-inorganic HNs following such systematic methods or strategies [21, 24].

Two different approaches are followed for the synthesis of HNs. In the first approach, the desired HNs are formed by the reaction between the pre-formed building blocks. In this case, the original integrity of the precursors is maintained at least partially. In the second approach, the starting materials are developed into a network structure to form novel one or two structural units. Both of these methodologies are widely used to prepare HNs and both methodologies have their advantages and disadvantages.

The approach involving the concept of building blocks involves a step-by-step fashion. While maintaining molecular integrity, simultaneous transmission of the typical qualities of the building blocks to the novel materials also takes place. Inorganic clusters may be modified or reactive organic groups may be attached to nanoparticles to have such well-defined building blocks. To achieve the best performance, they can be fabricated or designed in different ways and the best performance can be transferred to the newly formed materials. Some of the important strategies for synthesis along with their associated advantages and disadvantages are summarized in Table 1 [25].

Table 1. Strategies for synthesis of hybrid nanomaterials along with their associated advantages and disadvantages [25]

Method of synthesis	Definition	Advantages	Disadvantages
Building block approach	Molecular integrity partially maintained throughout the material formation	Appropriate design of the building blocks can be made to accomplish better performance	N/A
In situ formation	Precursors undergo chemical transformation throughout the formation of HNs	Simplicity along with commercial availability of the precursors; composition of precursors and the reaction conditions dictate the internal structure of HNs	N/A
Sol-gel process	Wet-chemical technique; colloidal and discrete materials dispersed in a solvent are brought back into solid-state	Versatility; well-defined structure due to homogeneous mixing at the molecular level; requirement of low energy since melting is not a requisite	Residual OH group and porosity; Concurrent precipitation of oxide when the formation of sol takes place; expensive preliminary set-up

Electrocrystallization	Formation of a solid phase at the electrode/solution interface	Production of superior quality single crystals in large size; selectivity of the reaction; excellent phase purity, facile kinetics and nice growth of crystals	N/A
Hydrothermal method	Crystallization from aqueous solution at elevated temperatures and pressurized conditions (> 100°C and >1 atm)	Simplicity; inexpensiveness; energy saving; better control of nucleation; free of pollution	Expensive due to the requirement of an autoclave; prolonged reaction; poor reproducibility in particular for achieving identical crystals
Wet chemistry approach	A range of conventional and simple methods involving direct experiments using liquids	Simplicity; inexpensiveness, almost no damage during processing; ease of manipulation, time-saving	Poor controllability with a few starting materials

4.1   In situ formation

In situ formation is distinctively different from the building-block approach. It is based on the chemical changes of the starting materials employed for synthesis and fabrication. The formation of organic polymers is carried out following this process; whereas the sol-gel process is followed in the formation of inorganic components. Multidimensional structures including one-, two-, and three-dimensional structures are developed from well-defined discrete molecules, which exhibit completely different properties from the original ones. This method is very advantageous for the achievement of marvelous properties. It is very simple and the precursor molecules are readily available. The precursor composition as well as the reaction environment determine the internal structure of the final product [20, 25].

4.2   Sol-gel process

The sol-gel process is associated with chemical condensation reactions, in particular, polycondensation reactions. In this wet-chemical process, discrete molecules in a liquid are brought back into a solid form. The reaction usually produces a three-dimensional crosslinked network. The HNs are formed by the use of small molecules as precursors, which offer several advantages. The method is excellent in terms of reproducibility and less energy consumption. This is a mostly investigated and critically acknowledged method compared to other synthetic strategies. There are many disadvantages as well. It is expensive and involves precipitation. Residual OH group and porosity are common issues and cannot be avoided [25].

4.3   Electrocrystallization

Electrocrystallization is a special type of crystallization process involving an electrochemical process. It occurs at the interface of the electrode in a solution and forms a phase nearly solid at the surface of the electrode. The method is so important for producing huge, good-quality single crystals. The selectivity, phase purity of the products, development of good crystals, and facile reaction are a few other important advantages. The greatest drawback is that it requires highly pure starting materials compared to other methods. Different parameters of the surrounding environment and the system like potential, current, supporting electrolyte, solvent, and the operating temperature also profoundly influence the fabrication of HNs [20].

4.4   Hydrothermal method

The hydrothermal method involves different crystallizing strategies using an aqueous solution at high temperatures and vapor pressures (> 100°C and > 1 atm). Simplicity, low cost, less energy consumption, superior control of nucleation, and pollution-free route are a few of the extraordinary points of interest in this strategy. On the other hand, this strategy needs the use of costly autoclaves and the reaction time is usually longer. The method also suffers from poor reproducibility, in particular for obtaining identical crystals [20].

4.5   Wet chemistry approach

The wet chemistry approach uses an assortment of conventional or basic strategies that include coordinated experimentation with fluids. It involves easy and simple steps and cheap precursor materials. The damage-free method with ease of manipulation and less time consumption has been increasingly popular. But the control of growth and structure is rather difficult especially due to the use of a little amount of starting materials [20].

4.6   Polymerization of organic monomers with preformed inorganic components

The process involves the preparation of organic polymers in the presence of inorganic components formed before polymerization. Different kinds of non-reactive or inert natural groups (for instance, an alkyl group), or moieties are used to adjust the inorganic surface.

The moieties containing receptive surface groups are generally used (such as polymerizable functionalities). For instance, surfactants or silane coupling agents are used to treat an inorganic surface to render it consistent with natural monomers. Especially, the monomers that are useful in responding to the surface of the inorganic material can be introduced. First-class HNs can be developed using this method. On the other hand, there are certain disadvantages of materials of this kind since no solid chemical interactions take place. The concerns of diffusion make the long-term stability of this type of material questionable [25].

Silica nanoparticles may be functionalized with alkyl chains and introduced into numerous polymers that are hydrophobic in nature. Furthermore, numerous metal nanoparticles may be more effectively introduced to the block copolymers containing poly(vinylpyridine), or hydroxyethyl methacrylates. Thus, the stability of the ultimate material depends on the extent to which the components interact, as in the second-class HNs. Cases of such materials are copolymers of organic monomers at the surface of polymers with certain polymerizable groups [25].

4.7   Simultaneous incorporation of components

This is a very special type of process and homogeneous interpenetrating networks can be formed through the formation of inorganic and organic polymers simultaneously [26-27]. There are different means to carry out the process. In general, monomers are assorted with the precursors for the sol-gel process, described earlier. The reactions take place at the same time to give homogeneous materials. They can be carried out with or without solvent. The process is associated with several advantages as well as disadvantages. One issue that emerges from this process is the sensitivity of numerous polymerization processes. The issue can be practically significant in the cases of ionic polymerization. To overcome the issue, free radical polymerization is recommended. Precursors containing alkoxides can be used in this particular process. Precursors can act as a monomer in the polymerization and can lead to HNs with decreased shrinkage and high homogeneity [25].

5.   Applications of hybrid nanomaterials

HNs are of many different kinds and their applications depend on the type of materials as well as their attainable properties. In the following sub-section, specific applications of a number of scientifically and technologically important HNs have been discussed.

5.1   Mesoporous silica based hybrid nanoparticles

Mesoporous silica nanoparticles are different types of silica nanoparticles. This distinctive form of silica nanoparticles contains pores that range in diameter from 2 to 50 nanometers [2-4, 28]. They have gained special attention after the invention of Mobil's M41S mesoporous molecular sieves, a porous material of silica. Later on, SBA-15, MCM-41, and MCM-48 were invented and are now referred to as the most common MSNs ranging from 2 to 10 nm with two-dimensional hexagonal and three-dimensional cubic structures

[28]. In a broader sense, the porous honeycomb-like structure of silica is defined as MSN. This material offers a number of advantages which include: a wide porous structure, changeable size, large specific surface area, biocompatibility, and desirable properties like controllable size and morphology, porosity, and high chemical stability. The high specific surface area of the pores is the biggest advantage, which allows different functional groups to be attached to the MSNs.

MSNs can be divided into different types based on their different characteristics. They can be mainly divided into three types: ordered, hollow or rattle type, and core/shell type. Ordered MSNs are characterized by a uniform distribution of pore size and a pore structure organized in an orderly fashion. The hollow or rattle type MSNs are preferred more over-ordered MSNs for application in medicine. Their porous structure allows the loading of drugs at an increased rate and they have the advantage of easy functionalization of the surface. Core/shell types of MSNs are generally synthesized where MSNs are coated over solid cores formed by metallic platinum. MSNs are thus functionalized with an aim toward the attainment of multifunctionality in a single material. Gold is also used for the functionalization of the surface of MSNs for various important applications [29].

MSNs have so far been used for various applications ranging from catalysis and environment   and from chemical removal to biomedicine. They are becoming increasingly important for drug delivery, diagnosis, catalysis, separation, and sensing [28].

Advanced HNs are being developed based on MSNs for exploring more enthralling properties and superior characteristics. The internal surfaces of mesoporous silica, upon functionalization through the integration of such organic-inorganic materials result in improved mesoporous silica-based HNs [5, 30]. As a consequence, HNs based on MSNs have been applied in the fields of biomedicine, drug carriers, and especially in cancer therapy [5, 31].

MSNs-based HNs have been crucial for the treatment of colon cancer. New MSNs based organic/inorganic HNs have been developed for this purpose [32]. These were covalently attached with poly(oligo(ethylene glycol) monomethyl ether methacrylate) (POEGMA) for improvement of stabilization and targeting peptide, arginyl glycyl aspartic acid (RGD). The aim has been to target the delivery of 5-fluorouracil (5-FU) for improving anticancer performance. The MSN-P(OEGMA-co-RGD) hybrid nanoparticles (Fig. 2) served as the drug carrier to treat colon cancer. 5-FU is known to be one of the most efficacious medications to treat colon cancer; although it is associated with serious side effects like mucositis and diarrhea. But there are serious concerns with the use of 5-FU. The therapeutic response rate has been reported to be very low for cancer tissues and it is poorly accumulated in tumor tissues. These altogether prolong the treatment and worsen cancer. According to current research, 5-FU loaded HNs based on MSNs show great results and improved efficacy for the treatment. The anticancer efficacy is enhanced and improved in vitro and in vivo, considering better medical applications as well as clinical applications. The internalization or penetration of this medication into the cancer cells has also been developed and the accumulation has also been enhanced in tumor tissues. Thus, the overall scenario indicates that the 5-FU could be effectively accumulated into the mesoporous structures of MSN-based nanoparticles. The integration or combination of RGD moieties played the most important role in reducing the side effects. The inclusion of the moieties on this specific nanomedicine showed tremendous results including improvement of internalization to target colon cancer cells and improvement of death of cancer cells, enhancement of accumulation in tumor tissues, reduction of tumor growth as well as lowering of side effects compared to 5-FU@MSN and free 5-FU [32].

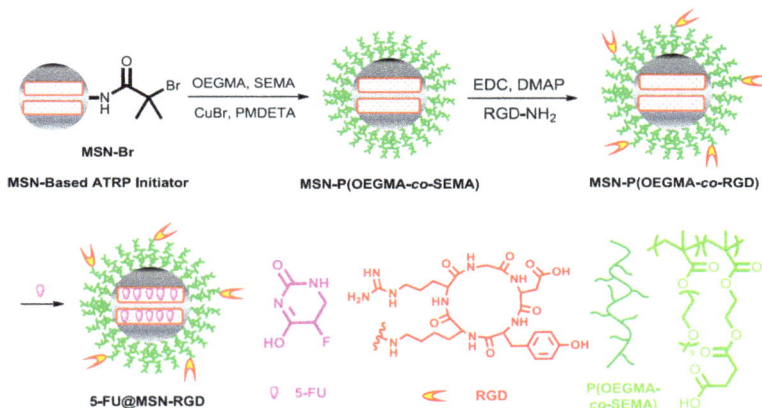

Figure 2. Route for synthesis for fabrication of MSN based tumour targeting nanocarriers, MSN-P(OEGMA-co-RGD), and 5-FU loaded MSN-P(OEGMA-co-RGD), 5-FU@MSN-RGD. DMAP: 4-dimethylaminopyridine; PMDTA: N,N,N',N'',N''- pentamethyl diethylenetriamine; EDC: 1-(3-dimethylaminopropyl)-3-ethylcarbodiimide hydrochloride; SEMA: 2-succinyloxyethyl methacrylate. Reproduced with permission [32]. Copyright 2014, Elsevier.

MSNs-based HNs have also been amazing for the selective delivery of drugs to certain tissues and cells such as bones, muscles, blood cells, and so on. HNs based on MSNs have recently been developed and used for bone-selective drug delivery (Fig. 3). The main drug delivery vehicle was MSNs with alendronate as the drug targeting tissues of bones. The drug delivery mechanism has been established by using a nonsteroidal anti-inflammatory drug, ibuprofen. The exterior surface of the particles is functionalized with a propionitrile derivative. It is hydrolyzed sequentially to transform into a carboxylic group. The diphosphonate drug, alendronate, is bound through electrostatic interaction to the carboxyl functional groups at the exterior of mesoporous silica. Hydroxyapatite, which mimics bone matrix, was prepared to assess the activity of targeting the resulting device. MSNs, thus,

Emerging Nanomaterials and Their Impact on Society in the 21st Century   Materials Research Forum LLC
Materials Research Foundations 135 (2023) 152-177   https://doi.org/10.21741/9781644902172-7

provided a solid foundation for the fabrication of multifunctional nanostructured drug delivery systems [33].

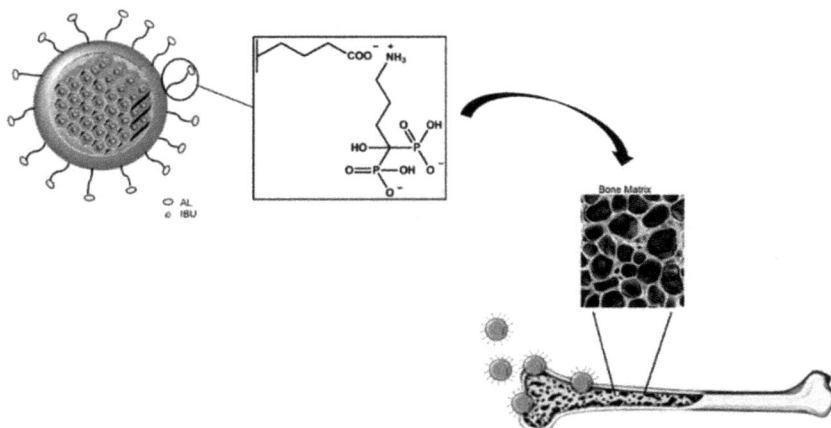

Figure 3. Schematics of bone-specific drug delivery of HNs based on mesoporous silica [33]. (Reprinted from [33]).

One of the most significant disadvantages of silica nano-drug delivery technology in cancer therapy is the possibility of bioaccumulation. In a recent development, a one-step technique was used to create an MSNs/hydroxyapatite (MSNs/HAP) hybrid drug carrier that improved silica biodegradability (Fig. 4). Ca^{2+} escapes from the skeleton and improves the degradability of nanoparticles in an acidic environment. The drug loading content of the MSNs/HAP sample is approximately five times that of MSNs. The MSNs/HAP were highly biocompatible and exhibited good anticancer activity. But more interestingly, it also significantly lowered the adverse effects of free doxorubicin (DOX). These, along with high efficiency for drug loading, and sensitivity for pH, have rendered the as-synthesized HNs a viable material for drug delivery technology [34].

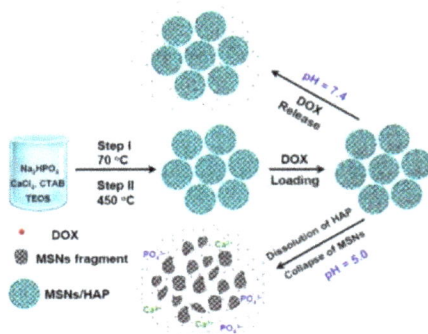

Figure 4. Synthesis MSNs/HAP composites and the mechanism of drug delivery. CTAB: Cetyltrimethylammonium bromide; TEOS: Tetraethyoxysilane. Reproduced with permission [34]. Copyright 2015, American Chemical Society.

5.2   Quantum dot based hybrid nanomaterials

Quantum dots are, in general, nanocrystals. They are colloidal, fluorescent semiconductors. Their sizes range from 1.5 to 10.0 nanometers [6, 35]. This size can be compared with the size between single molecules and bulk crystalline solids. The size and structural composition of quantum dots determine their properties and characteristics. As a consequence, the properties can be modified, controlled, and developed by tuning the size of QDs [6, 36].

QDs have size-dependent optical properties, photophysical properties, superior photostability, high extinction coefficient, and brightness, large Stokes shift, electronic properties, and excellent fluorescence properties [37]. They also exhibit catalytic and sensitive detection properties in many electrochemical applications. Their optical properties originate from quantum confinements. QDs also have highly tunable absorption and emission spectra. Due to such distinctive properties, QDs are applicable in a wide variety of applications. [6, 36].

QDs can be divided into various types based on their structure formation and composition. Core-type, core/shell type and alloyed QDs are the three main types of QDs. Core-type QDs have uniform internal compositions. They are the simplest of all types and their properties can easily be controlled or modified by a simple change in the size of the crystallite. Core/shell QDs are more complex compared to core-type QDs. QDs of this kind are developed with the integration of two semiconducting materials. Shells of a high band gap semiconducting material are grown around a semiconducting nanocrystal in this process. The QDs developed through this process are called core/shell QDs. Coating QDs with shells results in greater and improved quantum yield by passivizing non-radiative recombination sites. This also makes QDs more applicable and robust for different types of applications. Alloyed QDs help to tune the properties without changing or tuning the size. They allow the change of the optical and electronic properties by a change in the composition and internal structure. They are formed by alloying two semiconducting materials with different band gap energies. As a result, they possess better and additional composition-tunable properties.

Recently, QDs-based HNs have been developed by exploiting the enhancement of field and confinement of metal nanoparticles with fluorescent QDs. Such integration has opened a new avenue for nanophotonics with promises for biomedical applications, diagnosis, therapy, and energy-saving and green production processes. In fact, QDs-based hybrid nanomaterials are the emerging materials for future development [38].

QDs-based HNs are playing the most interesting role in bioimaging in modern technologies including intracellular labeling. The preparation of such type of material is rather simple (Fig. 5). HNs of monodisperse QDs were simply mixed with cholesterol-bearing pullulan (CHP) nanogels and was modified with amino groups (CHPNH2). The material, CHPNH2–QDs was successively inserted in different human cells. They also showed higher efficiency of cellular uptake in comparison with a normal carrier like cationic liposome. As a result, HNs like these could be a useful fluorescent probe for bioimaging [39].

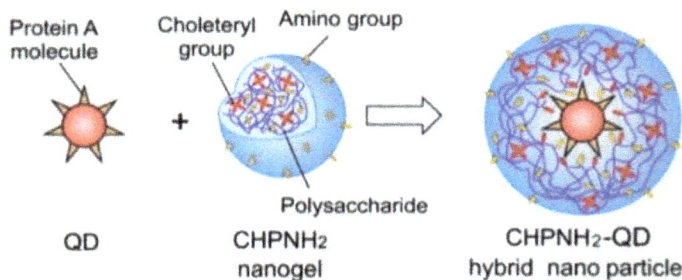

Figure 5. Formation of CHPNH2-QD hybrid nanomaterials. Reproduced with permission [39]. Copyright 2005, Elsevier.

Luminescent graphene QD-gold (GQD-Au) HNs have been developed for a variety of critical functional applications. They are prepared following a unique process. The GQDs are used in the conversion of chloroauric acid, $HAuCl_4.3H_2O$ to Au nanoparticles on their surface at ambient temperature (Fig. 6). GQDs with layers formed through self-passivation were created utilizing a microwave-assisted hydrothermal process with glucose as a precursor. The production of GQD and GQD-Au hybrid nanoparticles takes less than five minutes. The use of citrate to extract GQDs boosts the stability of the HNs for up to four weeks. The GQD-Au HNs were luminous under UV-A irradiation and had a size range of 5–100 nm and showed necessary promises for a variety of other applications, including luminous coatings for glass and automotive windowpanes, among others [40].

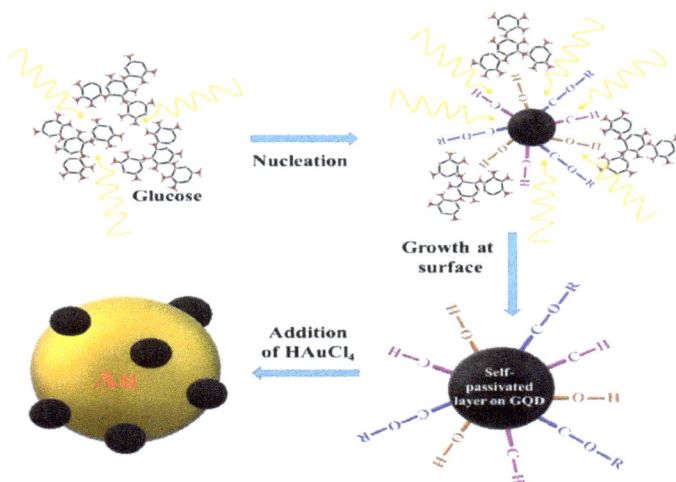

Figure 6. Schematic for the synthesis of GQD-Au hybrid nanoparticle [40]. (Reprinted from [40]).

Despite their wide *in vitro* uses, QDs have been limited in clinical applications due to their significant toxicity and dominating liver absorption. Therefore, the HNs based on QDs have been considered potential to settle the issues while enhancing the characteristic behaviors of QDs. For *in vivo* imaging, a new hybrid fluorophore was developed with good biocompatibility. The HNs comprised near-IR-emitting PbSe QDs encapsulated in solid fatty ester nanoparticles (QD-FEN) (Fig. 7). The headgroups of QD-encapsulating organic fatty acids were chemically modified to stop or remove inhibition of photoluminescence exhibited by PbSe nanocrystals. The addition of fatty esters increased the quantum yields of PbSe QDs in an aqueous solution by up to 45 percent, which was 50 percent greater than the aqueous quantum yields of water-soluble PbSe nanocrystals. QD-FEN demonstrated much lower short-term cytotoxicity compared to water-soluble QDs without encapsulation. More notably, there was decreased liver absorption, greater tumor retention, no toxic reaction, and almost complete clearance of QD-FEN from the studied animals. Near-IR spectral features and enhanced optical characteristics together with significant improvement in the biosafety profile have made the novel HNs-based fluorophore system very demanding. The HNs demonstrate real-time, deep-tissue fluorescent imaging of live animals and laid the foundation for the development of future clinical applications [41].

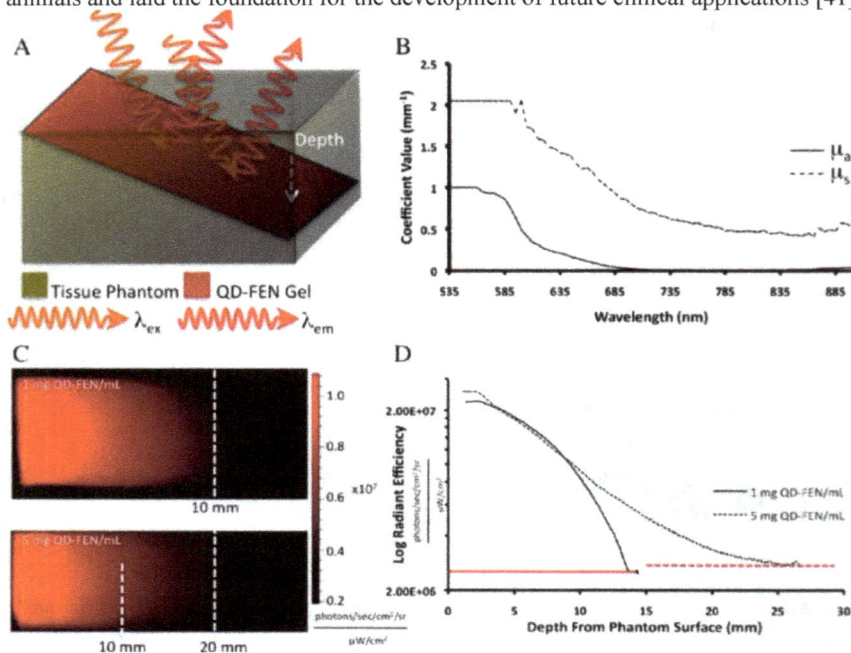

Figure 7. A tissue phantom model and measurement of fluorophore signal tissue depth penetration. Reproduced with permission [41]. Copyright 2011, American Chemical Society.

5.3   `Nanoscale metal-organic frameworks based hybrid nanomaterials

Metal-organic frameworks are hybrid solids belonging to the class of porous materials [10]. Metal ions or clusters of metal ions may be combined to have organic polydentate bridging linkers. The self-assembly process is followed in this regard. They are otherwise also called porous coordination polymers/networks. These frameworks can be prepared on the nanoscale as well [42] to give nanoscale metal-organic frameworks. The porous HNs of this kind is are characterized by diversified properties and have distinct advantages including facile synthesis, high storage capacity, excellent biocompatibility, ease of surface functionalization, and diverse compositions.

NMOFs have been ideal for different modern applications in particular for biomedical applications. They can be utilized in several ways. For instance, NMOFs can be used as carriers of drugs and preparation for therapeutics, biosensing, and imaging because of their high porosity, multifunctionality, and biocompatibility. NMOFs based diagnostic and therapeutic agents have been extensively developed and used for magnetic resonance imaging (MRI), computed tomography (CT), positron emission tomography (PET), and optical imaging [43].

Recent development shows the use of reverse microemulsions for the preparation of NMOFs by combining Gd^{3+} centers, benzenedicarboxylate and benzenetricarboxylate bridging ligands (Fig. 8). The presence of huge Gd^{3+} centers in each nanoparticle has offered NMOFs extremely high longitudinal and transverse relaxivities (R1 and R2 respectively). The effective time it takes for the MRI signal to decay in the longitudinal and transverse planes is referred to as T1 and T2 and NMOFs can serve as agents for contrast in MRI. Doping the NMOFs with Eu^{3+} or Tb^{3+} centers can also make them very luminous. The findings show that NMOFs might be employed in multimodal imaging as possible agents for contrasting [44].

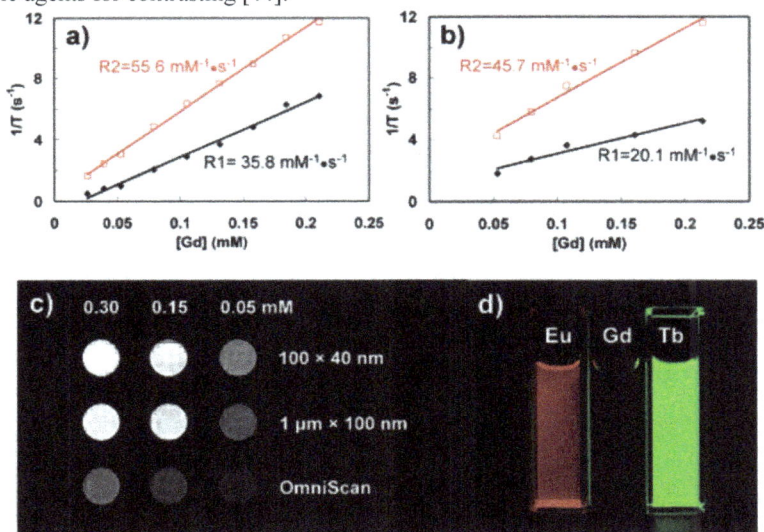

Figure 8. (a-b) Relaxivity curves of $Gd(BDC)_{1.5}(H_2O)_2$ nanorods (where BDC refers to BDC is 1,4-benzenedicarboxylate) (c) T1-weighted MR images of suspensions of $Gd(BDC)_{1.5}(H_2O)_2$ in water containing 0.1% xanthan gum. (d) Luminescence images of ethanolic suspensions of $Gd(BDC)_{1.5}(H_2O)_2$, and sample doped with 5 mol % of Eu^{3+}), and doped with 5 mol % of Tb^{3+}. Reproduced with permission [44]. Copyright 2006, American Chemical Society.

A recent finding also demonstrates the use of NMOFs for mitochondrial-targeted radiotherapy-radiodynamic treatment [45]. The first NMOFs was developed for cancer treatment by targeting mitochondria and triggering x-ray. The incorporation of tris(2,2'-bipyridyl)ruthenium(II) fluorophore, $Ru(bpy)_3^{2+}$ photosensitizer into Hf-DBB-Ru [Hf: Halfnium;   DBB-Ru   =   bis(2,2'-bipyridine)(5,5'-di(4-benzoato)-2,2'-bipyridine)ruthenium(II) chloride] resulted in the formation of a NMOF with high efficiency of targeting mitochondria (Fig. 9). In this process, Hf-DBB-Ru-enabled radiotherapy (RT) and radio dynamic therapy (RDT) were successively developed by hydroxyl radicals from the Hf_6 secondary building units (SBUs) and singlet oxygen from the DBB-Ru photosensitizers (PSs) when bombarded with modest doses of penetrating x-rays. The RT-RDT mechanism depolarized the mitochondrial membrane potential, released cytochrome c, and disrupted the respiratory chain, allowing apoptotic pathways for programmed cell death to be initiated. At very low x-ray dosages and with minimal side effects, the NMOF-enabled, mitochondria-targeted RT-RDT substantially reduced colorectal cancers in mice models [45].

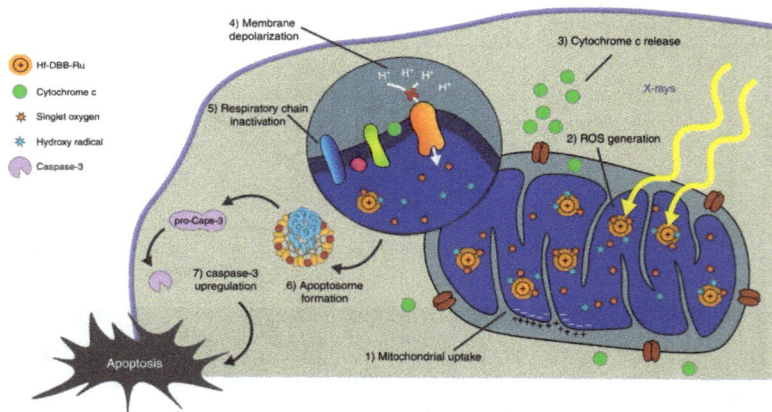

Figure 9. Mechanism of targeting mitochondria and RT-RDT mediated by Hf-DBB-Ru. [45]. (Reprinted from [45]).

5.4   Iron oxide nanoparticle based hybrid system

Iron oxides are chemical compounds formed by the combination of iron and oxygen. There is a total of 16 different forms in nature. Magnetite (Fe_3O_4), hematite (α-Fe_2O_3), and maghemite (γ-Fe_2O_3) are considered the three most common forms of iron oxides. Iron oxide nanoparticles are iron oxides developed at the nanoscale. They consist of γ-Fe_2O_3 and/or magnetite Fe_3O_4 particles, which cause them to be super magnetic. IONPs have been important since they are superparamagnetic, have a high surface-to-volume ratio and larger specific surface area, and can be easily separated [46].

HNs based on IONPs have also been attractive for biomedical applications because of their outstanding properties. They showcase exceptional properties like a response to external magnetic fields, high payload, good biocompatibility, and ease of surface engineering to avoid any biological interactions that are not favorable. They are now considered promising in cancer treatment and other biomedical applications [47].

The addition of gold to IONPs results in particles that are more stable and resilient. Magnetic properties of iron oxide and surface plasmon resonance of gold are synergistically combined to give them distinctive features for a multimodal platform suitable for usage as an agent for contrast in MRI and as a nano-heater. The seeding approach using a poly(ethylenimine) intermediate layer gives IONPs with a diameter of 30 nm for the core coated with gold. To improve biocompatibility and enhance retention durations *in vivo*, the fold-coated IONPS are coated further with poly(ethylene glycol). When the particles are treated with a human melanoma cell line, A375M cells, they appear to have no cytotoxic impact. The resulting HNs have the highest absorbance at 600 nm. After irradiation of the laser for just 90 seconds in an agar phantom, a temperature difference, ΔT of 32°C is attained. The HNs appear to reduce transverse relaxivity values (Fig. 10) [48].

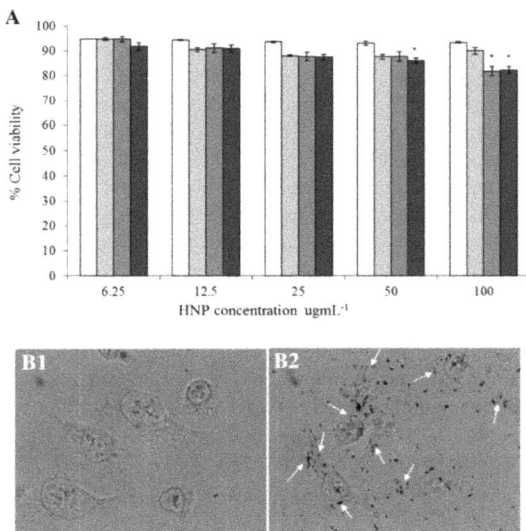

Figure 10. Interactions of hybrid nanoparticles (HNPs) with A375M cells. A) Cell viability estimated by Trypan blue exclusion assay. B) Cellular uptake of A375M Cells. Particles (50 μgmL^{-1}) were incubated for 24 h before fixation and silver enhanced staining. Arrows indicate nanoparticles bound to the cells (on the cell membrane or inside the cells) [48]. (Reprinted from [48]).

By modifying the surface layer coating, novel biocompatible Fe_3O_4 HNs could be realized for better uptake of target cells, improved R2 for sensitive MRI applications, dual-mode imaging, and image-guided cancer treatment. The active functional groups on Fe_3O_4 nanoparticles enable rational conjugation of medicinal and biological substances. Because of current and quick advances in the synthesizing and modifying surface of Fe_3O_4 nanoparticles, they may now be used for more effective combined therapeutics and diagnostics, in other words, theranostics applications. New generation molecular probes that combine computed tomography (CT), ultrasound scan (US), photoacoustic (PA), and positron emission tomography (PET) imaging into a single nanoprobe can provide more complete diagnostic information on the dynamics of disease development. Furthermore, Fe_3O_4 based nanosystems can be employed for imaging-guided and magnetic hyperthermia (MHT)-enhanced cancer therapy when combined with phototherapeutic, chemotherapeutic, and genetic agents. Despite tremendous development and encouraging results, many initiatives are still in a very rudimentary stage, and a number of difficulties need to be resolved and hurdles removed to warrant the use of nanoparticles as probes for diagnostic and therapeutic purposes [49].

Conclusions

At present, hybrid nanomaterials are the leading research topic worldwide. HNs are undoubtedly the future of newer, smarter, wearable, and multitasking technologies. These miniaturized technologies will experience further advancement and rapid development compared to other technologies. So far, HNs have performed a variety of functions in the evolution of contemporary technology, but their qualities and attributes are still insufficient. This is due to the fact that the characteristics and behavior of the HNs rely on the strategy of synthesis. Even a slight change in the protocol can have a significant impact on the properties of the HNs and the ultimate applications. Judicious selection of synthetic protocols is very essential. Intensive and systematic research should be performed to enhance the physical properties and mechanical characteristics of HNs and establish the means of tunability. The prospect of HNs of different kinds should be highlighted, and devices should be fabricated for biomedical applications with emphasis on performance as well as identification of barriers. Innovative ideas centered on improved performance and lifting impediments would broaden the scope of HNs to all major fields of applications. It can be envisaged that the world tomorrow is going to monitor a new wave of prosperity and development through the advancement of HNs.

Acknowledgement

M.A.B.H. Susan acknowledges the Centennial Research Grant from the University of Dhaka, Bangladesh for supporting the themed research.

Competing financial interests

There are no competing financial interests to declare.

References

[1] P. Gómez-Romero, C. Sanchez, Functional hybrid materials, John Wiley & Sons, 2006.

[2] M.R. Rashid, F. Afroze, S., Ahmed, M.S. Miran, M.A.B.H. Susan, Control of the porosity and morphology of ordered mesoporous silica by varying calcination conditions, Materials Today: Proceedings 15 (2019) 546-554. https://doi.org/10.1016/j.matpr.2019.04.119

[3] M. Khair, M.R. Rashid, S. Ahmed, M.A.B.H. Susan, Silica fillers for enhancement of dielectric properties of poly(vinylidene fluoride) and its copolymer, Materials Today: Proceedings 29 (2020) 1239-1245. https://doi.org/10.1016/j.matpr.2020.05.653

[4] M.S. Islam, M.S. Miran, M.Y.A. Mollah, M.M. Rahman, M.A.B.H. Susan, Polyaniline-silica composite materials: Influence of silica content on the thermal and thermodynamic properties, J. Nanostructured Polym. Nanocomposites 9 (2013) 83-89.

[5] K. Sayadi, A. Rahdar, M.R. Hajinezhad, S. Nikazar, M.A.B.H. Susan, Atorvastatin-loaded SBA-16 nanostructures: Synthesis, physical characterization, and biochemical alterations in hyperlipidemic rats, J. Mol Structure1202 (2020) 127296. https://doi.org/10.1016/j.molstruc.2019.127296

[6] A.M.M. Hasan, M.A. Hasan, A. Reza, M.M. Islam, M.A.B.H. Susan, Carbon dots as nano-modules for energy conversion and storage, Materials Today Commun. 29 (2021) 102732. https://doi.org/10.1016/j.mtcomm.2021.102732

[7] S.S., Satter, M. Hoque, M.M. Rahman, M.Y.A. Mollah, M.A.B.H. Susan, An approach towards synthesis and characterization of ZnO@Ag core@shell nanoparticles in water-in-oil microemulsion, RSC. Adv. 4 (2014) 20612-20615. https://doi.org/10.1039/C4RA01046A

[8] S. Mahmud, S.S. Satter, A.K. Singh, M.M. Rahman, M.Y.A., Mollah, M.A.B.H. Susan, Tailored engineering of bimetallic plasmonic Au@Ag core@shell nanoparticles, ACS Omega 4 (2019) 18061-18075. https://doi.org/10.1021/acsomega.9b01897

[9] M.K. Rahman, G. Aiba, M.A.B.H. Susan, Y. Sasaya, K. Ohta, M. Watanabe, Synthesis, characterization, and copolymerization of a series of novel acid monomers based on sulfonimides for proton conducting membranes, Macromolecules 37 (2004) 5572-5577. https://doi.org/10.1021/ma0498058

[10] A. Taher, M.A.B.H. Susan, N. Begum, I-M. Lee, Amine-functionalized metal-organic framework-based Pd nanoparticles: Highly efficient multifunctional catalysts for base-free aerobic oxidation of different alcohols, New J. Chem. 44 (2020) 19113-19121. https://doi.org/10.1039/D0NJ04138F

[11] M.S. Saveleva, K. Eftekhari, A. Abalymov, T.E.L. Douglas, D. Volodkin, B.V. Parakhonskiy, A.G. Skirtach, Hierarchy of hybrid materials- The place of inorganics-

in-organics in it, their composition and applications, Frontiers in Chemistry 7 (2019) 179. https://doi.org/10.3389/fchem.2019.00179

[12] C. Sanchez, P. Belleville, M. Popall, L. Nicole, Applications of advanced hybrid organic-inorganic nanomaterials: from laboratory to market, Chem. Soc. Rev. 40 (2011) 696-753. https://doi.org/10.1039/c0cs00136h

[13] P.N. Catalano, R.G. Chaudhary, M.F. Desimone, P.L. Santo, A Survey on Analytical methods for green synthesized nanomaterials. Curr. Pharmaceu Biotech, 22 (2021) 813-837. https://doi.org/10.2174/1389201022666210104122349

[14] C. Sanchez, B. Julián, P. Belleville, M. Popall, Applications of hybrid organic-inorganic nanocomposites, J. Mat. Chem. 15 (2005) 3559-3592. https://doi.org/10.1039/b509097k

[15] A. Taubert, F. Leroux, P. Rabu, V. de Zea Bermudez, Advanced hybrid nanomaterials, Beilstein J. Nanotechnology 10 (2019) 2563-2567. https://doi.org/10.3762/bjnano.10.247

[16] A. Taubert, F. Leroux, P. Rabu, V. de Zea Bermudez, Advanced hybrid nanomaterials, Beilstein-Institut, 2019. https://doi.org/10.3762/bjnano.10.247

[17] V.P. Ananikov, Organic-inorganic hybrid nanomaterials, Nanomaterials 9 (2019) 1197. https://doi.org/10.3390/nano9091197

[18] M. Aksit, V. Altstädt, Hybrid materials-historical perspective and current trends, COJ Rev Res. 2 (2020) .1-17

[19] S. Somiya, Handbook of advanced ceramics: materials, applications, processing, and properties, Academic Press, 2013. https://doi.org/10.1016/B978-0-12-385469-8.03001-X

[20] M.S. Umekar, A.K. Potbhare, G.S. Bhusari, M.F. Desimone, R.G. Chaudhary, Bioinspired reduced graphene oxide based nanohybrids for photocatalysis and antibacterial applications, Current Pharmaceutical Biotechnology, 22 (2021) 1759 - 1781. https://doi.org/10.2174/1389201022666201231115826

[21] R. Vargas-Bernal, Hybrid nanomaterials, hybrid nanomaterials- Flexible electronics materials, IntechOpen, 2020. https://doi.org/10.5772/intechopen.83326

[22] C. Auschra, R. Stadler, New ordered morphologies in ABC triblock copolymers, Macromolecules 26 (1993) 2171-2174. https://doi.org/10.1021/ma00061a005

[23] M. Faustini, L. Nicole, E. Ruiz-Hitzky, C. Sanchez, History of organic-inorganic hybrid materials: prehistory, art, science, and advanced applications, Advanced Functional Materials 28 (2018) 1704158. https://doi.org/10.1002/adfm.201704158

[24] M. Nanko, Definitions and categories of hybrid materials, AZojomo 6 (2009) 1-8.

[25] F. Tanasa, M. Zanoaga, Polymer based hybrid materials for aerospace applications, Scientific Research & Education in the Air Force-AFASES 1 (2012).

[26] G. Kickelbick, Introduction to hybrid materials, Hybrid Materials 1 (2007) 2. https://doi.org/10.1002/9783527610495.ch1

[27] J.D. Wright, N. Sommerdijk, Sol-Gel materials: Their chemistry and biological properties, Taylor & Francis Group, 2000.

[28] A. Mehmood, H. Ghafar, S. Yaqoob, D. Gohar, B. Ahmad, Mesoporous silica nanoparticles: A review, J. Developing Drugs 06 (2017) 174-188. https://doi.org/10.4172/2329-6631.1000174

[29] C. Rajani, P. Borisa, T. Karanwad, Y. Borade, V. Patel, K. Rajpoot, R.K. Tekade, 7 - Cancer-targeted chemotherapy: Emerging role of the folate anchored dendrimer as drug delivery nanocarrier, in: A. Chauhan, H. Kulhari (Eds.), Pharmaceutical Applications of Dendrimers, Elsevier 2020, pp. 151-198. https://doi.org/10.1016/B978-0-12-814527-2.00007-X

[30] E. Ilhan-Ayisigi, O. Yesil-Celiktas, Silica-based organic-inorganic hybrid nanoparticles and nanoconjugates for improved anticancer drug delivery, Engineering in Life Sciences 18 (2018) 882-892. https://doi.org/10.1002/elsc.201800038

[31] I. Miletto, E. Gianotti, M.-H. Delville, G. Berlier, Silica-based organic-inorganic hybrid nanomaterials for optical bioimaging, in: D. Marie-Hélène, T. Andreas (Eds.), Hybrid organic-inorganic interfaces : towards advanced functional materials, Wiley-VCH Verlag GmbH & Co. KGaA, Weinheim, Germany, 2018, pp. 729-765. https://doi.org/10.1002/9783527807130.ch17

[32] G. Pan, T.-T. Jia, Q.-X. Huang, Y.-Y. Qiu, J. Xu, P.-H. Yin, T. Liu, Mesoporous silica nanoparticles (MSNs)-based organic/inorganic hybrid nanocarriers loading 5-Fluorouracil for the treatment of colon cancer with improved anticancer efficacy, Colloids Surf B Biointerfaces 159 (2017) 375-385. https://doi.org/10.1016/j.colsurfb.2017.08.013

[33] L. Pasqua, I.E. De Napoli, M. De Santo, M. Greco, E. Catizzone, D. Lombardo, G. Montera, A. Comandè, A. Nigro, C. Morelli, A. Leggio, Mesoporous silica-based hybrid materials for bone-specific drug delivery, Nanoscale Advances 1 (2019) 3269-3278. https://doi.org/10.1039/C9NA00249A

[34] X. Hao, X. Hu, C. Zhang, S. Chen, Z. Li, X. Yang, H. Liu, G. Jia, D. Liu, K. Ge, X.-J. Liang, J. Zhang, Hybrid mesoporous silica-based drug carrier nanostructures with improved degradability by hydroxyapatite, ACS Nano 9 (2015) 9614-9625. https://doi.org/10.1021/nn507485j

[35] A.M. Wagner, J.M. Knipe, G. Orive, N.A. Peppas, Quantum dots in biomedical applications, Acta Biomaterialia 94 (2019) 44-63. https://doi.org/10.1016/j.actbio.2019.05.022

[36] N. Hong, Introduction to nanomaterials: Basic properties, synthesis, and characterization, 2019, pp. 1-19. https://doi.org/10.1016/B978-0-12-813934-9.00001-3

[37] N.K. Bakirhan, S.A. Ozkan, Chapter 28 - Quantum dots as a new generation nanomaterials and their electrochemical applications in pharmaceutical industry, in: C. M. Hussain (Ed.), Handbook of nanomaterials for industrial applications, Elsevier, 2018, pp. 520-529. https://doi.org/10.1016/B978-0-12-813351-4.00029-8

[38] M. Fernandez, A. Urvoas, P. Even-Hernandez, A. Burel, C. Mériadec, F. Artzner, T. Bouceba, P. Minard, E. Dujardin, V. Marchi, Hybrid gold nanoparticle-quantum dot self-assembled nanostructures driven by complementary artificial proteins, Nanoscale 12(2020) 4612-4621. https://doi.org/10.1039/C9NR09987E

[39] U. Hasegawa, S.M. Nomura, S.C. Kaul, T. Hirano, K. Akiyoshi, Nanogel-quantum dot hybrid nanoparticles for live cell imaging, Biochem. Biophys. Res. Commun. 331 (2005) 917-21. https://doi.org/10.1016/j.bbrc.2005.03.228

[40] S. Wadhwa, A.T. John, A. Mathur, M. Khanuja, G. Bhattacharya, S.S. Roy, S.C. Ray, Engineering of luminescent graphene quantum dot-gold (GQD-Au) hybrid nanoparticles for functional applications, MethodsX 7 (2020) 100963. https://doi.org/10.1016/j.mex.2020.100963

[41] A.J. Shuhendler, P. Prasad, H.-K.C. Chan, C.R. Gordijo, B. Soroushian, M. Kolios, K. Yu, P.J. O'Brien, A.M. Rauth, X.Y. Wu, Hybrid quantum dot−fatty ester stealth nanoparticles: Toward clinically relevant in vivo optical imaging of deep tissue, ACS Nano 5 (2011) 1958-1966. https://doi.org/10.1021/nn103024b

[42] F. Demir Duman, R.S. Forgan, Applications of nanoscale metal-organic frameworks as imaging agents in biology and medicine, J. Mat. Chem. B 9 (2021) 3423-3449. https://doi.org/10.1039/D1TB00358E

[43] D. Zhao, W. Zhang, Z.-H. Wu, H. Xu, Nanoscale metal−organic frameworks and their nanomedicine applications, Frontiers in Chemistry 9 (2022) 834171. https://doi.org/10.3389/fchem.2021.834171

[44] W.J. Rieter, K.M.L. Taylor, H. An, W. Lin, W. Lin, Nanoscale metal−organic frameworks as potential multimodal contrast enhancing agents, J. Am. Chem. Soc.128 (2006) 9024-9025. https://doi.org/10.1021/ja0627444

[45] K. Ni, G. Lan, S.S. Veroneau, X. Duan, Y. Song, W. Lin, Nanoscale metal-organic frameworks for mitochondria-targeted radiotherapy-radiodynamic therapy, Nature Commun. 9 (2018) 4321. https://doi.org/10.1038/s41467-018-06655-7

[46] A. Ali, H. Zafar, M. Zia, I. Ul Haq, A.R. Phull, J.S. Ali, A. Hussain, Synthesis, characterization, applications, and challenges of iron oxide nanoparticles, Nanotechnol. Sci. Appl. 9 (2016) 49-67. https://doi.org/10.2147/NSA.S99986

[47] M.P. Alvarez-Berríos, N. Sosa-Cintron, M. Rodriguez-Lugo, R. Juneja, J.L. Vivero-Escoto, Hybrid Nanomaterials Based on Iron Oxide Nanoparticles and Mesoporous Silica Nanoparticles: Overcoming Challenges in Current Cancer Treatments, J. Chem. 2016 (2016) 2672740. https://doi.org/10.1155/2016/2672740

[48] C. Hoskins, Y. Min, M. Gueorguieva, C. McDougall, A. Volovick, P. Prentice, Z. Wang, A. Melzer, A. Cuschieri, L. Wang, Hybrid gold-iron oxide nanoparticles as a multifunctional platform for biomedical application, J. Nanobiotech. 10(1) (2012) 27. https://doi.org/10.1186/1477-3155-10-27

[49] Y. Hu, S. Mignani, J.-P. Majoral, M. Shen, X. Shi, Construction of iron oxide nanoparticle-based hybrid platforms for tumor imaging and therapy, Chem. Soc. Rev. 47(5) (2018) 1874-1900. https://doi.org/10.1039/C7CS00657H

Emerging Nanomaterials and Their Impact on Society in the 21st Century   Materials Research Forum LLC
Materials Research Foundations 135 (2023) 178-199   https://doi.org/10.21741/9781644902172-8

Chapter 8

Carbon Nanomaterials for Efficient Perovskite Solar Cells

T. Swetha[1] and Surya Prakash Singh*[1,2]

[1]Polymers and Functional Materials Division, CSIR-Indian Institute of Chemical Technology, Uppal Road, Tarnaka, Hyderabad-500007, India

[2]Academy of Scientific and Innovative Research (AcSIR), Ghaziabad-201002, India

* spsingh@iict.res.in

Abstract

Metal halide perovskite solar cells (PSCs) have become the future candidates for the replacement of silicon solar cells with an efficiency exceeding 25%. PSCs have significant features like high absorption coefficient, better carrier mobility, and tunable bandgap. Despite these advantages, there are various challenges to reaching better efficiency, durability, and cost-effectiveness. To address these challenges, there is a necessity to develop new materials and modification of the conventional device architecture. In this scenario, Carbon nanotubes (CNTs) have emerged as the promising component for fabricating the PSCs. CNTs have attractive features that can offer unique advantages to enhance stability and device performance. In this chapter, we have discussed the utilization of CNTs in PSCs.

Keywords

Perovskite Solar Cells, Photovoltaic Applications, Electrodes, Hole-transporting Materials, Carbon Nanotubes

Contents

Carbon Nanomaterials for Efficient Perovskite Solar Cells.........................178

1.   Introduction...179

2.   Carbon nanotubes (CNTs) in various components of perovskite solar cells 181

        2.1   CNT in perovskite layer 181

        2.2   Carbon nanotubes in hole transport layer.........................188

Conclusions 192

References 192

1. Introduction

As the accomplishment of society is advancing, there is a huge demand for renewable and sustainable energy sources. Third-generation photovoltaics are prominent candidates to minimize the manufacturing cost of a traditional solar cell as well as enhancement of their potential optoelectronic applications. These solar cells consist of cost-effective light-harvesting materials like dyes [1], quantum dots [2], and conjugated polymers [3]. Nevertheless, third-generation solar cells (Dye-Sensitized Solar Cells (DSSCs), Organic photovoltaics (OPVs)) did not give competition in terms of efficiency to the conventional approaches. Organic and inorganic perovskite solar cells (PSCs) are the game changer candidate for the replacement of commercial Si Solar cells, within their little lifespan they have reached an efficiency of $\geq$ 25%. PSCs contain an efficient light absorber, i.e., perovskite, with better features, such as high carrier mobility and long carrier diffusion length [4-5], high molar extinction coefficient, and direct bandgap [6], and long balanced carrier diffusion length [7-9]. The perovskite has the ABX_3 formulae; A is the monovalent cation such as methylammonium [MA], formamidinium [FA], Cs, or Rb, B refers to a divalent cation and is the halide (Br, Cl, I), The first hybrid perovskite is invented in a dye-sensitized solar cell (DSSCs) using a mesoporous TiO_2 structure. Later, Miyasaka and the group achieved a PCE of 3.8%, in 2009 the perovskite ($CH_3NH_3PbI_3$ and $CH_3NH_3PbBr_3$) is used as light absorbers in the DSSCs structure [10]. Park and Grätzel et al. reported that the PSC device with improved the device stability with a solid hole transport layer [11]. After a lot of efforts were attempted by various researchers to optimization of device performance [6, 12-13], minimization of recombination rate, and optimization of the fabrication modes, currently, the certificated PCE of PSCs is 25.2% [14].

Despite of high PCE and notable progress in device architecture, PSCs need a lot of improvements in critical properties such as low cost, replacing many lead metals, and durability against degradation [15]. Stability is the most prominent issue in PSCs [16]. External and internal factors like thermal stress, UV light radiation, moisture ingress, selective charge transport layers, and metal electrodes affect their stability. A lot of research was done to improve the durability of the device. Apart from that cost of the fabrication process is also a vital parameter affecting the energy payback time of PSCs [17-18]. The general structure of a) perovskite and b) the energy level diagram of Perovskite Solar cells is shown in Fig. 1

Emerging Nanomaterials and Their Impact on Society in the 21st Century   Materials Research Forum LLC
Materials Research Foundations 135 (2023) 178-199   https://doi.org/10.21741/9781644902172-8

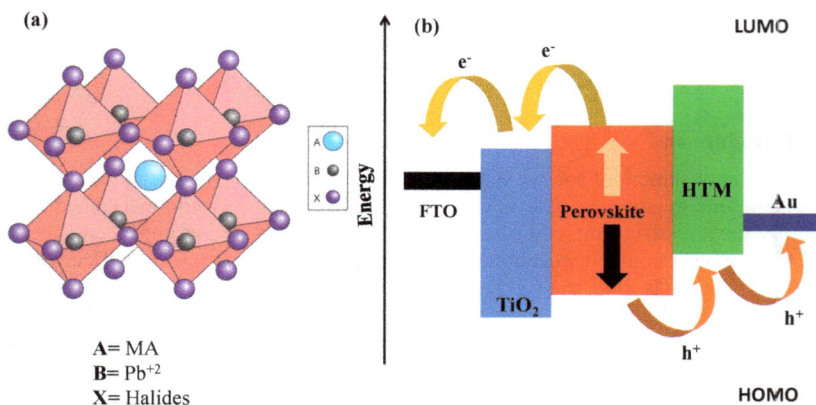

Figure 1. (a) Structure of perovskite (b) energy level diagram of Perovskite Solar cells

Generally, the standard configurations of PSCs are n-i-p or p-i-n (Fig. 2), containing the transparent conductive metal oxide electrode, an electron transport layer (ETL), and a perovskite light absorber layer, a hole transport layer (HTL), and a back counter electrode. The rigid or flexible substrates coated with transparent conductive metal oxide are the base for depositing the other layers. The Light-harvesting material is sandwiched between the ETL and HTL. Nowadays, in the PSCs device, one of the most crucial components is the light-harvesting material and hole transport layer. Apart from this transparent conductive oxide electrodes like indium tin oxide (ITO) or fluorine tin oxide (FTO) and metal back electrodes (Au, Ag) [19] are also important. The deposition of the metal back electrode evaporation will be done using the expensive energy-intensive high vacuum deposition, which led to poor mechanical flexibility and does not allow the PSCS as flexible PSCs. Apart from this, the unwanted recombination of electrons and holes will happen at the interface of the perovskite layers while device making, which leads to poor device performance. To address these features, we need to employ novel light-harvesting and HTL materials and modifications in the device components. Notable research is going on to prepare novel perovskite material and design new device structures to achieve better device performance [20-27].

Emerging Nanomaterials and Their Impact on Society in the 21st Century   Materials Research Forum LLC
Materials Research Foundations 135 (2023) 178-199   https://doi.org/10.21741/9781644902172-8

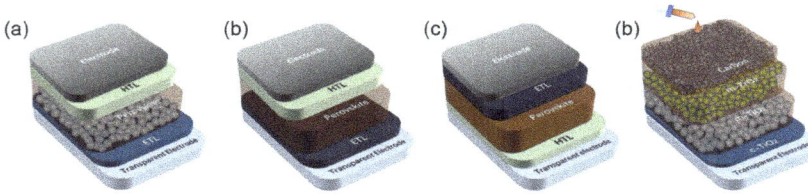

Figure 2. Schematic representation of (a) mesoporous n-i-p (b) planar n-i-p, (c) inverted p-i-n (d) mesoscopic HTL-free PSCs. Reproduced with permission [18].

Carbon materials are good candidates for addressing earlier challenges [28-29]. Carbon nanotubes (CNTs), as a representative 1D and 2D carbon nanomaterial, gained more interest in PSCs [30-31]. In the early 1990s, Iijima reported much work on CNTS in various optoelectronic applications [32-35]. Generally, CNTs are formed by the rolling-up single or multi-layered graphene sheet. The CNTS are classified into three types. They are i) single-walled CNTs (SWCNTs), ii) double-walled CNTs (SWCNTs), and iii) multi-walled CNTs (MWCNTs) [36]. SWCNTs have two categories, i.e., zigzag ($\theta = 0°$), chiral ($0° < \theta < 30°$), and armchair ($\theta = 30°$), depending on their chiral angles (q). The SWCNTs are either metallic or semiconducting based on their chiral angles and diameters [37-39]. This chapter discusses the importance of CNTs as perovskite layers and HTL in perovskite solar cell applications.

2.   Carbon nanotubes (CNTs) in various components of perovskite solar cells

2.1   CNT in perovskite layer

The active perovskite layer is the first layer in the PSC device; most of the device performance will depend on this layer. The quality of the active layer (charge transport capability, optical property, and morphology) will significantly impact PSC's durability and efficiency [40-42]. Because of the quick crystallization process, the light absorber layer is coated by the solution processing methods, which contain small grains with many grain boundaries. The grain boundaries have nonradiative recombination sites because they have many defects and traps and limit the device's performance [43-45]. However, stabilization of the perovskite films, passivation of grain boundaries, and increasing the grain size are the key points to getting high PSCs device performance. In this scenario, various composition management, additive strategy, and solvent engineering methods were discussed in multiple review articles [46-50].

Emerging Nanomaterials and Their Impact on Society in the 21st Century   Materials Research Forum LLC
Materials Research Foundations 135 (2023) 178-199   https://doi.org/10.21741/9781644902172-8

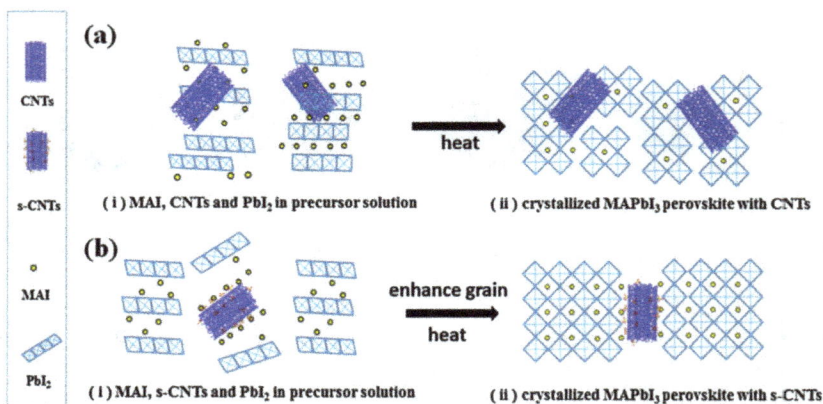

Figure 3. (a) Representation of CNT and s-CNT inserted perovskite films. Reproduced with permission [53].

From the literature, it was found that the incorporation of additives is an effective method to optimize the morphology of the perovskite film. Functionalized carbon nanotubes are one of the best additives to make the uniform film formation of perovskite. Combining the CNTs and perovskites led to hybrid hetero-junction structures with unique photovoltaic properties [51-52]. Chen et al. introduced the sulfonate carbon nanotubes (s-CNTs) while spin coating the $CH_3NH_3PbI_3$ perovskite material to enhance the grain size and reduce the grain boundary. By incorporating the s-CNT into the perovskite film, the process of the charge extraction was improved to obtain the large size and filled grain boundaries. The photoluminescence lifetime of perovskites was also enhanced. The fabricated device architecture is glass/FTO/c-TiO$_2$/mp-TiO$_2$-CH$_3$NH$_3$PbI$_3$ nanocomposite layer/ CH$_3$NH$_3$PbI$_3$ upper layer (320 nm)/spiro-MeOTAD/Ag) (Fig. 3). The device performance with the s-CNTs yielded a PCE of 15.1% and was compared with the pristine CNTs device (10.3%) [53]. Agbolaghi reported the incorporation of the grafted nanostructures with irregioregular (CNT-g-PDDT) and regioregular (CNT-g-P3HT) polymers into the perovskite film. To enhance the morphological, optical, and photovoltaic properties of PSCs. The grain size was increased with the CNT-g-PDDT and CNT-g- P3HT-based perovskite films compared to the pristine MWCNTs. The bare pristine device has shown a negative influence on grain size growth, and morphological and optical properties of active layers resulted in poor photovoltaic performance. The pristine CNTS gave poor absorbance, stability, and grain sizes. The grafted device CNT-g-P3HT-perovskite (50 kDa P3HT) exhibited the largest grain sizes. The overall PCE of PSCs containing CNT-g- PDDT, CNT-g-P3HT (5 kDa) and CNT-g-P3HT (50 kDa) is 14.5%, 16.3%, and 17.4%, respectively [54].

Emerging Nanomaterials and Their Impact on Society in the 21st Century   Materials Research Forum LLC
Materials Research Foundations 135 (2023) 178-199   https://doi.org/10.21741/9781644902172-8

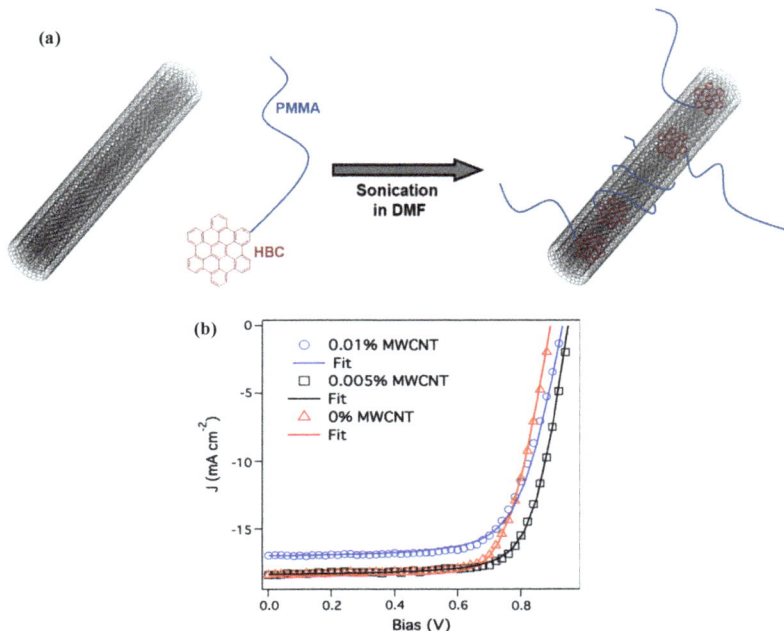

Figure 4. (a). Structure and (b). J–V curves of HBC-PMMA-incorporated MWCNT. Reproduce with permission [54].

Bag et al. introduced the functionalized MWCNT materials to improve device performance (Fig. 4). They have illustrated the feasible route for preparing perovskite/ poly (methyl methacrylate) end-functionalized hexabenzocoronene. They have also studied their effect on the charge recombination of the inverted p-i-n structured PSCs. The incorporation of the MWCNTs into perovskite results in the proper charge separation at the perovskite/MWCNTs interface. Consequently, reduced the charge carrier combination losses by >87%. The phenomenon was measured by impedance spectroscopy. The PCE of 0.0% CNT device is 12.07%, and the device with the 0.005 wt% MWCNTs shows a PCE of 12.91%. Nevertheless, the trap states of the perovskite layer were increased by incorporating the MWCNTS. So, the maintenance of optimal concentration of the MWCNTs in the Perovskite layer is mandatory to balance the trap state, charge recombination, and charge transport. Shigeo Maruyama introduced the small amount of surfactant (Sodium deoxycholate (DOC)) functionalized s-SWCNTs into the perovskite layer (Fig. 5). They behaved as a crystal growth template, grain boundary passivation, improved the mobility of perovskite, and reduced the charge traps, which resulted in the high J_{sc} and V_{oc}. Theoretical studies stated that it increased the nucleation energy of perovskites and slowed down the crystal growth rate happened when the Pb^{2+} Lewis acid in perovskite later interacted with the Lewis base in DOC. The PCE of planar n-i-p PSCs containing DOC-wrapped s-SWCNTs was 18.1% increased to 19.5% [55].

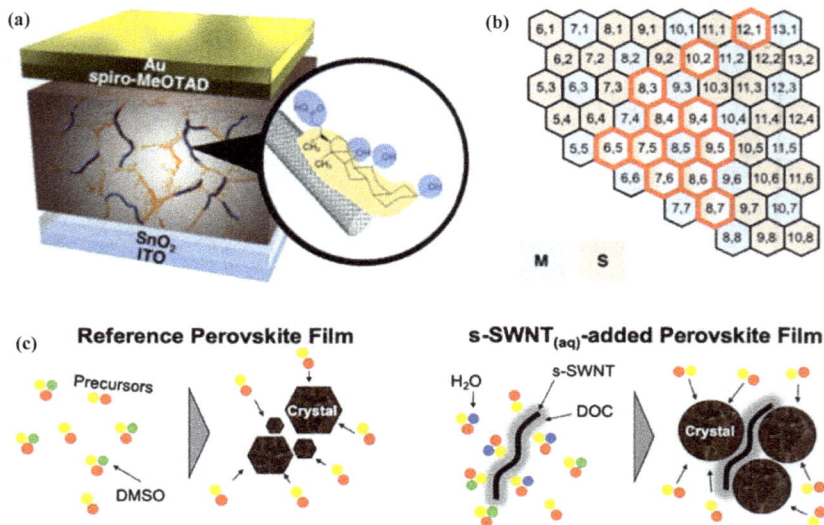

Figure 5. (a) Fabricated device, (b) Chirality index mapping (marked with red colour) c) schematic representation of the role of water and s-SWNT additives in the perovskite growth of the pristine and s-SWNT-added MAPbI₃ film. Reproduced with permission [55].

To replace the DOC surfactants on s-SWCNTs, the Lin group incorporated a polyaromatic nano tweezer surfactant 4,6-di(anthracen-9-yl)-1,3-phenylene bis-(dimethyl carbamate) (DPB) in perovskite layers (Fig. 6). The DPB-functionalized s-SWCNTs showed good dispersion in the perovskite solvents, DMF, and DMSO compared to the DOC-clenched s-SWCNTs. The PCE of DPB-clenched s-SWCNTs-based PSCs exhibited an increment from18.4% is 20.7%. The 0.05 wt % DPB-clenched s-SWNTs-added perovskite precursor solution enhanced the PCE, acting as both grain boundary passivator improved the perovskite crystal growth. Moreover, the DPB-attached s-SWNTs device has slightly better stability than conventional DOCs on SWNTs [56].

Emerging Nanomaterials and Their Impact on Society in the 21st Century   Materials Research Forum LLC
Materials Research Foundations 135 (2023) 178-199   https://doi.org/10.21741/9781644902172-8

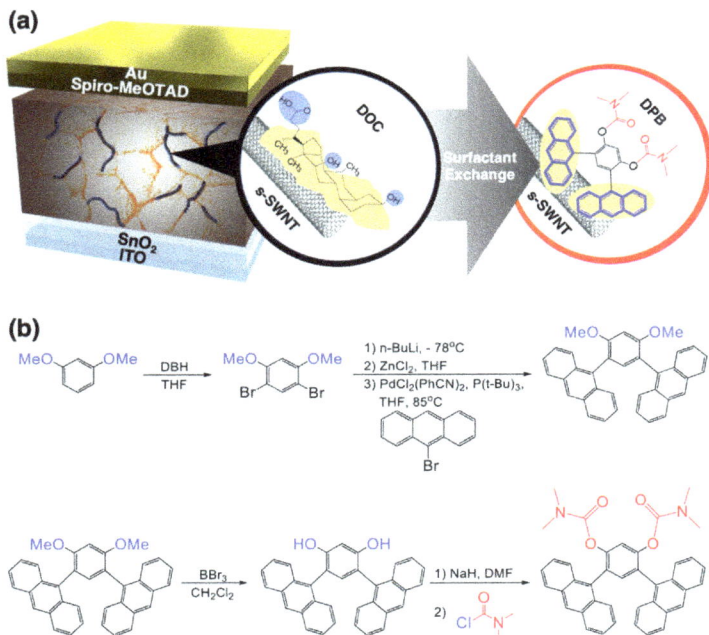

Figure 6. (a) Device architecture with surfactant-exchanged s-SWNTs from DOC to DPB. (b) Synthetic route of surfactant DPB, Reproduce with permission [56].

Wang et al. successfully incorporated the amino-functionalized CNTs (CNT-NH2) and methylammonium chloride (MACl) as additives into the crystalline uniaxial-orientated perovskite thin films (Fig. 7). The addition of MACl and CNT-NH$_2$ improved the morphology and crystallization of perovskite thin film and enhanced charge transport. The growth of the MA$_{0.85}$FA$_{0.15}$PbI$_3$ perovskite is observed as (110) and (220) crystallographic planes compared with MACl-added films, indicating that MACl and CNT-NH$_2$ can jointly manipulate the crystallization process. The planar PSCs device with this amino-functionalized CNTs (CNT-NH$_2$) and methylammonium chloride (MACl) showed the PCE of 21.02% and 19.3 %, under a similar environment, the pristine device exhibited the PCE of 17.7% [57]. The PSCs are facing the problem of an unfavourable interface contact because during the device making, usually perovskite layer will be deposited by a two-step spin coating process, which leads the structure of the perovskite to be noncuboidal and nanoparticles with a convex-concave rough surface. Cheng et al. introduced the multi-walled carbon nanotubes (MWCNTs) into the perovskite layer to enhance the charge transfer at the MAPbI$_3$/carbon electrode interface to address this issue. The MWCNTs

Emerging Nanomaterials and Their Impact on Society in the 21st Century   Materials Research Forum LLC
Materials Research Foundations 135 (2023) 178-199   https://doi.org/10.21741/9781644902172-8

behave as charge transport bridges between individual perovskite nanoparticles and the back carbon electrode to get a smooth flow of the photo-generated holes. The average PCE of MWCNT-based PSCs yielded a PCE of 11.6% under the optimized concentration of MWCNT [58].

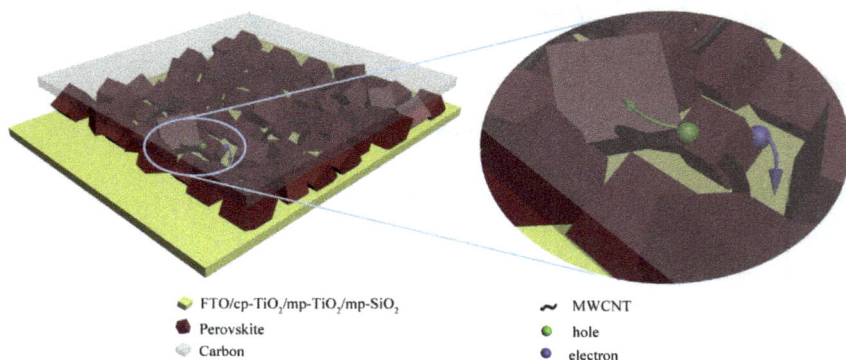

FTO/cp-TiO$_2$/mp-TiO$_2$/mp-SiO$_2$ ~ MWCNT
Perovskite   • hole
Carbon   • electron

Figure 7. Illustration of the charge transport mechanism in HTM-free-MWCNT PSCs Reproduced with permission [58].

Hosseine and Adelifard group employed both dopants (MWCNTs and RGO) in the silver bismuth iodide (SBI) perovskite thin films by spin-coating method. They cleared that all the absorber materials are in the R-3m space group symmetry and Ag_2BiI_5 phase with the hexagonal crystal structure. The resulting PCE doping of MWCNT and RGO is PCE to 1.61% and 1.32%, respectively. The Device stability test was done after 30 days and found that the PCE of SBI-0.3C andSBI-0.2G-based solar cells remains above 1.49% and 1.18%, respectively, which means that the efficiency remains the same (90%) under the environmental conditions [59]. Deposition of SBI films presented in Fig. 8.

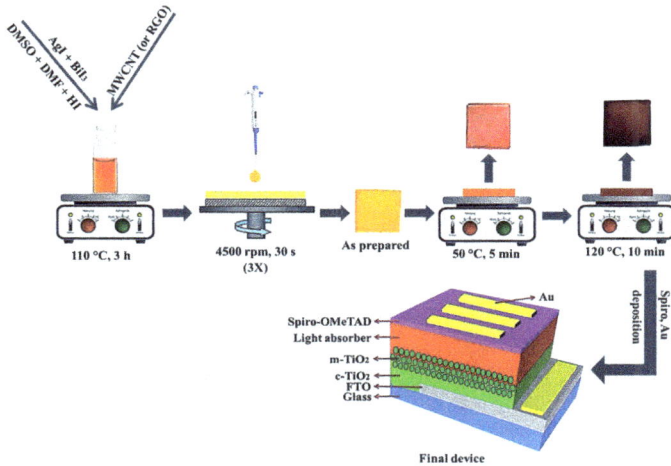

Figure 8. Fabrication Procedure for deposition of SBI films and final device. Reproduce with permission [59].

Hongxia Wang and their group developed the feasible method to produce the high-quality perovskite film with micrometer grain sizes by in situ incorporations of an ultrathin layer of octadecyl amine-functionalized SWCNTs (ODA-SWCNTs) (Fig.9a) before heat treatment. The ease of perovskite growth is due to the strong affinity of Pb^{2+} ions with functional groups (oxygen), which are present on the surface of ODA-SWCNTs. Moreover, the hydrophobic nature of the ODA-SWCNTs decreases the evaporation rate of the dimethyl sulfoxide and dimethylformamide solvents as well as the nucleation density of perovskites. The triple monovalent $(FA_{0.83}MA_{0.17})_{0.95}Cs_{0.05}Pb$ $(I_{0.83}Br_{0.17})_3$ $(CH(NH_2)_2,$ FA) perovskite as a light absorber showed PCE of 16.1% [60]. The SEM-EDS analysis is shown in Fig. 9b.

Emerging Nanomaterials and Their Impact on Society in the 21st Century   Materials Research Forum LLC
Materials Research Foundations 135 (2023) 178-199   https://doi.org/10.21741/9781644902172-8

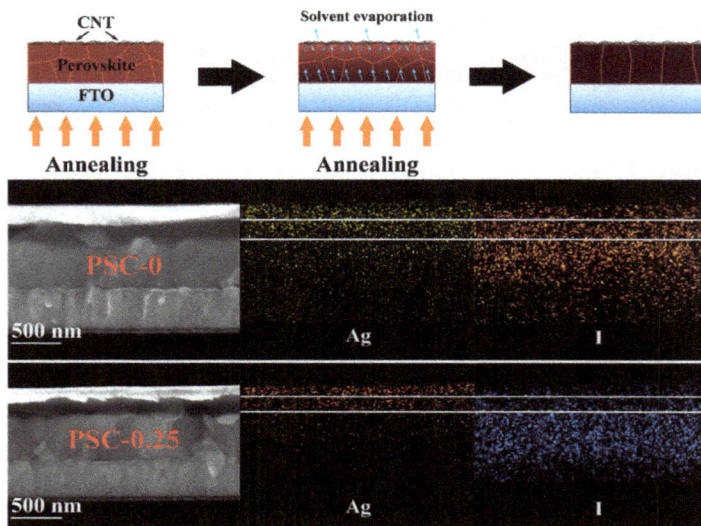

Figure 9. (a) Illustration of perovskite grain growth with the incorporation of SWCNTs and (b) SEM-EDS analysis of PSCs with (PSC-0.25) and without (PSC-0) ODA-SWCNT. Reproduced with permission [60].

2.2   Carbon nanotubes in hole transport layer

Apart from the other device layers, the hole transport layer (HTL) plays a crucial role. The HTL does the two main functions; it is good in the transfer of holes and blocks the transfer of the electrons which can be extracted into the circuit. Till now the most used HTL is the 2, 2', 7, 7'-tetrakis (N, N-di-p-methoxy-phenylamine)-9, 9'-spirobifluorene (Spiro-MeOTAD) [6, 61]. The HTL is necessary because of its effect on short circuit current density with hole mobility. The Spiro-OMeTAD has some disadvantages, such as a complicated synthetic route resulting in high cost and low intrinsic hole mobility [62]. To replace the commercial Spiro-MeOTAD, a variety of HTLs are reported. The efficient alternative HTM and back electrode material for the various PSCs [63-68], polymer solar cells [69], dye-sensitized solar cells [70-71], and silicon solar cells are the carbon nanomaterials like nanotubes (CNT), graphene flakes and nanoparticles. The desirable advantageous properties of carbon material attracted the researcher's interest in various filed such as photovoltaics [72], sensors [73-74], lithium-ion batteries [75], conductive ink [76-77], and redox flow batteries [78]. For the first to address the thermal and moisture stability of PSCs, Henry J. Snaith and his group reported the novel polymer-functionalized single-walled carbon nanotubes (SWNTs) for PSC applications with PCE up to 15.3% [79]. In this work, they have used the poly(3-hexylthiophene) (P3HT) wrapped poly(3-

hexylthiophene) (P3HT), which energy levels make the electron transfer from the light harvester to SWCNTs unfavorable. Further, it led to low device performance. To further improve the PCE of PSCs. Habisreutinger et al. replaced the PMMA with the HTL (Fig. 10). They tested the n-i-p device architecture with two HTL layers, i.e., undoped spiro-OMeTAD in a combination of P3HT/SWCNTs, and achieved the PCE OF 15.4%, with the same environment as the simple undoped spiro-OMeTAD has shown PCE of 6.8% [80].

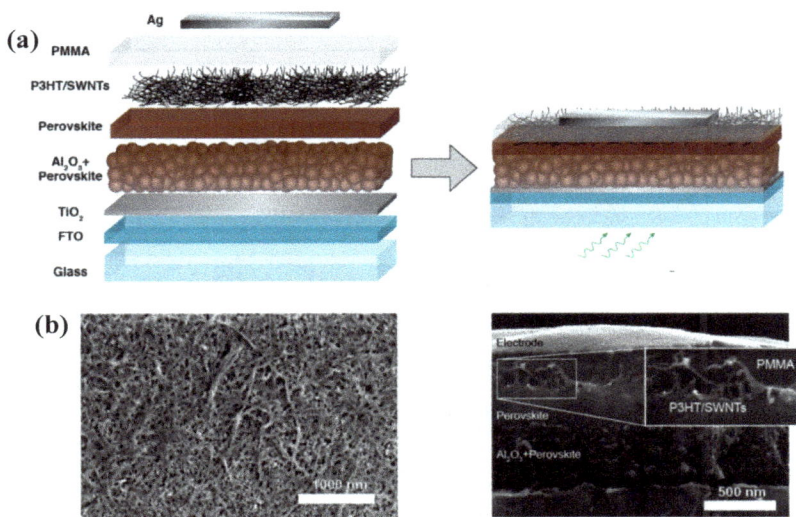

Figure 10. (a) Device Representation and (b) SEM images of mesoporous-superstructure PSCs. Reproduced with permission [81].

Mazzotta group replaced the P3HT/SWCNTs with polymer ethylene vinyl acetate wrapped CNTs in two-layer HTL and achieved the PCE of 17.1% [81]. In 2019, the same group investigated the combination of the charge transfer and recombination of two-layer HTL. They stated that to get low recombination of PSCs, the kinetics of the HTLS and charge extraction are crucial points. [82]. Wang and his group introduced the water/alcohol-soluble fullerene derivatives into perovskite film. The carbon materials C_{60} improved the electron diffusion length and PCE of PSCs. The BHJ Perovskite hybrid solar cells with fullerene derivative showed an enhancement of 22% in PCE with the best fill factor of 86.7% compared to pristine planar PSCs [83], J. Xu et al. incorporated graphene as additives in perovskite thin film, which yields suppressing of hysterics of PSCs (FF) [84].

Kertu Aitola and group reported the high-conductivity random network-type SWCNTs file by the drop-casting method of the Spiro-OMeTAD with the best PCE of 15.5%, the spin-coated Spiro-OMeTAD and a thermally evaporated counter electrode (Au) yielded the PCE of 18.8 % (Fig. 11). The authors tested the device with drop-casted SWCNT: Spiro-OMeTAD composite HTM-CE without an additional counter electrode (SWCNT: Spiro-OMeTAD cell). The device performance was compared with SWCNT film functioning as the sole HTM-CE (SWCNT cell), spin-coated Spiro-OMeTAD HTM and evaporated counter electrode (Au), Spiro-OMeTAD + gold cell, and the gold CE evaporated directly on the perovskite layer (Au cell) [85].

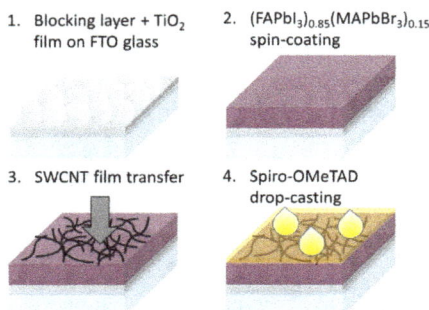

1. Blocking layer + TiO_2 film on FTO glass

2. $(FAPbI_3)_{0.85}(MAPbBr_3)_{0.15}$ spin-coating

3. SWCNT film transfer

4. Spiro-OMeTAD drop-casting

Figure 11. Stepwise layers deposition of SWCNT: Spiro-OMeTAD PSCs. Reproduced with permission [85].

In 2017 same group reported the mixed ion perovskite $Cs_5(MA_{0.17}FA_{0.83})_{95}Pb(I_{0.83}Br_{0.17})_3$ as the light absorber and long-term stable SWCNT for the PSCs. The architecture device is FTO glass substrate/C-TiO_2/m-TiO_2/perovskite as the active layer/Spiro-OMeTAD/back hole contact (Fig. 12). The reported PCE OF 18.4% and the linear efficiency loss during the 580 hrs, however the PSCS with gold contact showed the irreversible and drastic efficiency loss during the 140hrs [86].

To enhance the stability and hole extraction of PSCs, instead of the two HTLs incorporating, SWCNTs as an additive to HTL is the better phenomenon. Miletić et al. used the SWCNTs as an additive in place of the conventional lithium to enhance the hole mobility of the HTL. To modify (7,6)-enriched single-walled carbon nanotubes (SWCNTs) covalently, they have introduced the ligophenylenevinylene bearing alkoxy groups and amine groups (Fig. 13 a,b). The functionalized SWCNTs showed better stability and comparable efficiency with a standard doped HTL (spiro-MeOTAD) device. They achieved a PCE of 9.6 with the J_{sc} of 16.6 mACm^{-2}, V_{oc} 1.00V, and the FF is 1.00 [87]. In a similar way to reduce the charge recombination, which comes from the direct contact between the MWCNTS and light absorber, Lee and the group doped the spiro-OMeTAD with MWCNTs, which yielded the PCE of 15.1% and low series resistance of the PSCs. Fig. 13 c, d. [88].

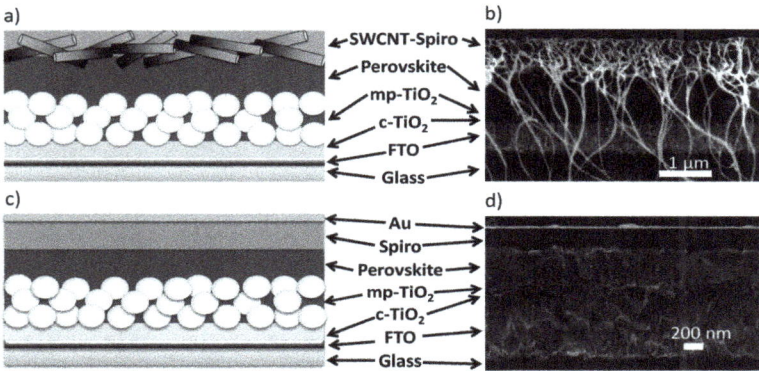

Figure 12. (a) Device profile and (b) cross-sectional SEM image of the SWCNT-incorporated PSC device. Reproduced with permission [86].

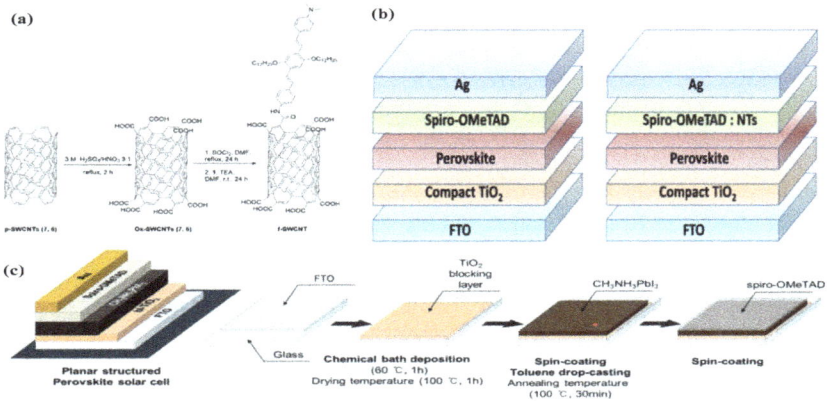

Figure 13. (a) Synthetic scheme of (7,6) (f-SWCNT) (b) device architecture. Reproduced with permission [87], (c) layers deposition of planar PSCs. Reproduced with permission [88].

Conclusions

In summary, incorporating carbon nanotubes (CNTs) is a good choice to improve the photovoltaic applications of metal halide-based perovskite solar cells (PSCs). Therefore, the functionalized CNTs are getting more interest from researchers to develop cost-effective and stable solar cell generation. The interesting application of CNT embedded perovskite or hole transport layer will mainly lead to the fibrous solar cells, which can be used in clothing and electronic devices. Stability and efficiency are the main factors in making efficient PSCs. These can address by CNTs, providing good optical, electronic, and mechanical properties. The CNTs can be incorporated in various PSCs device layers like perovskite and charge transport additives. Interface modifiers to the hole-transporting layer and charge-collecting electrodes make efficient, stable PSCs in the near future.

References

[1] M. Grätzel, Dye-sensitized solar cells J. Photochem. and Photobiol. C: Photochem. Rev. 4 (2003) 145-153. https://doi.org/10.1016/S1389-5567(03)00026-1

[2] S. Gunes, H. Neugebauer, N. S. Sariciftci, conjugated polymer-based organic solar cells, Chem. Rev. 107 (2007) 1324-1338. https://doi.org/10.1021/cr050149z

[3] I. J. Kramer, E. H. Sargent, Colloidal Quantum Dot Photovoltaics: A Path Forward, ACS Nano. 5 (2011) 8506-8514. https://doi.org/10.1021/nn203438u

[4] S. D. Stranks, G. E. Eperon, G. Grancini, C. Menelaou, M. J. P. Alcocer, T. Leijtens, L. M. Herz, A. Petrozza, H. J. Snaith, Electron-hole diffusion lengths exceeding 1 micrometer in an organometal trihalide perovskite absorber, Science. 342 (2013) 341-344. https://doi.org/10.1126/science.1243982

[5] G. C. Xing, N. Mathews, S. Y. Sun, S. S. Lim, Y. M. Lam, M. Grätzel, S. Mhaisalkar, T. C. Sum, Long-range balanced electron- and hole-transport lengths in organic-inorganic CH3NH3PbI3, Science. 342 (2013) 344-347. https://doi.org/10.1126/science.1243167

[6] H. S. Kim, C. R. Lee, J. H. Im, K. B. Lee, T. Moehl, A. Marchioro, S. J. Moon, R. Humphry-Baker, J. H. Yum, J. E. Moser, M. Grätzel, N-G, Park, Lead iodide perovskite sensitized all-solid-state submicron thin film mesoscopic solar cell with efficiency exceeding 9%, Sci. Rep. 2 (2012) 591. https://doi.org/10.1038/srep00591

[7] S. D. Stranks, Nonradiative losses in metal halide perovskites, ACS Energy Lett. 2 (2017) 1515-1525. https://doi.org/10.1021/acsenergylett.7b00239

[8] M. Liu, M. B. Johnston, H. J. Snaith, Efficient planar heterojunction perovskite solar cells by vapour deposition, Nature 501 (2013) 395-398. https://doi.org/10.1038/nature12509

[9] N. Aristidou, C. Eames, I. Sanchez-Molina, X. Bu, J. Kosco, M. S. Islam, S. A. Haque, Fast oxygen diffusion and iodide defects mediate oxygen-induced degradation of perovskite solar cells, Nat. Commun. 8 (2017), 15218. https://doi.org/10.1038/ncomms15218

[10] Q. Dong, Y. Fang, Y. Shao, P. Mulligan, J. Qiu, L. Cao, J. Huang, Electron-hole diffusion lengths > 175 μm in solution-grown CH3NH3PbI3 single crystals, Science. 347 (2015) 967-970. https://doi.org/10.1126/science.aaa5760

[11] A. Kojima, K. Teshima, Y. Shirai, T. Miyasaka, Organometal halide perovskites as visible-light sensitizers for photovoltaic cells. J. Am. Chem. Soc. 131 (2009) 6050-6051. https://doi.org/10.1021/ja809598r

[12] J.-H. Im, C. R. Lee, J. W. Lee, S. W. Park, N. G. Park, 6.5% efficient perovskite quantum-dot-sensitized solar cell, Nanoscale, 3 (2011) 4088-4093. https://doi.org/10.1039/c1nr10867k

[13] J. H.; Jang, I. H.; Pellet, N.; Grätzel, M.; Park, N. G. Growth of CH3NH3PbI3 cuboids with controlled size for high-efficiency perovskite solar cells Nat. Nanotechnol. 9 (2014) 927-932. https://doi.org/10.1038/nnano.2014.181

[14] H. Zhou, Q. Chen, G. Li, S. Luo, T. B. Song, H. Duan, Z. Hong, J. You, Y. Liu, Y. Yang, Interface engineering of highly efficient perovskite solar cells Science 345 (2014) 542-546. https://doi.org/10.1126/science.1254050

[15] J. Burschka, N. Pellet, S. J. Moon, R. Humphry-Baker, P. Gao, M. K. Nazeeruddin, M. Grätzel, Sequential deposition as a route to high-performance perovskite-sensitized solar cells, Nature, 499 (2013) 316-319. https://doi.org/10.1038/nature12340

[16] Y. Rong, Y. Hu, A. Mei, H. Tan, M. I. Saidaminov, S. I. Seok, M. D. McGehee, E. H. Sargent, H. Han, Challenges for commercializing perovskite solar cells Science, 361 (2018) 1214-1220. https://doi.org/10.1126/science.aat8235

[17] J. Correa-Baena, M. Saliba, T. Buonassisi, M. Grätzel, A. Abate, W. Tress, A. Hagfeldt, Promises and challenges of perovskite solar cells, Science. 358 (2017) 739-744. https://doi.org/10.1126/science.aam6323

[18] R. Wang, M. Mujahid, Y. Duan, Z. Wang, J. Xue, Y. Yang, A Review of Perovskites Solar Cell Stability, A Review of Perovskites Solar Cell Stability, Adv. Funct. Mater. 29 (2019) 1808843. https://doi.org/10.1002/adfm.201808843

[19] C. C. Boyd, R. Cheacharoen, T. Leijtens, M. D. Understanding degradation mechanisms and improving stability of perovskite photovoltaics, Chem. Rev. 119 (2019) 3418-3451. https://doi.org/10.1021/acs.chemrev.8b00336

[20] M. Cai, Y. Wu, H. Chen, X. Yang, Y. Qiang, L. Han, Cost-Performance Analysis of Perovskite Solar Modules, Cost-Performance Analysis of Perovskite Solar Modules, Adv. Sci. 4 (2017) 1600269. https://doi.org/10.1002/advs.201600269

[21] H. Tsai, W. Nie, J. Blancon, C. C. Stoumpos, R. Asadpour, B. Harutyunyan, A. J. Neukirch, R. Verduzco, J. J. Crochet, S. Tretiak, L. Pedesseau, J. Even, M. A. Alam, G. Gupta, J. Lou, P. M. Ajayan, M. J. Bedzyk, M. G. Kanatzidis, A. D. Mohite, High-efficiency two-dimensional Ruddlesden-Popper perovskite solar cells, Nature. 536 (2016) 312-316. https://doi.org/10.1038/nature18306

[22] H. Chen, F. Ye, W. Tang, J. He, M. Yin, Y. Wang, F. Xie, E. Bi, X. Yang, M. Grätzel, L. Han, A solvent- and vacuum-free route to large-area perovskite films for efficient solar modules, Nature. 550 (2017) 92-95. https://doi.org/10.1038/nature23877

[23] S. Turren-Cruz, A. Hagfeldt, M. Saliba, Methylammonium-free, high-performance, and stable perovskite solar cells on a planar architecture, Science. 362 (2018) 449-453. https://doi.org/10.1126/science.aat3583

[24] H. Min, M. Kim, S. Lee, H. Kim, G. Kim, K. Choi, J. H. Lee, S. I. Seok, Efficient, stable solar cells by using inherent bandgap of α-phase formamidinium lead iodide, Science. 366 (2019) 749-753. https://doi.org/10.1126/science.aay7044

[25] Y. Wang, M. I. Dar, L. K. Ono, T. Zhang, M. Kan, Y. Li, L. Zhang, X. Wang, Y. Yang, X. Gao, Y. Qi, M. Grätzel, Y. Zhao, Thermodynamically stabilized β-CsPbI 3-based perovskite solar cells with efficiencies >18, Science. 2019, 365, 591-595. https://doi.org/10.1126/science.aav8680

[26] Q. Jiang, L. Wang, C. Yan, C. Liu, Z. Guo, N. Wang, Nano-mesoporous TiO2 Vacancies Modification for Halide Perovskite Solar Cells, Eng Sci. 1 (2018) 64-68.

[27] E. H. Jung, N. J. Jeon, E. Y. Park, C. S. Moon, T. J. Shin, T. Yang, J. H. Noh, J. Seo, Efficient, stable and scalable perovskite solar cells using poly(3-hexylthiophene), Nature. 567 (2019) 511-515. https://doi.org/10.1038/s41586-019-1036-3

[28] R. G. Chaudhary, A. K. Potbhare, P. B. Chouke, A. R. Rai, R.P. Mishra, M. Desimone, A. Abdala, Graphene-Based Nanomaterials and their Nanocomposites with Metal Oxides: Biosynthesis, Electrochemical, Photocatalytic and Antimicrobial Applications, Magnetic Oxides and Composites II, Materials Research Forum, 83 (2020) 79-116. https://doi.org/10.21741/9781644900970-4

[29] A. U. Chaudhry, A. Abdala, S. P. Lonkar, R. G. Chudhary, A. Mabrouk, Thermal, electrical, and mechanical properties of highly filled HDPE/graphite nanoplatelets composites, Mater. Today: Proc, 29 (2020) 704-708. https://doi.org/10.1016/j.matpr.2020.04.168

[30] S. N. Habisreutinger, T. Leijtens, G. E. Eperon, S. D. Stranks, R. J. Nicholas, H. J. Snaith, Enhanced Hole Extraction in Perovskite Solar Cells Through Carbon Nanotubes J. Phys. Chem. Lett. 5 (2014) 4207-4212. https://doi.org/10.1021/jz5021795

[31] S. Iijima, Helical microtubules of graphitic carbon, Nature. 354 (1991) 56-58. https://doi.org/10.1038/354056a0

[32] S. Iijima, T. Ichihashi, Single-shell carbon nanotubes of 1-nm diameter, Nature. 363 (1993) 603-605. https://doi.org/10.1038/363603a0

[33] Z. Ku, Y. Rong, M. Xu, T. Liu, H. Han, Full printable processed mesoscopic CH3NH3PbI3/TiO2 heterojunction solar cells with carbon counter electrode, Sci. Rep. 3 (2013) 3132. https://doi.org/10.1038/srep03132

[34] R. K. Mishra, K. Verma, R. G. Chaudhary, T. Lambat, K. Joseph, An efficient fabrication of polypropylene hybrid nanocomposites using carbon nanotubes and PET

194

fibrils, Mater. Today: Proc, 29 (2020) 794-800.
https://doi.org/10.1016/j.matpr.2020.04.753

[35] H. Chen, S. Yang, Carbon-Based Perovskite Solar Cells without Hole Transport Materials: The Front Runner to the Market? Adv. Mater. 29 (2017) 1603994. https://doi.org/10.1002/adma.201603994

[36] L. Fagiolari, F. Bella, Carbon-based materials for stable, cheaper and large-scale processable perovskite solar cells, Energy Environ. Sci. 12 (2019) 3437-3472. https://doi.org/10.1039/C9EE02115A

[37] P. M. Ajayan, Nanotubes from Carbon, Chem. Rev. 99 (1999) 1787-1800. https://doi.org/10.1021/cr970102g

[38] T. Belin, F. Epron, Characterization methods of carbon nanotubes: a review. J. Mater. Sci. Eng. B 119 (2005) 105-118. https://doi.org/10.1016/j.mseb.2005.02.046

[39] M. S. Dresselhaus, G. Dresselhaus, A. Jorio, Unusual properties and structure of carbon nanotubes Annu. Rev. Mater. Res. 34 (2004) 247-278. https://doi.org/10.1146/annurev.matsci.34.040203.114607

[40] Y. Maeda, S. Kimura, M. Kanda, Y. Hirashima, T. Hasegawa, T. Wakahara, Y. Lian, T. Nakahodo, T. Tsuchiya, T. Akasaka, J. Lu, X. Zhang, Y. Yu, S. Nagase, S. Kazaoui, N. Minami, T. Shimizu, H. Tokumoto, R. Saito, Large-scale separation of metallic and semiconducting single-walled carbon nanotubes, J. Am. Chem. Soc. 127 (2005) 10287-10290. https://doi.org/10.1021/ja051774o

[41] Liu, Y. Cheng, Z. Ge, Understanding of perovskite crystal growth and film formation in scalable deposition processes, Chem. Soc. Rev. 49 (2020) 1653-1687. https://doi.org/10.1039/C9CS00711C

[42] J. Jiang, J. Xu, H. Walter, A. Kazi, D. Wang, G. Wangila, M. Mortazavi, C. Yan, Q. Jiang, The doping of alkali metal for halide perovskites, ES Mater. Manuf. 7 (2020) 25-33.

[43] E. Aydin, M. De Bastiani, S. De Wolf, Defect and Contact Passivation for Perovskite Solar Cells, Adv. Mater. 31 (2019) 1900428. https://doi.org/10.1002/adma.201900428

[44] H. D. Kim, H. Ohkita, H. Benten, S. Ito, Photovoltaic performance of perovskite solar cells with different grain sizes, Adv. Mater. 28 (2016) 917-922. https://doi.org/10.1002/adma.201504144

[45] W. Nie, H. Tsai, R. Asadpour, J. Blancon, A. J. Neukirch, G. Gupta, J. J. Crochet, M. Chhowalla, S. Tretiak, M. A. Alam, H. Wang, A. D. Mohite, High-efficiency solution-processed perovskite solar cells with millimeter-scale grains, Science. 347 (2015) 522-525. https://doi.org/10.1126/science.aaa0472

[46] D. W. de Quilettes, S. M. Vorpahl, S. D. Stranks, H. Nagaoka, G. E. Eperon, M. E. Ziffer, H. J. Snaith, D. S. Ginger, Impact of microstructure on local carrier lifetime in perovskite solar cells, Science 348 (2015) 683-686. https://doi.org/10.1126/science.aaa5333

[47] F. Zhang, K. Zhu, Additive engineering for efficient and stable perovskite solar cells, Adv. Energy Mater. 10 (2020) 1902579. https://doi.org/10.1002/aenm.201902579

[48] S. Liu, Y. Guan, Y. Sheng, Y. Hu, Y. Rong, A. Mei, H. Han, A review on additives for halide perovskite solar cells, Adv. Energy Mater. 10 (2020) 1902492. https://doi.org/10.1002/aenm.201902492

[49] R. Liu, K. Xu, Micro, Solvent engineering for perovskite solar cells: a review, Nano Lett. 2020, 15, 349-353. https://doi.org/10.1049/mnl.2019.0735

[50] A. Dubey, N. Adhikari, S. Mabrouk, F. Wu, K. Chen, S. Yang, Q. Qiao, A strategic review on processing routes towards highly efficient perovskite solar cells, J. Mater. Chem. A 6 (2018) 2406-2431. https://doi.org/10.1039/C7TA08277K

[51] Y. Chen, M. He, J. Peng, Y. Sun, Z. Liang, structure and growth control of organic-inorganic halide perovskites for optoelectronics: from polycrystalline films to single crystals, Adv. Sci. 3 (2016) 1500392. https://doi.org/10.1002/advs.201500392

[52] Li, H. Wang, D. Kufer, L. Liang, W. Yu, E. Alarousu, C. Ma, Y. Li, Z. Liu, C. Liu, N. Wei, F. Wang, L. Chen, O. F. Mohammed, A. Fratalocchi, X. Liu, G. Konstantatos, T. Wu, Ultrahigh carrier mobility achieved in photoresponsive hybrid perovskite films via coupling with single-walled carbon nanotubes, Adv. Mater. 29 (2017) 1602432. https://doi.org/10.1002/adma.201602432

[53] Y. Zhang, L. Tan, Q. Fu, L. Chen, T. Ji, X. Hua, Y. Chen, Enhancing the grain size of organic halide perovskites by sulfonate-carbon nanotube incorporation in high performance perovskite solar cells, Chem. Commun., 52 (2016) 5674-5677. https://doi.org/10.1039/C6CC00268D

[54] M. Bag, L. A.Renna, S. P. Jeong, X. Han ,C. L.Cutting, D.Maroudas, D.Venkataraman, Evidence for reduced charge recombination in carbon nanotube/perovskite-based active layers, Chem. Phys. Lett. 662 (2016) 35-41. https://doi.org/10.1016/j.cplett.2016.09.004

[55] S. Seo, Il Jeon, R. Xiang, C. Lee, H. Zhang, T. Tanaka, J-W Lee, D. Suh, T. Ogamoto, R. Nishikubo,e A. Saeki, S. Chiashi, J. Shiomi, H. Kataura, H. M. Lee, Y. Yang, Y. Matsuo, S. Maruyama , Semiconducting carbon nanotubes as crystal growth templates and grain bridges in perovskite solar cells, J. Mater. Chem. A, 7 (2019) 12987-12992 https://doi.org/10.1039/C9TA02629K

[56] H. Lin, S. Okawa, Y. Ma, S. Yotsumoto, C. Lee, S. Tan, S. Manzhos, M. Yoshizawa, S. Chiashi, H. M. Lee, T. Tanaka, H. Kataura, I. Jeon, Y. Matsuo, S. Maruyama, Polyaromatic nanotweezers on semiconducting carbon nanotubes for the growth and interfacing of lead halide perovskite crystal grains in solar cells, Chem. Mater. 32 (2020) 5125. https://doi.org/10.1021/acs.chemmater.0c01011

[57] Y. Wang, W. Li, T. Zhang, D. Li, M. Kan, X. Wang, X. Liu, T. Wang,Y. Zhao, Highly Efficient (110) Orientated FA-MA mixed cation perovskite solar cells via

functionalized carbon nanotube and methylammonium chloride additive, Small Methods, 4 (2020) 1900511. https://doi.org/10.1002/smtd.201900511

[58] N. Cheng, P. Liu, F. Qi, Y. Xiao, W. Yu, Z. Yu, W. Liu, S. Guo, X. Zhao, Effect of oxygen plasma treatment on the electrochemical performance of the rayon and polyacrylonitrile based carbon felt for the vanadium redox flow battery application, J. Power Sources 332 (2016) 24-29. https://doi.org/10.1016/j.jpowsour.2016.09.070

[59] S. S. Hosseini, M. Adelifard. The effect of multi-walled carbon nanotubes and reduced graphene oxide doping on the optical and photovoltaic performance of Ag2B1I5-based solar cells, J. Electron. Mater. 49 (2020) 5790-5800. https://doi.org/10.1007/s11664-020-08387-1

[60] V. T. Tiong, N. D. Pham, T. Wang, T. Zhu, X. Zhao, Y. Zhang, Q. Shen, J. Bell, L. Hu, S. Dai, H. Wang, Octadecylamine-functionalized single-walled carbon nanotubes for facilitating the formation of a monolithic perovskite layer and stable solar cells, Adv. Funct. Mater. 28 (2018) 1705545. https://doi.org/10.1002/adfm.201705545

[61] M. M. Lee, J. Teuscher, T. Miyasaka, T. N. Murakami, H. J. Snaith, Efficient hybrid solar cells based on meso-superstructured organometal halide perovskites., Science. 338 (2012) 643-647. https://doi.org/10.1126/science.1228604

[62] T. Leijtens, J. Lim, J. Teuscher, T. Park and H. J. Snaith, Charge Density Dependent mobility of organic hole-transporters and mesoporous tio2 determined by transient mobility spectroscopy: implications to dye-sensitized and organic solar cells, Adv. Mater., 25 (2013) 3227-3233. https://doi.org/10.1002/adma.201300947

[63] Z. Li, S. a Kulkarni, P. P. Boix, E. Shi, A. Cao, K. Fu, S. K. Batabyal, J. Zhang, Q. Xiong, L.H. Wong, N. Mathews and S. G. Mhaisalkar, Laminated carbon nanotube networks for metal electrode-free efficient perovskite solar cells, ACS Nano. 8 (2014) 6797-6804. https://doi.org/10.1021/nn501096h

[64] A. Mei, X. Li, L. Liu, Z. Ku, T. Liu, Y. Rong, M. Xu, M. Hu, J. Chen, Y. Yang, M. Grätzel and H. Han, S, A hole-conductor-free, fully printable mesoscopic perovskite solar cell with high stability, Science. 345 (2014) 295-298. https://doi.org/10.1126/science.1254763

[65] H. Zhou, Y. Shi, Q. Dong, H. Zhang, Y. Xing, K. Wang, Y. Du and T. Ma, Hole-conductor-free, metal-electrode-free TiO2/ CH3NH3PbI3 heterojunction solar cells based on a low-temperature carbon electrode, J. Phys. Chem. Lett. 5 (2014) 3241-3246. https://doi.org/10.1021/jz5017069

[66] X. Xu, Z. Liu, Z. Zuo, M. Zhang, Z. Zhao, Y. Shen, H. Zhou, Q. Chen, Y. Yang and M. Wang, Hole selective NiO contact for efficient perovskite solar cells with carbon electrode, Nano Lett. 15 (2015) 2402-2408. https://doi.org/10.1021/nl504701y

[67] S. N. Habisreutinger, T. Leijtens, G. E. Eperon, S. D. Stranks, R. J. Nicholas and H. J. Snaith, Carbon nanotube/polymer composites as a highly stable hole collection layer in perovskite solar cells, Nano Lett. 14 (2014) 5561-5568. https://doi.org/10.1021/nl501982b

[68] S. N. Habisreutinger, T. Leijtens, G. E. Eperon, S. D. Stranks, R. J. Nicholas and H. J. Snaith, Enhanced hole extraction in perovskite solar cells through carbon nanotubes, J. Phys. Chem. Lett. 5 (2014) 4207-4212. https://doi.org/10.1021/jz5021795

[69] A. Du Pasquier, H. E. Unalan, A. Kanwal, S. Miller, and M. Chhowalla, Conducting and transparent single-wall carbon nanotube electrodes for polymer-fullerene solar cells, Appl. Phys. Lett. 87 (2005) 1-3. https://doi.org/10.1063/1.2132065

[70] K. Aitola, A. Kaskela, J. Halme, V. Ruiz, A. G. Nasibulin, E. I. Kauppinen and P. D. Lund, Single-walled carbon nanotube thin-film counter electrodes for indium tin oxide-free plastic dye solar cells, J. Electrochem. Soc. 157 (2010) B1831-B1837. https://doi.org/10.1149/1.3500367

[71] K. Aitola, J. Halme, S. Feldt, P. Lohse, M. Borghei, A. Kaskela, A. G. Nasibulin, E. I. Kauppinen, P. D. Lund, G. Boschloo and A. Hagfeldt, Highly catalytic carbon nanotube counter electrode on plastic for dye solar cells utilizing cobalt-based redox mediator, Electrochim. Acta, 111 (2013) 206-209. https://doi.org/10.1016/j.electacta.2013.07.202

[72] Kalita, G.; Adhikari, S.; Aryal, H.R.; Afre, R.; Soga, T.; Sharon, M.; Umeno, M. Functionalization of multi-walled carbon nanotubes (MWCNTs) with nitrogen plasma for photovoltaic device application. Curr. Appl. Phys. 9 (2009) 346-351. https://doi.org/10.1016/j.cap.2008.03.007

[73] J. B. Raoof, R. Ojani, M. Baghayeri, M. Amiri-Aref, Application of a glassy carbon electrode modified with functionalized multi-walled carbon nanotubes as a sensor device for simultaneous determination of acetaminophen and tyramine, Anal. Methods. 4 (2012) 1579-1587. https://doi.org/10.1039/c2ay05494a

[74] E. Frackowiak, Carbon materials for supercapacitor application. Phys. Chem. Chem. Phys. 9 (2007) 1774-1785. https://doi.org/10.1039/b618139m

[75] T. Zhang, F. Ran, Design strategies of 3d carbon-based electrodes for charge/ion transport in lithium-ion battery and sodium ion battery, Adv. Funct. Mater. 31 (2021) 2010041. https://doi.org/10.1002/adfm.202010041

[76] C. Phillips, A. Al-Ahmadi, S.J. Potts, T. Claypole, D. Deganello, The effect of graphite and carbon black ratios on conductive ink performance. J. Mater. Sci. 52 (2017) 9520-9530. https://doi.org/10.1007/s10853-017-1114-6

[77] A. Ji, Y. Chen, X. Wang, C. Xu, Inkjet printed flexible electronics on paper substrate with reduced graphene oxide/carbon black ink, J. Mater. Sci. Mater. Electron. 29 (2018) 13032-13042. https://doi.org/10.1007/s10854-018-9425-1

[78] H. Zhang, N. Chen, C. Sun, X. Luo, Investigations on physicochemical properties and electrochemical performance of graphite felt and carbon felt for iron-chromium redox flow battery, Int. J. Energy Res. 44 (2020) 3839-3853. https://doi.org/10.1002/er.5179

[79] N. S.bisreutinger, T. Leijtens, G. E. Eperon, S. D. Stranks, R. J. Nicholas, H. J. Snaith, Carbon nanotube/ polymer composites as a highly stable hole collection layer

in perovskite solar cells, Nano Lett. 14 (2014) 5561-5568.
https://doi.org/10.1021/nl501982b

[80] S.N. Habisreutinger, T. Leijtens, G. E. Eperon, S. D. Stranks, R. J. Nicholas, H. J. Snaith, Research update: strategies for improving the stability of perovskite solar cells, J. Phys. Chem. Lett. 5 (2014) 4207-4212. https://doi.org/10.1021/jz5021795

[81] G, Mazzotta, M. Dollmann, S. N. Habisreutinger, M. G. Cristoforo, Z. Wang, H. J Snaith, M. K Riede, R.J Nicholas, Solubilization of carbon nanotubes with ethylene-vinyl acetate for solution-processed conductive films and charge extraction layers in perovskite solar cells, ACS Appl. Mater. Interfaces. 11 (2019) 1185-1191. https://doi.org/10.1021/acsami.8b15396

[82] S. N. Habisreutinger, N. K. Noel, B. W. Larson, O. G. Reid, J. L. Blackburn, Rapid charge-transfer cascade through SWCNT composites enabling low-voltage losses for perovskite solar cells, ACS Energy Lett. 4 (2019) 1872-1879. https://doi.org/10.1021/acsenergylett.9b01041

[83] K. Wang, C. Liu, P. Du, J. Zheng, and X. Gong, Bulk heterojunction perovskite hybrid solar cells with large fill factor, Energy Environ. Sci. 8 (2015) 1245-1255. https://doi.org/10.1039/C5EE00222B

[84] J. Xu, A. Buin, A. H. Ip, W. Li, O. Voznyy, R. Comin, M. Yuan, S. Jeon, Z. Ning, J. J. McDowell, P. Kanjanaboos, J-P Sun, X. Lan, L. N. Quan, D. Ha Kim, I. G. Hill, P. Maksymovych, E. H. Sargent, Perovskite-fullerene hybrid materials suppress hysteresis in planar diodes, Nature Comm. 8 (2015) 7081. https://doi.org/10.1038/ncomms8081

[85] K. Aitola, K. Sveinbjornsson, J. P. Correa Baena, A. Kaskela, A. Abate, Y. Tian, E. M. J. Johansson, M. Grätzel, E. Kauppinen, A. Hagfeldt, G. Boschloo, Carbon nanotube-based hybrid hole-transporting material and selective contact for high efficiency perovskite solar cells, Energy Environ. Sci. 9 (2016) 461-466. https://doi.org/10.1039/C5EE03394B

[86] K. Aitola, K. Domanski, J-P. C-Baena, K. Sveinbjörnsson, M. Saliba, A. Abate, M. Grätzel, E. Kauppinen, E. M. J. Johansson, W. Tress, Hagfeldt, G. Boschloo, High temperature-stable perovskite solar cell based on low-cost carbon nanotube hole contact, Adv. Mater. 29 (2017) 1606398 https://doi.org/10.1002/adma.201606398

[87] T. Miletić, E. Pavoni, V. Trifiletti, A. Rizzo, A. Listorti, S. Colella, N. Armaroli, D. Bonifazi, covalently functionalized SWCNTs as tailored p-type dopants for perovskite solar cells, ACS Appl. Mater. Interfaces. 8 (2016) 27966-27973. https://doi.org/10.1021/acsami.6b08398

[88] J. Lee, M. M. Menamparambath, J. Y. Hwang, S. Hierarchically structured hole transport layers of spiro-OMeTAD nd multiwalled carbon nanotubes for perovskite solar cells, Chem Sus Chem. 8 (2015) 2358. https://doi.org/10.1002/cssc.201403462

Emerging Nanomaterials and Their Impact on Society in the 21st Century   Materials Research Forum LLC
Materials Research Foundations 135 (2023) 200-225   https://doi.org/10.21741/9781644902172-9

Chapter 9

Nanoemulsions: Preparation, Properties and Applications

Preeti Gupta*, Vinita, S.S. Das

Department of Chemistry, DDU Gorakhpur University, Gorakhpur, India

*preeti17_nov@yahoo.com

Abstract

In the last few decades, Nanoemulsions (NE) have gained significant interest among researchers because of their improved functional properties in comparison to emulsions. These include significant properties such as optical and rheological properties, coalescence, flocculation etc. It also shows excellent kinetic and thermodynamic stability. These NEs have been used in a variety of applications such as food, cosmetic and oil industries, preservatives, antimicrobial agent, in different drug delivery systems, cell culture technology etc. The chapter focuses on the various synthesis methods of Nanoemulsions. This will also provide insight about the important and useful properties. Applications in a variety of disciplines have also been discussed in detail.

Keywords

Nanoemulsion, Drug Delivery, Stability, Cosmetic Industry, Ultrasonication

Contents

Nanoemulsions: Preparation, Properties and Applications **200**

1.   Introduction..**202**

2.   Preparation of nanoemulsion **203**

2.1   High energy method 204

2.1.1 High pressure homogenizer 204

2.1.2 Ultrasonication...205

2.1.3 Microfluidzation method 205

2.2   Low energy methods..206

2.2.1 Phase inversion temperature method (PIT) 206

2.2.2 Phase inversion composition method 206

2.2.3 Spontaneous emulsification..207

2.3   Vapour condensation method: new preparation technique
       for nanoemulsions...207

3.   Properties and characterization of nanoemulsion..............................207

3.1   Stability..207

3.2   Structure-function property 209

3.3   Rheology..209

3.4   Optical property 210

3.5   Mechanical and barrier properties 211

3.6   Release property 212

4.   Applications of nanoemulsions 212

4.1   Food 213

4.2   Cosmetics..213

4.3   Cell culture technology...214

4.4   Non-toxic disinfectant 214

4.5   Drug delivery 214

4.5.1 Oral delivery 214

4.5.2 Permanent drug delivery..214

4.5.3 Pulmonary drug delivery 214

4.5.4 Intranasal drug delivery 215

4.5.5 Ocular drug delivery system..215

4.5.6 Dermal and transdermal drug delivery system 215

4.5.7 Vaccine delivery 216

4.5.8 Cancer therapy 216

4.5.9 Gene therapy 216

4.6   Nanoemulsion in agriculture 216

Conclusions..216

References 217

Emerging Nanomaterials and Their Impact on Society in the 21st Century   Materials Research Forum LLC
Materials Research Foundations 135 (2023) 200-225   https://doi.org/10.21741/9781644902172-9

1.   Introduction

Over the past decade, a considerable interest has been drawn to develop and improve new technologies to accomplish the demand of society. Among the new developing techniques, nanotechnology has achieved an important place in various industries viz; food packaging, cosmetics, drug delivery [1,2]. Presently, it is much more significant to make new nanomaterials which have enhanced surface properties, antioxidant and antimicrobial properties etc.

Nanoemulsions are one of most significant nanomaterials which has widely been used in food and cosmetics industry, in drug and drug delivery etc. [3-30]. Specifically, nanoemulsions are submicron sized colloidal system and sizes ranges from 10 to 1000 nm. It is also known as miniemulsions/submicron emulsions/ultrafine emulsions. These nanoemulsions consist of two immiscible liquids. The dispersed phase forms the spherical droplets and the dispersed medium is a liquid surrounding it [21]. Another name for dispersed phase is discontinuous or internal phase while the dispersed medium is called as continuous or external phase. An interfacial film of surfactant and co-surfactant acts as stabilizer of nanoemulsions. The droplet size of these surfactants lies between 20-600 nm [10].

According to the diameter of dispersed phase nanoemulsions can broadly be classified in two types (i) Biphasic i.e. oil in water (O/W) and water in oil (W/O), (ii) multiple NEs (W/O/W) (Figure 1) [17].

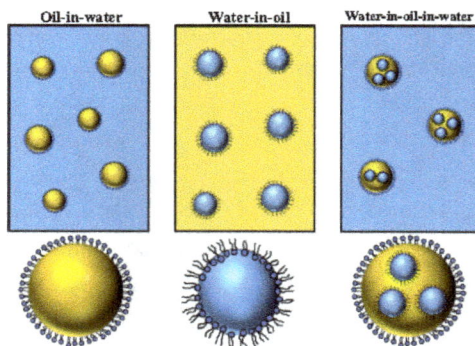

Figure 1 Nanoemulsion (i) O/W (ii) W/O (iii) W/O/W.

In O/W system, oil is dispersed phase and water is dispersed medium whereas in W/O system water is dispersed phase and oil is dispersion medium. The phase:volume ratio (ϕ), relative volumes of internal and external liquids, governs the stability of the system and also the number of droplets of nanoemulsions. The volume of internal phase (dispersed phase) is generally less than the external phase (dispersed medium). On increasing the phase: volume ratio ($\phi>40\%$), there will be phase inversion from W/O NE to an O/W system. In the case of multiple NE, containing a dispersed phase, will also incorporate other immiscible liquid phase droplets. Polar solvents i.e. water and cosolvents (example carbohydrates, alcohol, proteins and polyols) form the aqueous phase of nanoemulsions [22] and this defines its phase behavior, polarity, rheology, interfacial tension, and ionic strength. Moreover, the oil phase is made up of monoacylglycerol, di- and triacylglycerol and also free fatty acids. Different lipophilic substance, mineral oil, non-polar essential oil, waxes lipids etc are also served as oil phase. Different functional properties, stability as well as the formation of NEs are significantly affected by interfacial tension, refractive index, viscosity, phase behavior, density of the oil phase [23].

Apart from immiscible liquids, Emulsifying agents are the materials which are used to stabilize the mixture of aqueous solution and lipids [27]. This has been achieved by lessening its interfacial tension and form stable and small droplets. Nature of these emulsifiers can be nonionic, anionic, cationic and zwitterionic [28]. Suitable surfactant should inhibit coalescence by forming a coherent film around dispersed phase. In order to form droplets of desired size, the concentration should be fairly low and viscosity and zeta potential of the system should be optimized. Few surfactants are Gelucire 44/14, 50/13., Capryol 90, Imwitor 191, 742, 780 k, 928, 988, Poloxamer 124 and 188, Softigen 701, 767, Cremophor RH 40, Labrafil CS, M, 2125 CS, PEG MW, 4000, Lauroglycol 90, Labrasol, Plurol Oleique CC 497, Cremophor EL, Tween 20, 60 and 80 [26]. Nanoemulsions have large oil/water interface and small droplet size. The stability of such system can be achieved by adding cosurfactant along with surfactant. Propylene and ethylene glycol, transcutol P, PEG 400, glycerin, ethanol, propanol, isopropyl alcohol, n-butanol and carbitol have extensively been used as cosurfactants [8].

In view of huge significance in various fields, this chapter compiles the important methods of preparation of nanoemulsions, properties and application of these materials.

2.   Preparation of nanoemulsion

Because of non-equilibrated system, preparation of nanoemulsions requires ample energy or surfactant or sometime both energy and surfactant are essential. Thus, two groups of methods can be used to prepare nanoemulsions i.e. High energy method and Low energy method. The high energy method is an important traditional method. However, the low energy methods have drawn great attention because of their extensive uses and benefits (Figure 2).

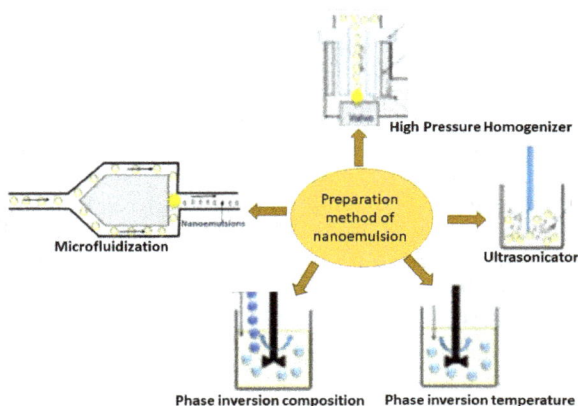

Figure 2 Different Preparation methods of Nanoemulsions

2.1 High energy method

This method requires high intense force and high mechanical forces. This method makes use of heavy mechanical devices which produce disruptive forces that break the oil droplets. It includes several techniques and can be categorized into three major groups which is based on the type of devices used.

2.1.1 High pressure homogenizer

It is the most commonly used method to homogenize nanoemulsions with particle size up to 1 nm. This high-pressure valve homogenizer comprised of a pressure valve, positive displacement pump and chambers for homogenization. The suction stroke of the pump helps to suck the mixture of oil, water and surfactant into the homogenization chamber. This mixture emulsion is subjected to high pressure (up to 300 MPa) generated in the chamber which pushed emulsion through small gap of micrometric size (up to 5-10 μm) using the homogenizer valve. In this method the emulsification takes place in two steps; (i) The size of emulsion droplets is reduced to finer and smaller size because of the turbulence, shear stress, and cavitation. This increases the surface area of dispersed phase in the homogenizing chamber. Due to complete coverage of oil and water with sufficient amounts of surfactant, droplet diameters of the emulsion (~ 100 nm) will be produced. The droplet coalescence is prevented by the high adsorption kinetics is high [30,31]

(ii) In the interaction chamber, the stabilization of emulsion droplets caused by the addition of emulsifier and its adsorption at the freshly formed interfaces.

There are numerous other reasons for the reduction of the size of the droplets i.e, an increase in concentration of emulsifier, reduction in interfacial tension and also conserve the low viscosity ratio of dispersed and continuous phase [32]. In conclusion, nanoemulsions were obtained by high-pressure homogenization under 1000–1200 bar at 25°C for 10 times. The obtained droplet size was under 200 nm. [6, 33]. High pressure homogenization is extensively used to synthesis nanoemulsions of lemon, Carvacrol, mandarin oil, lycopene nanodispersions for beverages, lentil nanoemulsions and many more [36,37]

2.1.2 Ultrasonication

Ultrasonic emulsification is a very efficient technique to reduce the size of droplets of emulsion and to produce homogenous emulsions. There are two mechanisms proposed for ultrasonic action;(a) Interfacial waves are produced acoustic field which results in the breakdown of the oil phase into the continuous phase, (b) acoustic cavitation is produced by ultrasound. This results in the subsequent breakdown of microemulsion into nanosized droplets because of the fluctuation of pressure [6].

Device made up of an ultrasonic chamber along with an ultrasonic probe. The nanoemulsion is formed in the ultrasonic chamber because cavitation, turbulence, and interfacial waves produces disruptive forces. Factors that affect nanoemulsion formation are the amplitudes of applied waves, concentration of emulsifier, viscosity ratio of dispersed and continuous phase [38].

There are various nanoemulsion based on different essential oil which have been preprared by the ultrasonication technique.

2.1.3 Microfluidzation method

The fundamental principle of mircofluidization is somewhat parallel to high pressure homogenizer. The applied pressure in this method is about 270 MPa. The basic difference between high pressure valve homogenizer and mircofluidization is that in the mircofluidization, the chamber is made up of a series of special microchannels with dimension 50–300 μm. In the series of microchannels, interaction among the microemusions takes place through the force applied which results the collisions (velocity ~400 ms^{-1}). Because of this process, fine nanoemulsions are formed whose particle size may vary with applied pressure. It also depends upon on the number of passes through the microchannels.

During the interaction of streams in the channel, strong impulsive force has been produced which is enough for disruption of droplets and as a result fine emulsions are formed. This process is repeated for more than two cycles to increase emulsification time and pressure [21]. The schematic diagram of different methods has been shown in Figure 1.

2.2   Low energy methods

According to the name, this technique employs low energy for manufacturing of nanoemulsions. In order to yield nano-sized emulsion, these methods are influenced by the physicochemical characteristics of the surfactants, emulsifiers and oil and modification in the phase transitions. Because of this phase transitions, it is also named as the condensation process of nanoemulsification [8]. Nanoemulsions are formed by the use of energy input using chemical potential of the components.

The formation of Nanoemulsion governed by the composition, temperature and solubility and it can be controlled by two methods (a) First methods is to alter the temperature without changing composition, (b) second method is to varying the compositions and interfacial properties at constant temperature [39].

It incorporates phase inversion transition (PIT), phase inversion composition (PIC), and spontaneous emulsification process [40, 41]. The intrinsic physical and chemical characteristics and interfacial behaviour of the ternary phase played crucial role in nanoemulsion production [42,43].

2.2.1   Phase inversion temperature method (PIT)

This is most widely used method for the preparation of nanoemulsions in industries. It has been observed that the PIT method is based on the variation of solubility of non-ionic surfactants i.e. polyoxyethylene at a given temperature [44]. Due to the dehydration of polyoxyethylene chains, these surfactants behave as lipophilic or hydrophilic with the rise of temperature.

Generally, in this method, water, oil along with non-ionic surfactant are mixed at room temperature. It is made up of oil-water emulsion in which oil exists in excess and surfactant monolayer exhibit a positive curvature. At intermediate temperature, the surfactant monolayer exhibit zero curvature and non-ionic surfactant has same affinity with both the oil and water layer. Hence, it forms lamellar liquid crystalline or continuous system. When the temperature is gradually increases, due to lipophilicity of surfactant, it gets solubilized in the oil layer and the oil- water emulsion will turn in to water-oil emulsion.

Because of the rise of surface tension with an increase of temperature formation of small droplets emulsion take place. However, at phase inversion temperature, there small droplets get coalesce and form macroemulsion. Liew et al. [44] suggest that to form oil-water nanoemulsions preservation of temperature below PIT and one need to cool the system. With rapid cooling and heating, emulsion of small droplet size with the alteration of conductivity, viscosity and turbidity, phase inversion temperature can be determined.

2.2.2   Phase inversion composition method

This method is also known as self nanoemulsification method. Lipophilic or hydrophilic behaviour of emulsification has been changed according to the variation in the composition of constituents. The nanoemulsion of oil in a water bath can be changed to water in oil by the addition of salt as an emulsifier. On the other hand, dilution with water, water-in-oil emulsion with high content of salt convert into oil-in water nanoemulsions. It is a highly thermodynamically stable technique with low cost. It also does not require an organic solvent and heat.

2.2.3 Spontaneous emulsification

Spontaneous emulsification process has been used to prepare nanoemulsions by mixing of oil, water and surfactant which is followed by stirring. The process of stirring allowed an emulsifier to enter inside the water layer and increased the oil-water interface and subsequently oil droplets are produced. On applying the reduced pressure, the extra oil is evaporated. There are some limitations of this method i.e. requirement of high amount of surfactant, there is low oil content and difficult removal of solvent. Hence, this is not regularly used in industries.

This method has been used to prepare vitamin E acetate nanoemulsions with droplet size <50 nm, vitamin D nanoemulsions (droplet size < 200 nm, fish oil nanoemulsions, cinnamaldehyde nanoemulsions etc.

2.3   Vapour condensation method: new preparation technique for nanoemulsions

Vapor condensation method is an advance technique to produce nanoemulsions with low-cost production. This method has been used to prepare water in oil emulsion. In this process, the nucleation of water droplet takes place at the interface of oil and air and there is spontaneous dispersion within the oil. The concentration of oil-surfactant deeply affects the spreading of oil on water. The droplet size of nanoemulsions reduce to 100 nm by this condensation method [21].

3.   Properties and characterization of nanoemulsion

Nanoemulsions exhibit number of physiochemical and physiological properties such as stability, optical, mechanical, interfacial, rheological properties along with bioactivity, antimicrobial, antifungal properties etc. Characteristic properties of droplets viz; size, appearance, composition, texture and stability of droplets affect significantly the various properties [45-52].

3.1   Stability

There are variety of physiochemical process (gravitational separation, creaming, coalescence, flocculation and phase separation) which greatly influence the stability of nanoemulsions and it disintegrate over time. In general, it is said that nanoemulsions system is thermodynamically stable.

Gravitational force is the force which is responsible for either sedimentation or creaming process. It usually depends upon the densities of surrounding liquids i.e. dispersed and continuous phase. Sedimentation is a process in which droplet move in downward direction because of droplets have higher density than the surrounding liquid. However, creaming is just an opposite process. In this, droplets move in upward direction as the density of droplet is lower than the surrounding liquid. Sedimentation process is more frequently observed in water-oil nanoemulsions and creaming is in oil-water system. This is due to lower densities of oil than water and oil have tendency to move upward and water to downward. Stokes' law has defined the creaming velocity but for nanoemulsions, the law is somewhat modified and noticed that the Brownian movement is also essential for droplets [53]. It dominates over the gravitational force when the droplet size is relatively small, $d<100nm$. Droplet aggregation is another significant process in nanoemulsions. This takes place when the droplets continuously collide with each other because of mechanical agitation/ gravitational force/ Brownian motion. There are two major types of aggregation i.e. coalescence and flocculation. In Coalescence, the droplets get combine and form a single large droplet whereas in flocculation number of small aggregates are formed. There are many interactions works between the colliding droplets. The attractive interactions are bridging, depletion, van der Waals and hydrophobic whereas repulsive interactions are electrostatic and steric interaction. Strength of both the attractive and repulsive interaction increase with the increase of the size of droplets.

Flocculation is a phenomenon which allows the emulsion drops to aggregate and the stabilizing layer at the interface is not disturbed. When the attractive force among the droplets is sufficient the Flocculation mechanism has also been considered as a destabilizer of nanoemulsion [37]. Flocculation is the process that emulsion drops aggregate, without disrupting of the stabilizing layer at the interface. This process takes place, when the is a net attractive force between nanodroplets is fair enough to prevail over the persistent aggregation and thermal agitation.

Oswald ripening is another important process in order to understand the stability of nanoemulsions. In this phenomenon, oil molecules from small droplets get diffuse to the larger droplets via the surrounding liquid in oil-water nanoemulsion and form large droplet. Laplace pressure, a difference in pressure across a curved interface, is driving force for Ostwald ripening.

The process of Oswald ripening can be inhibited by using oil with very low aqueous solubility. It is described in literature that oil phase of nanoemulsions made up of long chain triacylglycerol viz; olive, plam, fish peanut or sunflower oil is stable towards Ostwald ripening. Kabalnov and Shchukin [54] have reported that when inhibitors, water insoluble oil, mixed water-soluble oil then nanoemulsions are formed. These inhibitors create entropy of mixing effect to compensate the effect of curvature. Chen et.al. [55] have explained that the Oswald ripening phenomenon has been retarded by adding >30% bean oil (hydrophobic)in eugenol O/W nanoemulsion.

Droplet size and polydispersity index (PI) is determined by number of instruments viz; Dynamic light scattering (DLS) or quasielastic light scattering (QELS), laser diffraction (LD), X-ray diffraction peak broadening etc. In Dynamic light scattering (DLS) or quasielastic light scattering (QELS) measurement, the particles of nanoemulsions get interact with the light, which is at 90^0 observation angle (dos Santos et al., 2018). It also analyses the variations in scattering intensity by the Brownian motion of droplets. PI gives an idea about the homogeneity of dispersion. LD is a technique to evaluate the particle size which is based on volume and can be expressed as mass mean diameter and volume of equivalent spheres [8,52]

3.2   Structure-function property

Structure-function relationship deals with composition, concentration, charge, particle size distribution and physical state of droplets. Composition of droplets of nanoemulsion is dependent upon the hydrophobic (oils) and emulsifiers. The hydrophobic core with radius 'r' is coated by a thin layer with thickness 'δ_s'. Nanoemulsion have almost similar dimensions of r and δ_s ($r \approx \delta_s$). Hence, shell layer equally contributes to overall composition of droplet. The particle composition has a very good impact on density and refractive index and largely affect creaming and optical property.

Concentration of droplets represents number, mass, volume, surface area per unit volume. In general volume fraction (ϕ) is commercially used to present the concentration of droplets. It can be altered by the proportion of oil which is used to create the nanoemulsions. Another term effective volume fraction ($\phi_{effective}$) has been introduced to explain the concentration part. It is mentioned above in this section, that $r \approx \delta_s$, thus $\phi_{effective}$ will be higher than ϕ.

Another important parameter that largely effect stability, rheology and release properties of nanoemulsion is particle size distribution (PSD). It can be defined by fraction of droplets within a range of discrete size. It is more convenient to represent PSD as polydispersity index. PSD can be controlled by the variation of homogenizing conditions. Dimensions of hydrophobic core and thin layer shell collectively form an effective radius of droplet i.e. $r_{effective} = r + \delta_S$.

Thus, $r_{effective}$ depends upon the thickness of layer of encapsulation. It can be enhanced by adsorption of charged particle on the surface of the droplets.

Charge of droplets plays another vital role the explain various properties such as stability, theology, surface adhesion and ingredient interaction. The electrical charge produce on the droplet by the adsorption of charge material for example mineral ion, biopolymers, ionized emulsifiers etc. This is specially characterized by surface potential, charge density and Zeta- or ζ-potential. Among the three electrical properties, ζ-potential is the best over two other properties because it is quiet easy to measure it and is independently responsible for adsorption of any charged counter ions. The values of ζ-potential ranges from -60 mV to +60 mV and hinge on type and nature of emulsifier [56].

3.3   Rheology

Rheology deals with flow behaviour of matter as well as its deformation. This is an significant parameters that affect the functional behaviour of materials. Rheology of nanoemulsions is very important to understand because it influences the viscosity of two immiscible phases influences the droplet disruption inside the homogenizer, shelf-life of products based on nanoemulsion, manufacturing process and the sensory behaviour of nanoemulsions.

Shear stress and shear strain measurement is characteristic of rheological properties and Shear viscosity is the most important parameter which is generally measured. Droplet concentration and droplet interaction are the parameters which decides the rheology of nanoemulsion. Albert Einstein [57] has given an approximation for the viscosity of nanoemulsions

$$\eta = \eta_0(1 + 2.5\phi) \tag{1}$$

Where η = viscosity of nanoemulsion, η_0= viscosity of continuous phase, ϕ = volume fraction of dispersed phase. This equation clearly indicates that the viscosity of dilute nanoemulsion will linearly increase with an increase of concentration of droplets. However, in the case of concentrated nanoemulsions where the volume fraction is greater than 5% , the droplets start interacting and the impact of colloidal and hydrodynamic interaction has greater on the shear viscosity. Because of hydrodynamic interaction, the flow of the fluid of a droplet can be influenced by the other droplet and large amount of energy is dissipated. This is also another reason why shear viscosity is lager for concentrated system. It is also noticed that at low shear stress Brownian motion advances random movement of droplets and the shear viscosity is raised.

There are two another droplet interaction i.e. repulsive and attractive interaction which greatly influence the rheological property. During repulsive interaction, the effective volume fraction is increased because it inhibits the droplets to come closer. When the shell thickness (and droplet diameter (r) is approximately equal the impact of repulsive interaction is considerable.

Another interaction is attractive interaction which can overcome the repulsive interaction. Due to attractive interaction, flocculation takes place. The viscosity of non-flocculated nanoemulsion is less than the flocculated nanoemulsions. This is because the continuous phase is surrounded by aggregated droplets.

The flow behaviour of the emulsions was evaluated in a controlled shear rate rheometer (Anton Paar, ReolabQC, Austria) with a concentric cylinder geometry conforming to DIN 53018. [58]

3.4   Optical property

Nanoemulsions with enhanced optical property has extensively been used in many applications specially emulsion-based food and beverage products. There are food products which is slightly turbid or transparent for examples jam, soft drinks and fruit beverages while few products are opaque in nature such as sauces, mayonnaise. If any oil droplets are added in transparent product, there will be no change in opacities but change may take in the overall appearance on the addition of oil droplets due to scattering of light.

In general, the light waves are either transmitted or reflected when it encounters the nanoemuslions' surface. Scattering of light and absorption also takes place by the droplets of nanoemulsions. UV-Visible spectrophotometers used to measure the optical properties. This determines the transmittance and reflection of light waves in the visible region and the obtained spectra is used to depict the optical properties. By using trichromatic colorimeters, tristimulus color coordinates such as L*a*b* system can be illustrated. In this, L* shows the lightness and the opacity of nanoemulsion is described whereas a* red-green and b* yellow-blue represents color, thus the color intensity is described by this. In transparent or slightly turbid system, transmission measurement is considered and opacity is explained in term of turbidity. However, for opaque system, reflectance measurement will be taken in account. The optical properties of nanoemulsion broadly effected by droplet concentration, relative refractive index and absorption spectrum.

Size of the droplet (r) and wave length of light (λ) determines the scattering pattern. The nanoemulsion system is opaque when the r≈ λ. Nonetheless, the system looks clear, when r<<λ. It is experimentally proved by Wooster et.al. [59] that the nanoemulsion system is clear when the droplet size is less than 20-30nm. Refractive index explains that how light move via medium and transparency of nanoemulsions. The refractive index increases with the increase in scattering of light and hence explains the opacity/ turbidity of nanoemulsions. Largely, refractive index depends upon the structure and composition of droplets. As the thickness of emulsifier increase on interfacial layer increases, the refractive index also increases.

Abbes type refractometer is used to measure the refractive index of the nanoemulsion at 25±0.5° A drop of nanoemulsion was placed on slide and comparing it with refractive index of water (1.333). If refractive index of nanoemulsion has equal refractive index as that of water, then the nanoemulsion is considered to have transparent nature [60].

3.5 Mechanical and barrier properties

Mechanical property includes Young's modulus, tensile strength and elongation at break. It is very essential property to study in context to the packaging materials. Young's modulus explains stiffness of film, tensile strength shows the resistance of film to external mechanical forces and the stretching capacity of material has been explained by elongation at break. An extensive research work has done in the field of nanoemulsion based food packaging [61]. Table 2 shows mechanical properties of nanoemulsions. [61-65].

Barrier property for food packaging application represents the extent of water and vapor permeability. The water vapor permeability parameter is important as the food product with low-water activity absorb moisture from the surrounding. The influence of nanoemulsions on barrier properties determined by factors which includes concentration and size of droplets, chemical composition etc. Table 1 defines the different mechanical strength of ginger essential oil, gelatin and MMT.

Table 1. Mechanical strength of gelatin, gelatin/GEO, gelatin/MMT

System	Tensile strength (MPa)	Elongation at break	Young's Modulus (MPa)
Gelatin	30.3 ±3.3	48.2±4.9	61±0.4
Ginger Essential oil (GEO)	29.7±1.4	56.0±5.9	7.6±0.8
Montmorillonite (MMT)	33.8±2.4	45.7±5.4	8.2±0.5
MMT+ GEO	32.4±2.3	58.7±4.6	8.0±0.3

3.6   Release property

There are number of bioactive agents (colours, flavours, preservatives, vitamins, pharmaceuticals, nuterticals that are potential applications in nanoemulsions. These agents are hydrophilic, hydrophobic or amphiphilic in nature. In drug delivery system based on nanoemulsions, release and retention properties of the bioactive component play an important role. It depends upon the composition and microstructure. Droplet size of emulsion is inversely proportional to solubility of component and it is important for the stability of nanoemulsion.

4.   Applications of nanoemulsions

Nanoemulsions have been applied in various aspects such as Food, Cosmetics, Drug delivery system, cell culture technology, agriculture, nontoxic disinfectant cleaner etc (Figure 3).

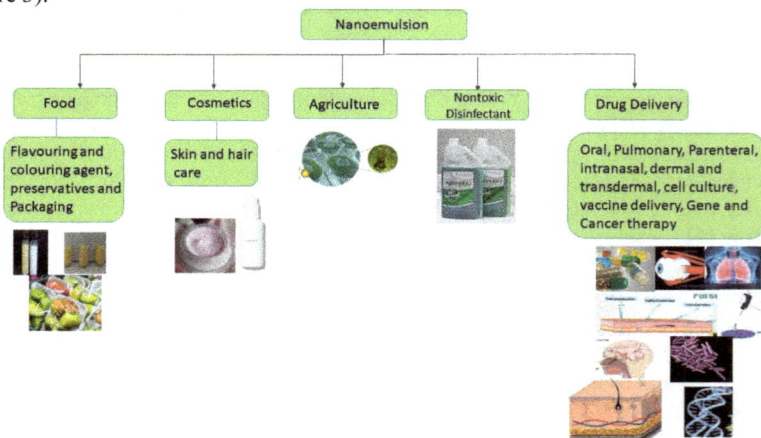

Figure 3 Applications of nanoemulsions

4.1   Food

Nanoemulsion in food application have been extensively studied. Development of functional food deals with certain limitations viz; stability, low solubility and bioavailability of bioactive compounds. These challenges have been overcome by the incorporation of bioactive compounds in the oil phase or surfactants. This enhances its stability and bioavailability and also protect it from degradation, pH, temperature and light. It prevents the bioactive compound from oxidation during storage. Nanoemulsions loaded with essential oils in the application of food have been broadly reviewed by researchers [45, 66]. Nanoemulsion food includes flavour and coloring agents, preservatives, Nutraceuticals and food packaging.

Flavor and coloring agent: Generally, ketones, aldehydes and easter functional groups containing compounds are used as flavoring and coloring agents in food. This makes these agents sensitive to photolytic and oxidative degradation. When the bioactive compounds are incorporated in nanoemulsion, the shelf life is enhanced and deleterious effects is checked [67].

Literature review shows that the oil water NEs of citral with antioxidant β-carotene, black tea extract have large chemical stability during its storage. Akin to that, the same nanoemulsions with uniquinol also shows elevated chemical stability. Oil in water nanoemulsion with β-carotene exhibit good physical stability and physiochemical properties. Wei and Gao [68] have reported the enhanced physiochemical properties of β-carotene coated with starch, caseinate and chitosan epigallocalectin-3-gallate conjugates. Preservatives: Essential oil (thyme, clove, orange, oregano) have excellent antimicrobial activity against pathogens. But its hydrophobic nature of the components limits the usage in food application. Formulation of NEs of essential oils break the limitation and they are extensively used as preservatives and food packaging [69].

Oregano oil nanoemulsion inhibit the growth of bacteria [66], Cinnamaldehyde NEs encapsulated in pectin [70], orange oil NEs [71], NEs of oregano and clove oil with droplet size 180-250 nm [72] shows excellent antimicrobial activity.

Food packaging: In the food packaging application, nanoemulsion plays and important role. Essential oil nanoemulsion film have been prepared by the dispersion of bioactive ingredients in the continuous phase. Cinnamaldehyde and clove oil nanoemulsions with pectin has been used as edible films because it has less vapour and water permeability. Lemongrass, thyme, sage oil with sodium alginate, lemon, bergamot, mandarin, carvacrol oil with chitosan collectively form NE film with antimicrobial activity.

4.2   Cosmetics

Nanoemulsions are excellent delivery systems. Thus, nowadays, it is extensively utilized in the formulation of cosmetic products. They are very good ingredients for cosmetics because of their small droplet size with high surface area, lipophilic interior and bioactive affects. These NEs are also protected from sedimentation, flocculation, creaming and coalescence. Polygycol-4 laurate and dilauroyl citrate NE (99 from anna cosmetics), potassium lauroyl wheat amino acid, palm glycerides, capryloyl glycine oil water NEs are used for skin care [73-75].

On the addition of emulsifier, Surbiton oleate, tween 80, polysorbate 80 in the silicon oil, oil in water nanoemulsion in silicon oil is formed. This has enhanced absorption property on the surface of hair [76].

4.3   Cell culture technology

Nanoemulsions has been introduced to this cell culture technology to improve the growth of the cells. Earlier, it is cumbersome to provide a medium with high bioavailability. With the use of nanoemulsions, oil soluble components quickly absorb the cell in the cell culture.

4.4   Non-toxic disinfectant

There are numbers of areas where nontoxic disinfectants cleaners are used. These disinfectants are utilized in travel, health care, military application and in food applications. Nanoemulsions also falls in category of disinfectant cleaners. They are formed by oil droplets with the size <100 nm and is suspended in water. The Nanoemulsions has surface charge which penetrate the microorganism and the electrical boundary is disturbed.

4.5   Drug delivery

4.5.1   Oral delivery

Numbers of drugs, such as hormones, steroids, antifungal, antibiotics, are lipophilic in nature. The nanoemulsions are an appropriate means for oral delivery. This delivery system protects drugs from pH and other conditions. It is reported in the literature that curcumin is drug for breast cancer when it is encapsulated in castor oil dispersed phase and Tween 80, cremphor RH-40 as surfactant. The oral bioavailability has been increased with improved anticancer activity [8, 77].

4.5.2   Permanent drug delivery

In parenteral drug delivery system, therapeutic agents are encapsulated by parenteral nanoemulsions. It shows variety of advances such as enhance bioavailability, effective and fine therapeutic index. Because of their enhanced properties, these nanoemulsions release the drugs in a controlled way, thus the dosages may be reduced. There are several drugs Feinstein, Insulin, docetaxel, carbamazepine, primaquine etc. are described in the literature and their parenteral formulation has been done. [8, 78].

4.5.3   Pulmonary drug delivery

Lungs are the important target for drug delivery. This is because it has a large surface area for the absorption of the drug and the direct drug delivery can takes place at the site of action. Pulmonary aspergillosis has been treated by amphotericin B. Nasr et al [11,79] has

Emerging Nanomaterials and Their Impact on Society in the 21st Century   Materials Research Forum LLC
Materials Research Foundations 135 (2023) 200-225   https://doi.org/10.21741/9781644902172-9

prepared the nanoemulsions of amphotericin B and reported the improved lung deposition, first pass metabolism and pulmonary retention of drugs.

4.5.4   Intranasal drug delivery

Besides parenteral and oral routes, this drug delivery system is a well-known route for this delivery purpose. Olfactory veins in the nasal mucosa appear as a good way from nasal tract to brain. Due to reduced activity of enzymes, epithelial layer with moderate permeability and high immunoactivity sites, nasal cavity is the best tract and a good effective site. Many drugs (for example: risperidone) have shown less bioavailability.

In order to enhance the bioavailability, risperidone forms oil in water nanoemulsion by mixing it into compound MCM, transcutol, tween 80 and polyethylene glycol. It is reported that due to ultra-fine droplet size (15.5-16.7 nm). There is effective and rapid dilution of drug to brain via intranasal process. Fluoxetine HCl, olanzapine, saquinavir mesylate and quetiapine drugs are mixed with capmul MCM and tween 80 to form nanoemulsion.

4.5.5   Ocular drug delivery system

It is well defined route to treat the eye disease. The drugs used for ocular treatment have low bioavailability thus, the nanoemulsions are introduced in the ocular drug delivery system. Nanoemulsions of dorzolamide HCl, and antiglaucoma drug has been synthesized [8] with prolong effect and substantial therapeutic efficacy. Oil in water nanoemulsions with timolol maleate is employed as ocular drug delivery system [80,81].

4.5.6   Dermal and transdermal drug delivery system

Delivery of drugs through skin is of great interest because it possesses a huge surface area. Nanoemulsions based drugs has been introduced in this field and drawn prominent attraction as it has small droplet size, high stability, mild nature and with simple formulation method. Aciclovir incorporated water/oil/water nanoemulsion have shown excellent penetration of drug in the skin [82]. Pentyl gallate encapsulated in self emulsifying nanoemulsions represents good anti therapeutic activity and skin penetration. Alvardo et al. [83] and Isailovic [84] have explored antiflammatory activity by using nanoemulsion. Acecbfenac, olerimacic acid, ursolic acid, an antiflammatory and anticancer drug have high capacity for penetration. Amphotericin B an antifungal drug [85] administrated via nanoemulsion to cure candidiasis and aspergillosis as an application of dermal drug delivery.

Transdermal drug delivery can be explained as nanoemulsions are absorbed through skin and reach to the target site by systematic circulation. Nanoemulsions for transdermal activity shows an important area of research. It improves the bioavailability of drug and therapeutic efficacy. Aquil et al. [86] demonstrated olmestron nanoemulsion with clove oil and reported sound pharmacokinetic properties and better bioavailability. Caffeine in water-oil nanoemulsion is appropriate for skin cancer treatment [87]. Gua et al. [88] has reported 10,11 methylenedioxycamptothecin loaded with hylauric acid nanoemulsion for keloid therapy

4.5.7 Vaccine delivery

Nanoemulsions has good use in vaccine delivery due to the presence of surfactant which act as enhancer for the penetration of drugs. It has chance to incorporate other substance to improve transdermal absorption [89]. Ahmad et al. [90] reported that intentlukin-17 nanoemulsion shows better ability against hypervirulent Mycobacterium tuberculosis.

4.5.8 Cancer therapy

Cancer therapy includes new approaches such as photodynamic and theranostic approach to malignant and non-malignant disease. One can see the cancer cell via fluorescent imaging. In recent years, number of researchers have worked in this field [91-93]. Chloro aluminium phthalocyanine loaded nanoemulsion is employed for the therapy of prostate cancer, cervical adenocarcinoma and glioblastoma. Camptothecin loaded nanoemulsion with chitosan is used to treat breast cancer [93]

4.5.9 Gene therapy

Considerable interest has been attained in gene therapy because it treats many of hereditary diseases. Cationic nanoemulsions has investigated a prominent candidate for gene delivery because it has capacity to pass over the barrier that is set by nucleic acid. Siliva et. al [94] suggested that the ultrasonic steryl amine loaded nanoemulsion is applicable for gene therapy.

4.6   Nanoemulsion in agriculture

Pesticidal nanoemulsions are an active candidate as its application on crop hinder the spreading of disease and increase the agricultural yield. In order to achieve, the maximum ability of pesticides, nanoemulsions play a significant role in delivering agrochemical bioactive compounds to the target site in the plant. Due to small droplets size and large surface area, they are able to uptake, release and accumulate the ingredients. When these ingredients are encapsulated in nanoemulsions, it shows good wettability, low surface tension, better kinetic stability and solubility. Coating of nanoemulsion on pesticides prevents the photo degradation. There are several nanoemulsion based pesticides which helps to prevent several diseases and also help them to enhance agrochemical application [95].

Conclusions

Nanoemulsions are system with two liquid phases and are immiscible in nature. Because of their unique properties they are used in variety of applications. This chapter clearly explain the method of high energy and low energy methods of the preparation of nanoemulsions in detail. Properties of nanoemulsions droplet size, surface tension, rheology, optical, release properties define its application in a variety of fields.

References

[1] J. Ashaolu, Nanoemulsions for health, food, and cosmetics: a review, Tolulope, Environ. Chem. Lett., 19 (2021),3381-3395. https://doi.org/10.1007/s10311-021-01216-9. https://doi.org/10.1007/s10311-021-01216-9

[2] P. Chowdhury, M. Gogoi, S. Borchetia et al, Nanotechnology applications and intellectual property rights in agriculture. Environ Chem Lett 15(3) (2017) 413-419. https:// doi. org/ 10. 1007/s10311- 017- 0632-4. https://doi.org/10.1007/s10311-017-0632-4

[3] A. M. Rodríguez , A. T.Ibarz , L. S.Trujillo , O. M. Belloso, Incorporation of antimicrobial nanoemulsions into complex foods: A case study in an apple juice-based beverage, LWT-Food Sci Tech., 141 (2021) 110926. https://doi.org/10.1016/j.lwt.2021.110926

[4] J. S. Franklynea , P. M. Gopinatha , A. Mukherjee, N. Chandrasekaran, Nanoemulsions: The rising star of antiviral therapeutics and nanodelivery system-current status and prospects, Current Opinion Colloid & Inter. Sci., 2021, 54:101458. https://doi.org/10.1016/j.cocis.2021.101458

[5] S. M. El-Sayed, H. S. El-Sayed, Antimicrobial nanoemulsion formulation based on thyme (Thymus vulgaris) essential oil for UF labneh preservation journal of materials research and technology,;10 (2021)1029-1041. https://doi.org/10.1016/j.jmrt.2020.12.073

[6] P. Norouzia, A. Rastegarib, F. Mottaghitalabc , M. Farokhid , P.Zarrintaje , M.R.Saeb, Nanoemulsions for intravenous drug delivery in: Nanoengineered Biomaterials for Advanced Drug Delivery, Elsevier Ltd (2020). https://doi.org/10.1016/B978-0-08-102985-5.00024-3. https://doi.org/10.1016/B978-0-08-102985-5.00024-3

[7] M.L. Zambrano-Zaragoza, D. Quintanar-Guerrero, N. Mendoza-Mun~oz, G. Leyva-Go'mez, Nanoemulsions and nanosized ingredients for food formulations in: Handbook of Food Nanotechnology Elsevier ltd. (2020). DOI: https://doi.org/10.1016/B978-0-12-815866-1.00006-6 ©. https://doi.org/10.1016/B978-0-12-815866-1.00006-6

[8] S.A. Chime, F.C. Kenechukwu and A.A. Attama, Nanoemulsions - Advances in Formulation, Characterization and Applications in Drug Delivery, in: A.D.Sezer (Ed.), Applications in Drug Delivery, Intech Open, London. 2014. https://doi.org/10.5772/58673

[9] A.C. Faria-Silva, A. M. Costa, A. Ascenso, H. M. Ribeiro, J. Marto, L. Maria Gonc¸ alves, M. Carvalheiro, S.Simo~es, Nanoemulsions for cosmetic products,

Nanocosmetics Fundamen.Appl. Toxic. Micro Nano Techn., (2020) 59-77
https://doi.org/10.1016/B978-0-12-822286-7.00004-8

[10] V.Pandey, R. Shukla, A. Garg, M.L.Kori, G.Rai, Nanoemulsions for cosmetic
products: From laborartory to market, DOI: https://doi.org/10.1016/B978-0-12-
822286-7.00004-8 2020 Elsevier Inc. https://doi.org/10.1016/B978-0-12-822286-
7.00021-8

[11] R. K. Harwansha, R. Deshmukha , Md Akhlaquer Rahman , Nanoemulsion:
Promising nanocarrier system for delivery of herbal bioactives Journal of Drug
Delivery Science and Technology 51 (2019) 224-233
https://doi.org/10.1016/j.jddst.2019.03.006

[12] Z. Karami, M. R. S. Zanjani, M. Hamidi, Nanoemulsions in CNS drug delivery:
recent developments, impacts and challenges, Drug Discovery Today, 24, (2019)
1104-1115. https://doi.org/10.1016/j.drudis.2019.03.021

[13] N. Kumar, A. Verma, A. M. Formation, Characteristics and oil industry applications
of nanoemulsions: A review, J. Petro. Sci. Eng., 206, (2021) 109042.
https://doi.org/10.1016/j.petrol.2021.109042

[14] Y.Singh, J.G.Meher, K..Raval, F. A.Khan, M. Chaurasia, N. K. Jain, M. K.
Chourasia, Nanoemulsion: Concepts, development and applications in drug delivery, J.
Controlled Release, 252 (2017) 28-49. https://doi.org/10.1016/j.jconrel.2017.03.008

[15] N. M. Aljabri, N. Shi, A.Cavazos, Nanoemulsion: An emerging technology for
oilfield applications between limitations and potentials, J. Petroleum Sci. Engg., 208
(2022) 109306. https://doi.org/10.1016/j.petrol.2021.109306

[16] M. M. Badr, M. E.I. Badawy, N. E.M. Taktak, Characterization, antimicrobial
activity, and antioxidant activity of the nanoemulsions of Lavandula spica essential oil
and its main monoterpenes, J. Drug Delivery Sci. Tech., 65 (2021) 102732.
https://doi.org/10.1016/j.jddst.2021.102732

[17] L. Almasi, M. Radi, S. Amiri, D. J. McClements, Fabrication and characterization of
antimicrobial biopolymer films containing essential oil-loaded microemulsions or
nanoemulsions, Food Hydrocolloids, 117 (2021) 106733.
https://doi.org/10.1016/j.foodhyd.2021.106733

[18] D. J. McClements, Advances in edible nanoemulsions: Digestion, bioavailability,
and potential toxicity, Progress in Lipid Res,, 81 (2021) 101081.
https://doi.org/10.1016/j.plipres.2020.101081

[19] P. Nirale, A. Paul, K. S. Yadav, Nanoemulsions for targeting the neurodegenerative
diseases: Alzheimer's, Parkinson's and Prion's, Life Sci., 245 (2020) 117394.
https://doi.org/10.1016/j.lfs.2020.117394

[20] Y. Ozogula, E. K. Boğaa, I. Akyolb , M.Durmusa , Y. Ucarc , J. M. Regensteind ,
A.R. Köşker, Antimicrobial activity of thyme essential oil nanoemulsions on spoilage
bacteria of fish and food-borne pathogens, Food Biosci., 36 (2020) 10063.
https://doi.org/10.1016/j.fbio.2020.100635

[21] J. B.Aswathanarayan, R. R. Vittal, Nanoemulsions and Their Potential Applications in Food Industry, Frontiers in sustainable Food systems, 3 (2019) 1-21. https://doi.org/10.3389/fsufs.2019.00095

[22] A. Saxena, T.Maity, A.Paliwal, S. Wadhwa, Technological aspects of nanoemulsions and their applications in the food sector, in: Nanotechnology Applications in Food, Cambridge, MA: Academic Press), (2017). 129- 152. https://doi.org/10.1016/B978-0-12-811942-6.00007-8

[23] D. J. McClements, J. Rao, Food-grade nanoemulsions: formulation, fabrication, properties, performance, biological fate, and potential toxicity., Crit. Rev. Food Sci. Nutr., 51 (2011). 285-330. https://doi.org/10.1080/10408398.2011.559558

[24] S. Kaul, N. Gulati, D. Verma, S. Mukherjee, U. Nagaich, Role of nanotechnology in cosmeceuticals: a review of recent advances. J Pharm, (2018) 3420204. https://doi.org/10.1155/2018/3420204. https://doi.org/10.1155/2018/3420204

[25] Feng J, Zhang Q, Liu Q et al (2018) Application of nanoemulsions in formulation of pesticides. In:Nanoemulsions. Academic Press, pp 379-413. https://doi.org/10.1016/B978-0-12-811838-2. 00012-6 https://doi.org/10.1016/B978-0-12-811838-2.00012-6

[26] ZAA Aziz, Mohd- H Nasir, A. Ahmad, Mohd.S. H.Setapar, W. L. Peng4, S. C.Chuo, A. Khatoon, K. Umar, A. A. Yaqoob, Mohd. N. Mohamad Ibrahim, Role of nanotechnology for design and development of cosmeceutical: application in makeup and skin care. Front Chem 7 (2019) 739. https://doi.org/10.3389/fchem.2019.00739

[27] S. Sharma, M. Hegde, V. Sadananda, B. Matthews, In vitro study to evaluate laser fluorescence device for monitoring the effect of aluminum gallium arsenide laser on noncavitated enamel lesions, J. Dent. Lasers, 11 (2017) 26. https://doi.org/10.4103/jdl.jdl_17_16

[28] S., Talegaonkar, L.M., Negi, Nanoemulsion in drug targeting. In: Devarajan, P.V., Jain, S. (Eds.), Targeted Drug Delivery: Concepts and Design. Springer International Publishing, Cham, 2015, pp. 433459. https://doi.org/10.1007/978-3-319-11355-5_14

[29] N. Bainun, N. Hashimah, S. Shatir, S. Hassan, Nanoemulsion: Formation, Characterization, Properties and Applications- A review, Adv. Mater. Res., 1113 (2015) 147-152. https://doi.org/10.4028/www.scientific.net/AMR.1113.147

[30] M. Quintanilla-Carvajal, B. Camacho-Diaz, L. Meraz-Torres, J. Chanona-Perez, L. Alamilla -Beltran, A. Jimenez-Aparicio, et al., Nanoencapsulation: A new trend in food engineering processing, Food Eng. Rev. 2 (2010) 39-50., https://doi.org/10.1007/s12393-009-9012-6

[31] S. M. Afari, E. Assadpoor, Y. H. He, and B. Bhandari, Re-coalescence of emulsion droplets during high-energy emulsification, Food Hydrocoll. 22 (2008) 1191-1202. doi: 10.1016/j.foodhyd.2007.09.006 https://doi.org/10.1016/j.foodhyd.2007.09.006

[32] S. J. Lee, and D. J. McClements, Fabrication of protein-stabilized nanoemulsions using a combined homogenization and amphiphilic solvent dissolution/evaporation

approach, Food Hydrocoll. 24 (2010) 560-569. doi: 10.1016/j.foodhyd.2010.02.002. https://doi.org/10.1016/j.foodhyd.2010.02.002

[33] R. Cappellani, M. ,Perinelli, D. Romano, P. Laura, S. Aurélie C. Marco , C. Riccardo, B. Paolo, Injectable nanoemulsions prepared by high pressure homogenization: processing, sterilization, and size evolution, Appl. Nanosci. 8 (6) (2018) 1483-14919 https://doi.org/10.1007/s13204-018-0829-2

[34] H. Schubert, and R. Engel, Product and formulation engineering of emulsions. Chem. Eng. Res. Des. 82 (2004) 1137-1143. doi: 10.1205/cerd.82.9.1137.44154Liu et al., 2019 https://doi.org/10.1205/cerd.82.9.1137.44154

[35] R. Severino, G. Ferrari, K. D. Vu, F. Donsi, S. Salmieri, and M. Lacroix, Antimicrobial effects of modified chitosan based coating containing nanoemulsion of essential oils, modified atmosphere packaging and gamma irradiation against Escherichia coli O157:H7 and Salmonella Typhimurium on green beans, Food Control, 50 (2015) 215-222. doi: 10.1016/j.foodcont.2014 https://doi.org/10.1016/j.foodcont.2014.08.029

[36] Y. N. Shariffa, T. B. Tan, U. Uthumporn, F.Abas, H. Mirhosseini, I. A. Nehdi, et al. Producing a lycopene nanodispersion: formulation development and the effects of high pressure homogenization, Food Res. Int. 101 (2017) 165-172. https://doi.org/10.1016/j.foodres.2017.09.005

a. doi: 10.1016/j.foodres.2017.09.005

[37] A. Håkansson, Can high-pressure homogenization cause thermal degradation to nutrients? J. Food Eng. 240 (2019) 133-144. https://doi.org/10.1016/j.jfoodeng.2018.07.024

a. doi: 10.1016/j.jfoodeng.2018.07.024

[38] K. Nakabayashi, F. Amemiya, T. Fuchigami, K. Machida, S. Takeda, K. Tamamitsub, et al. Highly clear and transparent nanoemulsion preparation under surfactant-free conditions using tandem acoustic emulsification. Chem. Commun. 47 (2011) 5765-5767. doi: 10.1039/c1cc10558b https://doi.org/10.1039/c1cc10558b

[39] N. Anton, and T. F. Vandamme, Nano-emulsions and microemulsions: clarifications of the critical differences, Pharm. Res. 28 (2011) 978-985. doi: 10.1007/s11095-010-0309-1 https://doi.org/10.1007/s11095-010-0309-1

[40] R. K. Harwansha, R. Deshmukha, M. A. Rahman, Nanoemulsion: Promising nanocarrier system for delivery of herbal bioactives, J. Drug Delivery Sci. Tech., 51 (2019) 224 -233. https://doi.org/10.1016/j.jddst.2019.03.006

[41] L. Wang, X. Li, G. Zhang, J. Dong, J. Eastoe, Oil-in-water nanoemulsions for pesticide formulations, J. Colloid Interface Sci. 314 (2007) 230-235. https://doi.org/10.1016/j.jcis.2007.04.079

[42] S. Lamaallam, H. Bataller, C. Dicharry, J. Lachaise, Formation and stability of miniemulsions produced by dispersion of water/oil/surfactants concentrates in a large

amount of water, Colloid. Surf. Physicochem. Eng. Asp. 270 (2005) 44-51.
https://doi.org/10.1016/j.colsurfa.2005.05.035

[43] C. Solans, J. Esquena, A.M. Forgiarini, N. Uson, D. Morales, P. Izquierdo,
Nanoemulsions: formation and properties, in Solution: Fundamentals and
Applications, Surfactants Marcel Dekker, New York, (2002) 525-554.
https://doi.org/10.1201/9780203910573.ch25

[44] J. C. L. Liew, Q. D. Nguyen, and Y. Ngothai, Effect of sodium chloride on the
formation and stability of n-dodecane nanoemulsions by the PIT method. Asia Pac. J.
Chem. Eng. 5 (2010) 570-576. doi: 10.1002/apj.445. https://doi.org/10.1002/apj.445

[45] G. R. Deen, J. Skovgaard, J. S. Pedersen, FORMATION AND PROPERTIES OF
NANOEMULSIONS, Emulsions, (2016) 193-226. http://dx.doi.org/10.1016/B978-0-
12-804306-6.00006-4. https://doi.org/10.1016/B978-0-12-804306-6.00006-4

[46] A. D. Gadhave, Nanoemulsions: Formation, Stability and Applications International
Journal for Research in Science & Advanced Technologies Issue-3, Volume-2 (2014)
038-043, 2014.

[47] G. P. Malode, S. A. Chavhan, S. A. Bartare, L. L. Malode, J. V. Manwar, R. L.
Baka, A Critical Review on Nanoemulsion: Advantages, Techniques and
Characterization Journal of Applied Pharmaceutical Sciences and Research, 4(3)
(2021). https://doi.org/10.31069/japsr.v4i3.2

[48] Development and characterization of nanoemulsion as carrier for the enhancement
of bioavailability of artemether Moksha Laxmi, Ankur Bhardwaj, Shuchi Mehta &
Abhinav Mehta, Artificial Cells, Nanomedicine, and Biotechnology, 43 (2015) 334-
344. https://doi.org/10.3109/21691401.2014.887018

[49] D. J. McClements, Fabrication, characterization and properties of food, nano
emulsions, University of Massachusetts, USA, Woodhead Publishing Limited, (2012).
https://doi.org/10.1533/9780857095657.2.293

[50] W. Jin, W. Xu, H. Liang, Y. Li, S. Liu, B. Li, NANOEMULSIONS FOR FOOD:
PROPERTIES, PRODUCTION, CHARACTERIZATION, AND APPLICATIONS,
Emulsions. (2016) http://dx.doi.org/10.1016/B978-0-12-804306-6.00001.
https://doi.org/10.1016/B978-0-12-804306-6.00001-5

[51] G. R. Deen, J. Skovgaard, J. S. Pedersen, FORMATION AND PROPERTIES OF
NANOEMULSIONS, Emulsions, (2016) http://dx.doi.org/10.1016/B978-0-12-
804306-6.00006-4. https://doi.org/10.1016/B978-0-12-804306-6.00006-4

[52] Z. Zhang and D. J. McClements, Overview of Nanoemulsion Properties: Stability,
Rheology, and Appearance, Nanoemulsions (2018) https://doi.org/10.1016/B978-0-12-
https://doi.org/10.1016/B978-0-12-811838-2.00002-3

[53] D.J. McClements, Food Emulsions: Principles, Practices, and Techniques. CRC
Press, Boca Raton, FL. (2015). https://doi.org/10.1201/b18868

[54] A.S. Kabalnov, E.D. Shchukin, Ostwald ripening theory: applications to fluorocarbon emulsion stability. Adv. Colloid Interf. Sci. 38 (1992) 69-97. https://doi.org/10.1016/0001-8686(92)80043-W

[55] J. Chen, J.R. Stokes, Rheology and tribology: two distinctive regimes of food texture sensation. Trends Food Sci. Technol. 25 (2012) 4-12. https://doi.org/10.1016/j.tifs.2011.11.006

[56] D.J. McClements, Emulsion design to improve the delivery of functional lipophilic components. Annu. Rev. Food Sci. Technol. 1 (2010) 241-269. https://doi.org/10.1146/annurev.food.080708.100722

[57] R.J. Hunter, Foundations of Colloid Science. Oxford University Press, New York, NY (2001).

[58] K. Fuentes, C. Matamala, N. Martínez, R. N. Zúñiga, and E. Troncoso, Comparative Study of Physicochemical Properties of Nanoemulsions Fabricated with Natural and Synthetic Surfactants Processes 9 (2021) 2002. https://doi.org/10.3390/pr9112002. https://doi.org/10.3390/pr9112002

[59] T.J. Wooster, D. Labbett, P. Sanguansri, H. Andrews, Impact of microemulsion inspired approaches on the formation and destabilisation mechanisms of triglyceride nanoemulsions. Soft Matter 12 (2016) 1425-1435. https://doi.org/10.1039/C5SM02303C

[60] K. GURPREET AND S. K. SINGH, Review of Nanoemulsion Formulation and Characterization Techniques, Indian Journal of Pharmaceutical Science, J Pharm Sci 80(5) (2018) 781-789. https://doi.org/10.4172/pharmaceutical-sciences.1000422

[61] P. J. P. Espitia , C. A. Fuenmayor , and C. G. Otoni, Nanoemulsions: Synthesis, Characterization, and Application in Bio-Based Active Food Packaging, Comprehensive Reviews in Food Science and Food Safety, (2018) doi: 10.1111/1541-4337.12405. https://doi.org/10.1111/1541-4337.12405

[62] N. Robledo, L. López, A. Bunger, et al. Effects of antimicrobial edible coating of thymol nanoemulsion/quinoa protein/chitosan on the safety, sensorial properties, and quality of refrigerated strawberries (Fragaria × ananassa) under commercial storage environment. Food and Bioprocess Technology, 11(8) (2008) 1566-1574. https://doi.org/10.1007/s11947-018-2124-3, https://doi.org/10.1007/s11947-018-2124-3

[63] Y. A. Oh, Y. J. Oh, A. Y. Song, J. S. Won, K. B. Song, & S. C. Min, Comparison of effectiveness of edible coatings using emulsions containing lemongrass oil of different size droplets on grape berry safety and preservation. LWT, 75 (2017) 742-750. https://doi.org/10.1016/j.lwt.2016.10.033, https://doi.org/10.1016/j.lwt.2016.10.033

[64] I. Dammak, R. A. d. Carvalho, C. S. F. Trindade, R. V. Lourenc, P. J. d. A. Sobral, Properties of active gelatin films incorporated with rutin-loaded nanoemulsions, International Journal of Biological Macromolecules 98 (2017) 39-49. https://doi.org/10.1016/j.ijbiomac.2017.01.094

[65] E. M. C. Alexandre1, R. V. Lourenço, A. M. Q. B. Bittante, I. C. F. Moraes, P. J. d. A. Sobral, Gelatin-based films reinforced with montmorillonite and activated with nanoemulsion of ginger essential oil for food packaging applications , Food Packaging and Shelf Life 10 (2016) 87-96 . https://doi.org/10.1016/j.fpsl.2016.10.004

[66] K. Bhargava, D. S. Conti, S. R. P. da Rocha, and Y. Zhang, Application of an oregano oil nanoemulsion to the control of foodborne bacteria on fresh lettuce. Food Microbiol. 47 (2015) 69-73. doi: 10.1016/j.fm.2014.11.007. https://doi.org/10.1016/j.fm.2014.11.007

[67] S. Goindi, A. Kaur, R. Kaur, A. Kalra, and P. Chauhan, Nanoemulsions: an emerging technology in the food industry. Emulsions. 3 (2016) 651-688. doi: 10.1016/B978-0-12-804306-6.00019-2. https://doi.org/10.1016/B978-0-12-804306-6.00019-2

[68] Z. Wei, and Y. Gao, Physicochemical properties of β-carotene bilayer emulsions coated by milk proteins and chitosan-EGCG conjugates. Food Hydrocoll. 52 (2016) 590-599. doi: 10.1016/j.foodhyd.2015.08.002. https://doi.org/10.1016/j.foodhyd.2015.08.002

[69] E. M. C. Alexandre, R. V. Lourenço, A. M. Q. B. Bittante, I. C. F. Moraes, and P. J. A. Sobral, Gelatine based films reinforced with montmorillonite and activated with nanoemulsion of ginger essential oil for food packaging applications. Food Pack. Shelf Life. 10 (2016) 87-96. doi: 10.1016/j.fpsl.2016.10.004. https://doi.org/10.1016/j.fpsl.2016.10.004

[70] C. G. Otoni, M. R. de Moura, F. A. Aouada, G. P. Camilloto, R. S. Cruz, M. V. Lorevice, et al. Antimicrobial and physicalmechanical properties of pectin/papaya puree/cinnamaldehyde nanoemulsion edible composite films, Food Hydrocoll. 41 (2014) 188-194. doi: 10.1016/j.foodhyd.2014.04.013 . https://doi.org/10.1016/j.foodhyd.2014.04.013

[71] S. Sugumar, S. Singh, A. Mukherjee, and N. Chandrasekaran, Nanoemulsion of orange oil with non ionic surfactant produced emulsion using ultrasonication technique: evaluating against food spoilage yeast. Appl. Nanosci. 6 (2015) 113-120. doi: 10.1007/s13204-015-0412-. https://doi.org/10.1007/s13204-015-0412-z

[72] C. G. Otoni, S. F. Pontes, E. A., Medeiros, and N. F. Soares, Edible films from methylcellulose and nanoemulsions of clovebud (Syzygium aromaticum) and oregano (Origanum vulgare) essential oils as shelf life extenders for sliced bread. J. Agric. Food Chem. 62 (2014) 5214-5219. doi: 10.1021/jf501055f. https://doi.org/10.1021/jf501055f

[73] J. Meyer, G. Polak, R. Scheuermann, Preparing PIC emulsions with a very fine particle size. Cosmet Toilet, 122 (2007) 6170.

[74] P. Heunnemann, S. Pre'vost, I. Grillo, C.M. Marino, J. Meyer, M. Gradzielski, Formation and structure of slightly anionically charged nanoemulsions obtained by the

phase inversion concentration (PIC) method. Soft Matter 7(12) (2011) 5697710. https://doi.org/10.1039/C0SM01556C. https://doi.org/10.1039/c0sm01556c

[75] Sinerga. NANOCREAMs-Fine particle-size emulsifier, ,https://www.sinerga.it/files/materie-prime/ nanocream/nanocream-flyer.pdf.; [accessed 16.10. 19].

[76] Z, Hu M, Liao Y, Chen Y, Cai L, Meng Y, Liu et al. A novel preparation method for silicone oil nanoemulsions and its application for coating hair with silicone, Int J Nanomedicine.,7 (2012) 5719-5724. https://doi.org/10.2147/IJN.S37277

[77] K.H. Ranjit, C.P. Kartik, K.P. Surendra, S. Jagadish, A.R. Mohammed, Nanoemulsions as vehicles for transdermal delivery of glycyrrhizin. Braz. J. Pharm. Sci. 47(4) (2011) 13-19. https://doi.org/10.1590/S1984-82502011000400014

[78] D. Mou, H. Chen, D. Du, C. Mao, J. Wan, H. Xu, X. Yang, Hydrogel-thickened nanoemulsion system for topical delivery of lipophilic drugs, Int. J. Pharm. 353 (2008) 270-276. https://doi.org/10.1016/j.ijpharm.2007.11.051

[79] O.A., Hussein, H.A. Salama, M. Ghorab, A.A. Mahmoud, Nanoemulsion as a po-tential ophthalmic delivery system for dorzolamide hydrochloride. AAPS PharmSci-Tech. 10(3) (2009) 808-819. https://doi.org/10.1208/s12249-009-9268-4

[80] C. Chircov and A. M. Grumezescu, Nanoemulsion preparation, characterization, and application in the field of biomedicine, Nanoarchitectonics in Biomedicine. (2019) DOI: https://doi.org/10.1016/B978-0-12-816200-2.00019-0 https://doi.org/10.1016/B978-0-12-816200-2.00019-0

[81] M. Gallarate, D. Chirio, R. Bussano, E. Peira, L. Battaglia, F. Baratta, et al., Development of O/W nanoemulsions for ophthalmic administration of timolol. Int. J. Pharm. 440 (2) (2013) 126134. https://doi.org/10.1016/j.ijpharm.2012.10.015

[82] J.C. Schwarz, V. Klang, S. Karall, D. Mahrhauser, G.P. Resch, C. Valenta, Optimisation of multiple W/O/W nanoemulsions for dermal delivery of aciclovir. Int. J. Pharm. 435 (1) (2012) 6975. https://doi.org/10.1016/j.ijpharm.2011.11.038

[83] H.L. Alvarado, G. Abrego, E.B. Souto, M.L. Gardun˜o-Ramirez, B. Clares, M. L. Garcı'a, et al., Nanoemulsions for dermal controlled release of oleanolic and ursolic acids: in vitro, ex vivo and in vivo characterization. Colloids Surf., B: Biointerfaces 130 (2015) 4047. https://doi.org/10.1016/j.colsurfb.2015.03.062

[84] T. Isailovic, S. Đorđevi, B. Markovi D. Ranđelovi N. Ceki M. Luki, et al., Biocompatible nanoemulsions for improved aceclofenac skin delivery: formulation approach using combined mixture-process experimental design. J. Pharm. Sci. 105 (1) (2016) 308323. https://doi.org/10.1002/jps.24706

[85] L. Sosa, B. Clares, H.L. Alvarado, N. Bozal, O. Domenech, A.C. Calpena, Amphotericin B releasing topical nanoemulsion for the treatment of candidiasis and aspergillosis. Nanomed. Nanotechnol. Biol. Med. 13 (7) (2017) 23032312. https://doi.org/10.1016/j.nano.2017.06.021

[86] M. Aqil, M. Kamran,A. Ahad, S.S. Imam, Development of clove oil based nanoemulsion of olmesartan for transdermal delivery: BoxBehnken design optimization and pharmacokinetic evaluation. J. Mol. Liq. 214 (2016) 238248 https://doi.org/10.1016/j.molliq.2015.12.077

[87] F. Shakeel, W. Ramadan, Transdermal delivery of anticancer drug caffeine from water-in-oil nanoemulsions. Colloids Surf., B: Biointerfaces 75 (1) (2010) 356362. https://doi.org/10.1016/j.colsurfb.2009.09.010

[88] Y. Gao, X. Cheng, Z. Wang, J. Wang, T. Gao, P. Li, et al., Transdermal delivery of 10,11-methylenedioxycamptothecin by hyaluronic acid based nanoemulsion for inhibition of keloid fibroblast. Carbohydr. Polym. 112 (2014) 376386. https://doi.org/10.1016/j.carbpol.2014.05.026

[89] I. Tamayo, C. Gamazo, J. de Souza Rebouc, J.M. Irache, Topical immunization using a nanoemulsion containing bacterial membrane antigens. J. Drug Delivery Sci. Technol. 42 (2017) 207214. https://doi.org/10.1016/j.jddst.2017.02.009

[90] G. Ahmad, R. El Sadda, G. Botchkina, I. Ojima, J. Egan, M. Amiji, Nanoemulsion formulation of a novel taxoid DHA-SBT-1214 inhibits prostate cancer stem cell-induced tumor growth. Cancer Lett. 406 (2017) 7180. https://doi.org/10.1016/j.canlet.2017.08.004

[91] Leandro, F.Z., Martins, J., Fontes, A.M., Tedesco, A.C., 2017. Evaluation of theranostic nanocarriers for near-infrared imaging and photodynamic therapy on human prostate cancer cells. Colloids Surf., B: Biointerfaces 154, 341349. https://doi.org/10.1016/j.colsurfb.2017.03.042

[92] L.P. Franchi, C.F Amantino, M.T. Melo, A.P. de Lima Montaldi, F.L. Primo, A.C. Tedesco, In vitro effects of photodynamic therapy induced by chloroaluminum phthalocyanine nanoemulsion. Photodiagn. Photodyn. Ther. 16 (2016) 100105. https://doi.org/10.1016/j.pdpdt.2016.09.003

[93] S. Natesan, A. Sugumaran, C. Ponnusamy, V. Thiagarajan, R. Palanichamy, R. Kandasamy, Chitosan stabilized camptothecin nanoemulsions: development, evaluation and biodistribution in preclinical breast cancer animal mode. Int. J. Biol. Macromol. 104 (2017) 18461852. https://doi.org/10.1016/j.ijbiomac.2017.05.127

[94] A.L. Silva, H.R. Marcelino, L.M. Verissimo, I.B. Araujo, L.F. Agnez-Lima, E.S. do Egito, Stearylamine-containing cationic nanoemulsion as a promising carrier for gene delivery. J. Nanosci. Nanotechnol. 16 (2) (2016) 13391345. https://doi.org/10.1166/jnn.2016.11671

[95] I. F. Mustafa and M. Z. Hussein, Synthesis and Technology of Nanoemulsion-Based Pesticide Formulation Nanomaterials 10 (2020) 1608. https://doi.org/10.3390/nano10081608

Materials Research Foundations 135 (2023) 226-264   https://doi.org/10.21741/9781644902172-10

Chapter 10

Effect of Nanomaterials on the Properties of Binding Materials in the Construction Industry

N.B. Singh[1], B. Middendorf[2] and Richa Tomar[1*]

[1]Department of Chemistry and Biochemistry, Sharda University, Greater Noida, India

[2]Department of Structural Materials and Construction Chemistry, University of Kassel, Kassel, Germany

* richa.tomar@sharda.ac.in

Abstract

The construction industry uses a lot of binding materials. One of the most important binders is cement and concrete. Most used cement in concrete is Portland cement (OPC). Its manufacture consumes lot of raw materials and energy and emits huge amounts of hazardous gases particularly CO_2 into the atmosphere. Attempts are being made to overcome the above problems by using blended cements, geopolymer and LC^3 cements. However, there are a number of limitations in using alternatives to OPC. In recent years researches are being carried out to use nanomaterials such as nano silica, nano titanium oxide, nano alumina, nano $CaCO_3$, nano clay, nano carbons, etc. in cement and concrete to improve the properties. The basic objectives of using nanomaterials in binding materials are to increase mechanical properties and durability. Nanomaterials, if added in appropriate amounts, accelerate the hydration, increase mechanical properties and decrease the porosity. In addition to improving physical properties, nanomaterials particularly nano TiO_2 acts as a cleaning agent. Nano embedded phase change materials perform better. Since it is a new field, little work has been done so far. In the present chapter the effect of different nanomaterials on the properties of OPC, concrete, geopolymer cement and LC^3 cement have been discussed at length.

Keywords

Nanomaterials, Portland Cement, Concrete, Geopolymer, LC3 Cement, Phase Change Materials

Contents

Effect of Nanomaterials on the Properties of Binding Materials in the Construction Industry 226

Materials Research Foundations 135 (2023) 226-264   https://doi.org/10.21741/9781644902172-10

1.   **Introduction**..**228**

2.   **Different type of binding materials in construction industry** **228**

     2.1   Portland cement and concrete...228

     2.2   Geopolymer cement and concrete 231

     2.3 LC^3 233

3.   **Nanomaterials** **234**

4.   **Nanoscience and nanotechnology in the building sector** **237**

5.   **NMS in cement and concrete**...**237**

     5.1   Effect of nano silica...239

     5.2   Effect of nano-Fe_2O_3...240

     5.3   Effect of nano-$CaCO_3$ (NC)..240

     5.4   Effect of nano Al_2O_3 (NA) 241

     5.5   Effect of nano ZnO(NZ) 241

     5.6   Effect of nano TiO_2 (NT)..242

     5.7   Effect of carbon nanotubes (CNT) 244

     5.8   Effect of graphene based NMs 245

     5.9   Effect of nano-clay 247

     5.10   Effect of nano-enhanced phase change materials.............................247

6.   **Nanomaterials and their impact on the properties of geopolymer cement concrete** **249**

     6.1   Effect of nanosilica (NS) 250

     6.2   Effect of nano TiO_2 (NT)..253

     6.3   Effect of nanoclay..253

     6.4   Effect of nanocarbons 254

     6.5   Effect of nanoalumina (NA)...254

7.   **LC^3 in presence of nano silica**...**255**

Conclusions **256**

References **257**

1. Introduction

Due to population growth, industrialization and modernization, a lot of constructions are going on all over the world for which huge amounts of binding materials are needed. Cement and concrete are the main components of binders in the construction industry. Portland cement (OPC), a main binding component in concrete, consumes lot of raw materials, energy and emits huge amount of CO_2 gas during its production. Concrete, the most important construction materials, is produced at about 27.3 billion tonnes in 2015 [1-7]. It is expected that OPC production may increase by nearly 200% by 2050 with 6,000 million tonnes [8]. Now the major challenges before us are the production of concrete with smaller amounts of raw materials clinkers with better durability [9,10]. There is a need to minimise energy consumption, CO_2 gas emissions and raw materials in the production of OPC and concrete. This is possible by using certain additives. It is expected that nanotechnology may be of great help in improving the quality of cement and concrete [11-20]. Nanotechnology may revolutionize the construction industry making OPC a high-tech material but the use of nanotechnology in the construction industry, particularly for binding materials, is not used much. Nanomaterials (NMs) improve the hydration reactions and enhance the strength and mechanical properties of cement mortar and concrete.

Out of different construction materials, binders particularly OPC, Geopolymer cement and concretes made from them are very important. In recent years one new cement LC^3 has also emerged. Strength and durability are improved considerably in the presence of NMs. In this article, properties of cement and concrete in the presence of NMs have been discussed. However, it is essential to know about binding materials such as Portland cement, Geopolymer cement, concretes and NMs used in the construction industries.

2. Different type of binding materials in construction industry

2.1 Portland cement and concrete

Portland cement (OPC), when mixed with water, hydrates giving strength. It contains tricalcium aluminate, calcium silicates, and tetra calcium alumino ferrite and calcium sulphate as gypsum. In presence of water, it has properties capable of binding together mineral fragments so as to produce a continuous compact mass of masonry. The production and consumption of OPC, the most important binding materials, is an index of development of a country. Raw materials when mixed in appropriate amounts and fired at about 1450°C in a rotary kiln forms OPC clinkers. These clinkers when mixed with 3-5% gypsum results in OPC. The whole process of OPC manufacture is shown in Fig.1a. Cement, paste, mortar and concrete have been defined in Fig.1b.

Emerging Nanomaterials and Their Impact on Society in the 21st Century   Materials Research Forum LLC
Materials Research Foundations 135 (2023) 226-264   https://doi.org/10.21741/9781644902172-10

(a)   (b)

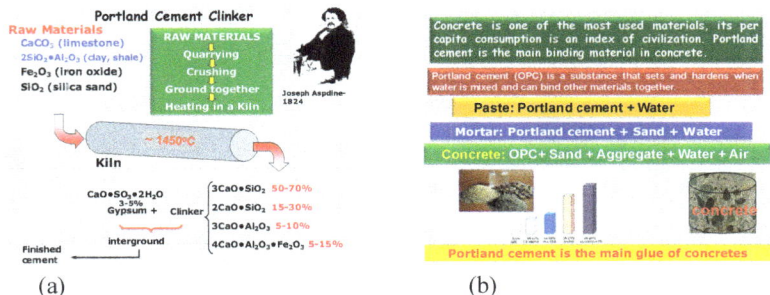

Figure1. (a) OPC manufacture process (b) OPC, paste, mortar and concrete

In cement chemistry shorthand notations are used (Fig.2).

Shorthand notation in cement chemistry

Compound	Symbol
SiO_2	S
Al_2O_3	A
Fe_2O_3	F
CaO	C
MgO	M
K_2O	K
Na_2O	N
SO_3	$\bar{S}$
H_2O	H

Chemical name	Chemical Formula	Oxide Formula	Cement Notation	Mineral Name
Tricalcium Silicate	Ca_3SiO_5	$3CaO.SiO_2$	C_3S	Alite
Dicalcium Silicate	Ca_2SiO_4	$2CaO.SiO_2$	C_2S	Belite
Trium Alumiateicalc	$Ca_3Al_2O_6$	$3CaO.Al_2O_3$	C_3A	Aluminate
Tetracalcium Alumino Ferrite	$2(Ca_2AlFeO_5)$	$4CaO.Al_2O_3.Fe_2O_3.Fe_2O_3$	C_4AF	Ferrite
Calcium Hydroxide	$Ca(OH)_2$		CH	Portlandite
Calcium sulphate dihydrate	$CaSO_4.2H_2O$	$CaO.SO_3.2H_2O$	CSH_2	Gypsum

Figure 2. Shorthand notation and nomenclature in cement chemistry

When water is added to OPC, different phases hydrate in different way giving different hydration products. The hydration reactions are given below.

$2C_3S + 7H \Rightarrow C_3S_2H_8 + 3CH$, $H = -500J/g$

$2C_2S + 5H \Rightarrow C_3S_2H_8 + 3CH$, $H = -250J/g$

C_3A hydration under different conditions

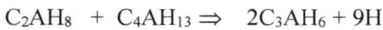

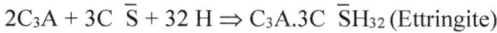

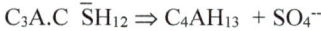

$2C_3A + 21H \Rightarrow C_2AH_8 + C_4AH_{13}$

$C_2AH_8 + C_4AH_{13} \Rightarrow 2C_3AH_6 + 9H$

$C_3A + CH + 12H \Rightarrow C_4AH_{13}$

$2C_3A + 3C\ \bar{S} + 32\ H \Rightarrow C_3A.3C\ \bar{S}H_{32}$ (Ettringite)

$2C_3A + C_3A.3C\ \bar{S}H_{32} + 4H \Rightarrow 3C_3A.C\ \bar{S}H_{12}$ (monosulphoaluminate)

$C_3A.C\ \bar{S}H_{12} \Rightarrow C_4AH_{13} + SO_4^{--}$

$C_3A + CH + 12H \Rightarrow C_4AH_{13}$

$C_4AH_{13} \Rightarrow C_3AH_6 + CH + 6H$

$C_4AF + 2CH + 14H \Rightarrow C_4(A,F)H_{13} + (A,F)H_{13}$

The speed of hydration of different phases of OPC along with major hydration products are given Fig.3.

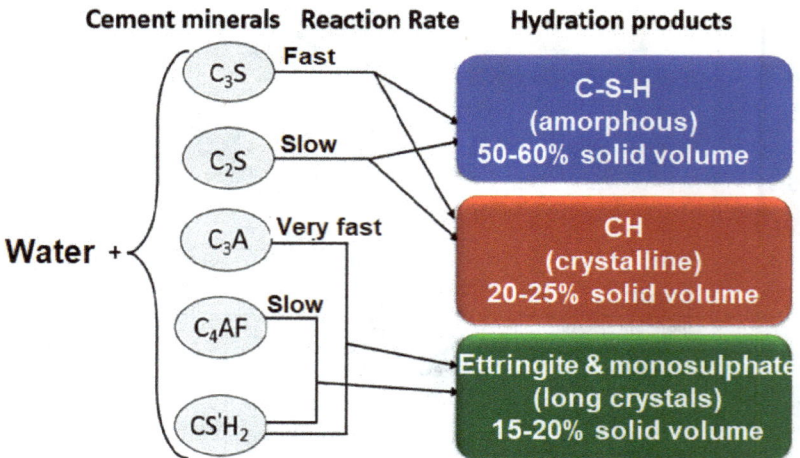

Figure 3. Heat of hydration of different phases of OPC along with hydration products

In presence of nanomaterials, above reactions are affected considerably, leading to changes in various properties responsible for strength development.

2.2   Geopolymer cement and concrete

Davidovits in 1978, for the first time, proposed the concept of geopolymers [21]. These are obtained by the reaction of industrial by-products containing aluminosilicate materials (generally slag, ,fly ash, metakaolin, rice husk ash, palm oil fuel ash, etc.) with alkali solutions (NaOH/KOH) at room or high-temperature [22,23]. The precursors as mentioned, when mixed with appropriate amount of alkali of appropriate concentration, geopolymerization occurs. During the process if sand and aggregates are mixed, geopolymer concrete is formed. The overall process of synthesis of geopolymers is given in Fig.4. [24].

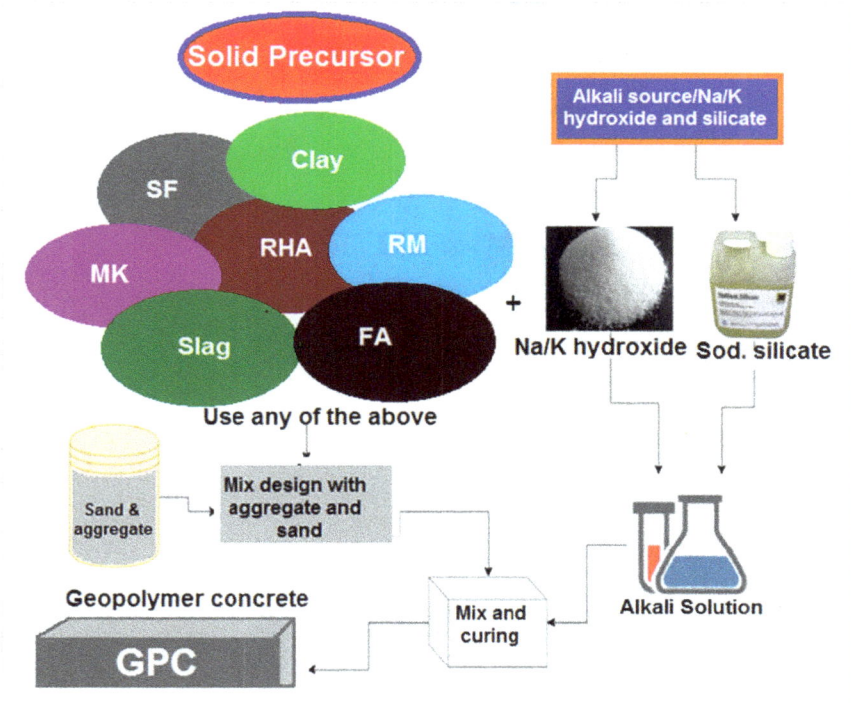

Figure 4. Geopolymer cement-concrete formation

When aluminosilicate minerals are mixed with alkalis, geopolymers are formed and considered to be an inorganic polymer. Aluminosilicate minerals on dissolutions in solutions, give free SiO_4 and AlO_4 tetrahedral units, which react by sharing oxygen atoms. Finally, the polymeric Si–O–Al–O bonds which are similar to amorphous feldspar are formed [25]. These have superior mechanical properties, low shrinkage, and lower material cost [26,27]. In the process of geopolymerization, water acts as the reaction medium and promotes the dissolution of aluminosilicate [28].

Alumino silicate reacts with alkali, forming first the geopolymer precursor, which then converted to geopolymer backbone. Reactions occurring during geopolymerization, are shown in Fig. 5 [24].

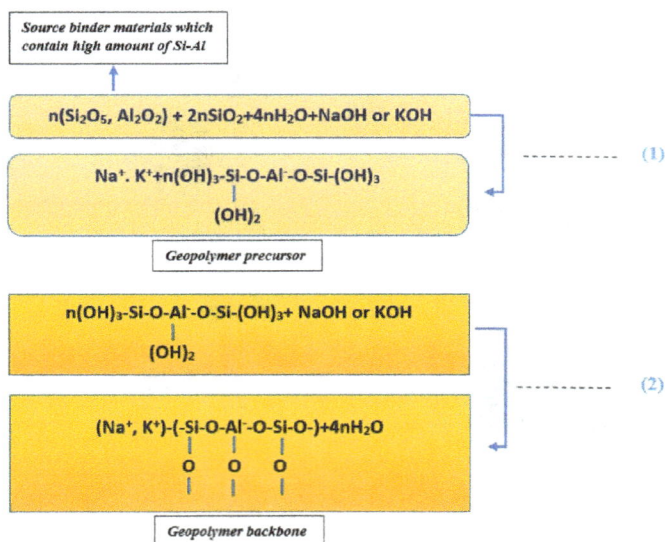

Figure 5. Geopolymerization reaction [24]

Geopolymer cement and concrete have a number of advantages as given in Fig.6 [24].

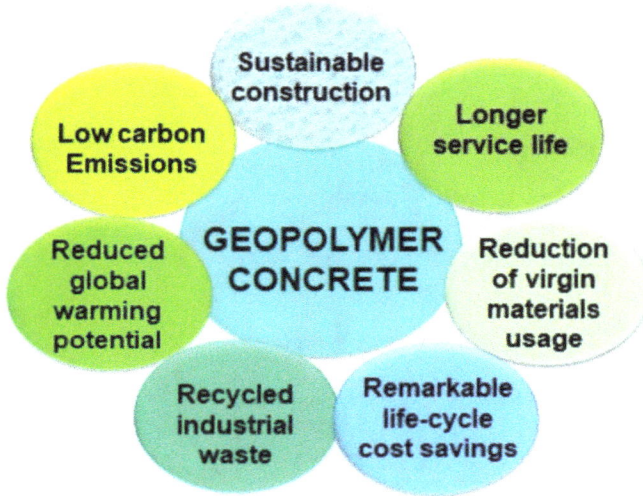

Figure 6. Importance of geopolymer concrete [24]

2.3   LC³

On blending limestone, calcined clay, clinker and gypsum in appropriate amount (Limestone-15%, Calcined clay-30%, Clinker-50%, Gypsum-5%), a new type of cement known as limestone – calcined clay-clinker-gypsum (LC³) cement is formed. It is cost effective and reduces CO_2 emissions [29]. LC³ has a lot of advantages (Fig.7). Further, due to the lower calcination energy requirements, LC³ is more economical than OPC for similar performance. Especially at locations with shortage of high-quality fly ash, low limestone quality or excess reserves of waste limestone, LC³ is economical as compared to PPC.

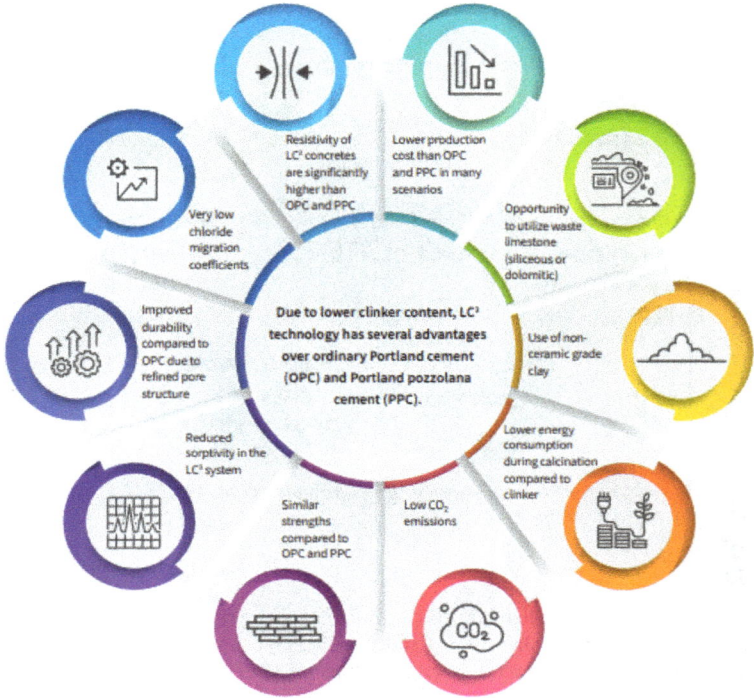

Fig. 7 Advantages of LC³

3.   Nanomaterials

Materials in the small size range (nanorange) were known long before but not in terms of nanoscience and nanotechnology [30]. The first concept came by Richard Feynman lecture of 1959 [31] and the nanotechnology was used for the first time by Norio Taniguchi [32]. Now nanoscience and nanotechnology is deal at nanoscale with dimension about 1–100 nm in at least one dimension [33].

The United States National Nanotechnology Initiative (NNI) define nanotechnology as the *"understanding and control of matter at the nanoscale, at dimensions between approximately 1 and 100 nm, where unique phenomena enable novel applications"* [34]

Emerging Nanomaterials and Their Impact on Society in the 21st Century   Materials Research Forum LLC
Materials Research Foundations 135 (2023) 226-264   https://doi.org/10.21741/9781644902172-10

NMs may be of three categories-Inorganic, organic and carbon based. (Fig.8) [35] and can be classified based on dimensionality, morphology, state and chemical composition (Fig.9) [36]. NMs in general are prepared by two approaches: top down approach - physical methods and bottom up approach - chemical and biological methods (Fig.10) [37].

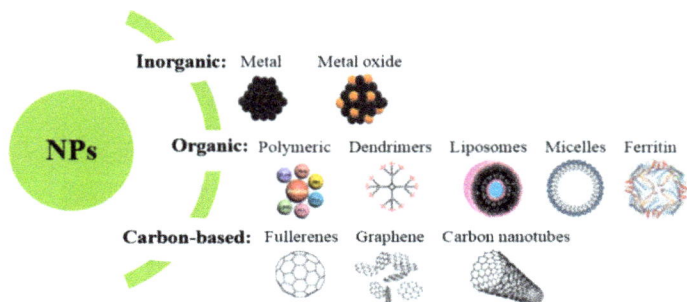

Figure 8 Type of NMs [35]

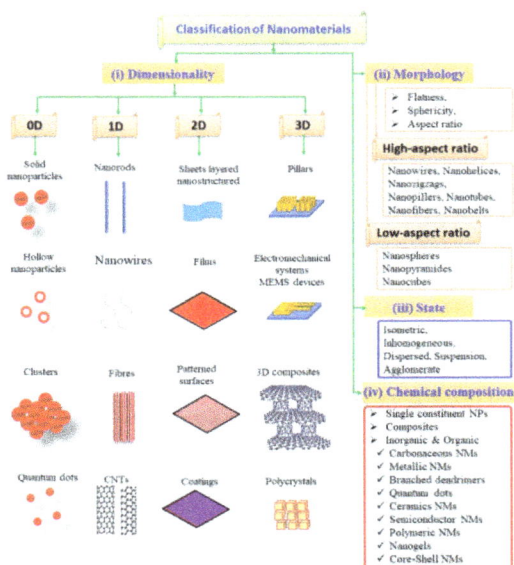

Figure 9. Classification of NMs [36]

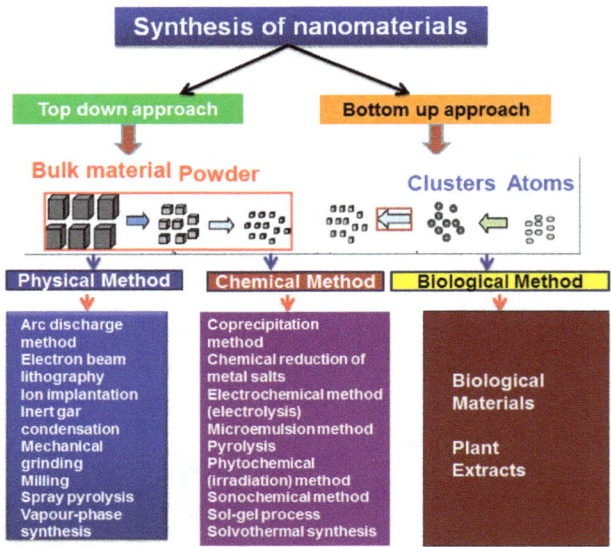

Figure 10 Synthesis of NMs [37]

NMs can be used in different sectors as shown in Fig.11 [38].

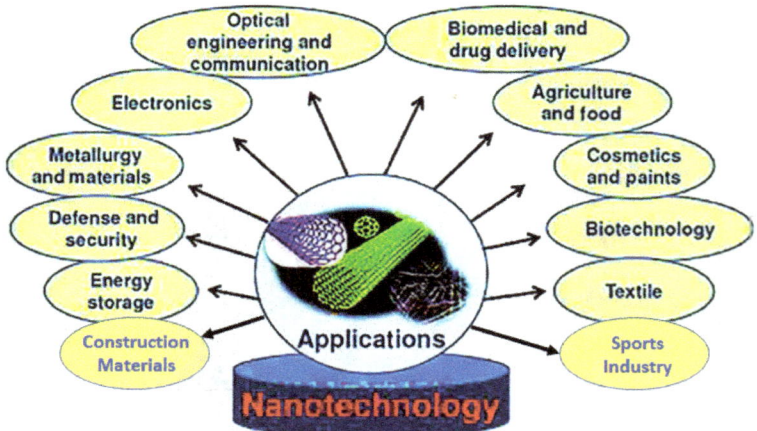

Figure 11. Applications of NMs in different areas [38]

4.   Nanoscience and nanotechnology in the building sector

The building sector, traditionally has not been adopting any innovative change but in recent years new and emerging technologies are also being adopted making a revolutionary change [39]. Out of different technologies, nanotechnology is now being adopted in the construction industry particularly in improving the properties of binding materials. In addition to improving the properties of binders, nanoparticles are used for mortar's repair, self-healing applications, air purification, self-cleaning surfaces, water or germ repellents, thermal insulation materials, sensors, etc. Other applications of nanoparticles in buildings are: nano self-colour changing thin films, energy saving and energy production [39]. Nanoparticles are also being used for crack recovery in buildings. Nano metal oxides are used as photocatalysts, self-cleaning, antibacterial, and water or germ repellents at the surface of buildings [40]. Technologies for novel smart windows are also being developed for energy saving and indoor air quality control. For solar energy generation in buildings, quantum dot technologies can be useful [41]. Some of the applications of NMs in building constructions other than strength improvements and durability are shown in Fig. 12 [42].

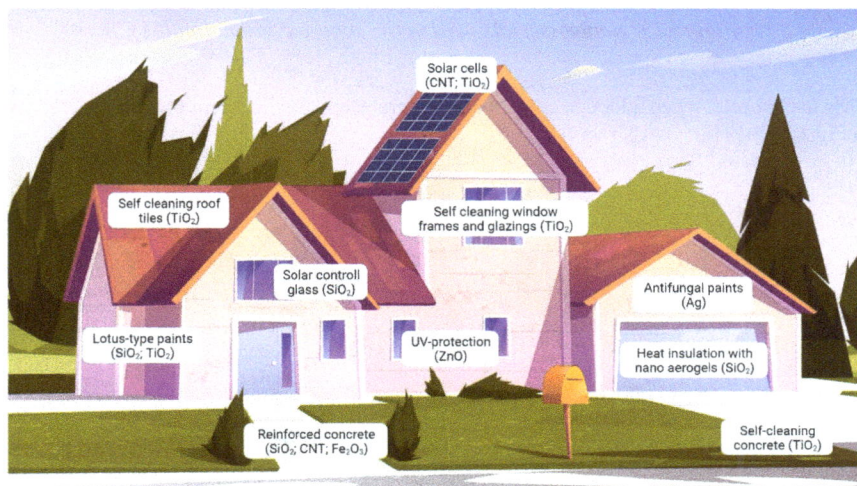

Figure 12. Nanotechnology enabled functionalities in construction [42]

5.   NMS in cement and concrete

Number of NMs such as SiO_2, Al_2O_3, TiO_2, Fe_2O_3, etc., CNTs and carbon nanofibers, nanoplatelets such as nanoclays, graphene and graphite oxide are being frequently used in cementitious systems for different applications [43,44]. Size of the constituents matters a lot in cement chemistry. Sizes of constituents in concrete incorporating NMs are given in Fig. 13 [45].

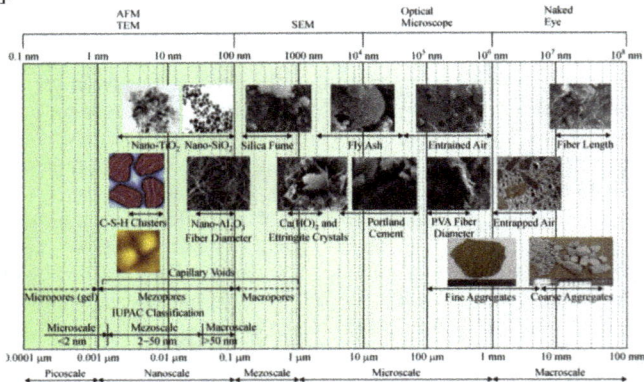

Figure 13. Nanomaterials used in the concrete formation [45]

NMs in concrete, when added in cement and concrete, modify different properties as shown in Fig.14 [46].

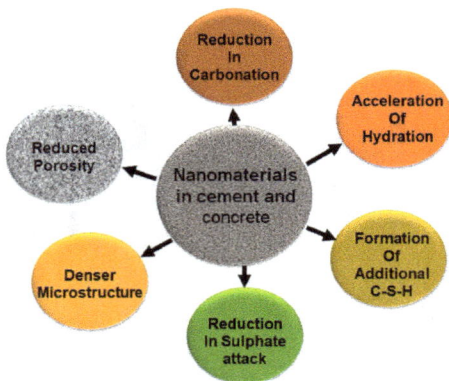

Fig.14 Effect of NMs in cement and concrete [46]

In the following section, effect of different NMs on the properties of cement and concrete has been discussed in length.

5.1   Effect of nano silica

NMs when added to cement and concrete, enhance the microstructures and bulk properties [47]. Due to high specific surface area-to-volume ratio, nano silica accelerates the pozzolanic reactions during hydration and increases mechanical strength of concrete. It changes microstructures [48]. Results have shown improvement in bulk density, better durability indices and more compactness. Furthermore, it helps in the decrease of leaching of calcium hydroxide, responsible for concrete deterioration and as a result extending the service life of the concrete [35]. Different processes such as nucleation and chemical reactions combining $Ca(OH)_2$ and forming C–S–H gel are involved during hydration of cement in presence of NS (Fig.15)[49]. These processes accelerate the hydration.

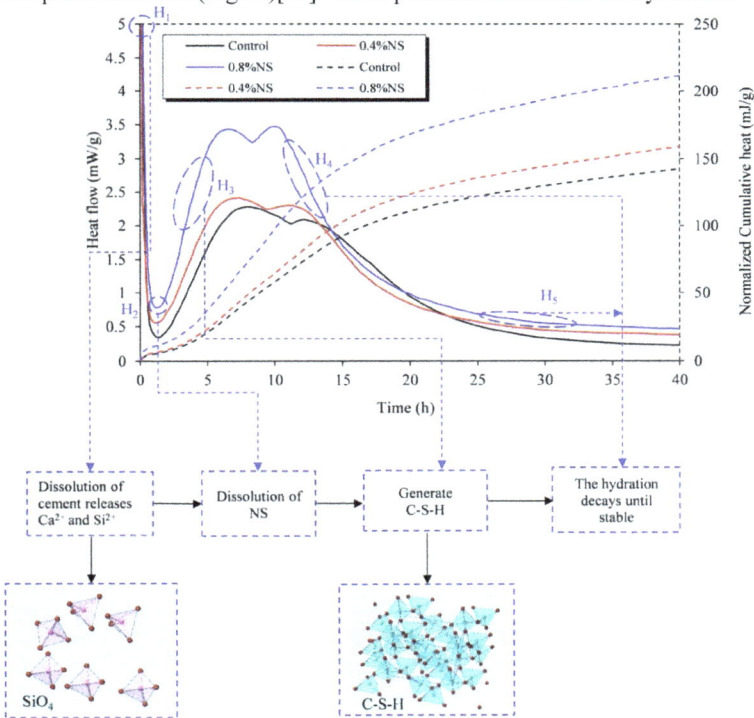

Figure 15. Hydration of cement in presence of NS [49]

NS being of smaller size, fills the pores, reduces the pore volume, optimizes the pore size distribution and increases the compactness. As a result, NS blocks the migration channels

Emerging Nanomaterials and Their Impact on Society in the 21st Century   Materials Research Forum LLC
Materials Research Foundations 135 (2023) 226-264   https://doi.org/10.21741/9781644902172-10

of water molecules and reduce cracking and damage [49]. This also optimizes the microstructure. Overall, NS improves the quality of the binding material.

In recent years, a lot of work is being done on the use of recycled concrete powder (RCP) into concrete as a substitute for cement. However, RCP has a negative effect and decrease the mechanical properties of cement-based materials. In presence of RCP, the hydration products of cement-based materials decreased but on addition of NS, properties are improved. NS improves the properties by filling the press, increasing the nucleation of hydration products and increasing pozzolanic reaction [50]. The whole processes are given in Fig.16 [50].

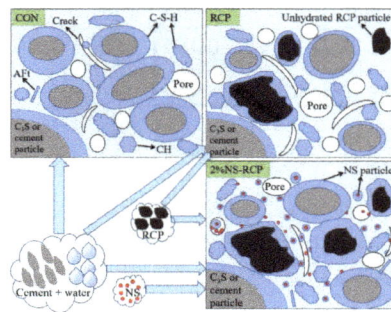

Figure 16. Illustration of the gelling process in presence of RCP and NS [50]

5.2   Effect of nano-Fe₂O₃

Nano-Fe_2O_3 (NF) when mixed with cementitious system refines the pore structure and the viscosity of the cement matrix is increased. It generates compact microstructure and reduces microcracks. It improves workability of concrete, mechanical properties and electrical resistivity. At the same time it reduces chloride penetration, water absorption, and sulfate attack [51].

5.3   Effect of nano-CaCO₃ (NC)

NC accelerates cement hydration. Due to nucleation effect, NC particles disturb the silicon-rich layer on the surface of cement particles and weaken the silicon-rich layer and NC particles interaction. NC can also act as a nucleation site outside the silicon-rich layer, resulting in acceleration of cement hydration by shortening the induction period of cement hydration [52]. High surface energy of NC promote adsorption of Ca^{2+} dissolved in the solution and accelerate ion migration. This decreases the induction period of hydration and as a result accelerate the heat of hydration. These, in turn, shift the cement hydration peak towards lower hydration time indicating acceleration of hydration (Fig17) [52].

Emerging Nanomaterials and Their Impact on Society in the 21st Century   Materials Research Forum LLC
Materials Research Foundations 135 (2023) 226-264   https://doi.org/10.21741/9781644902172-10

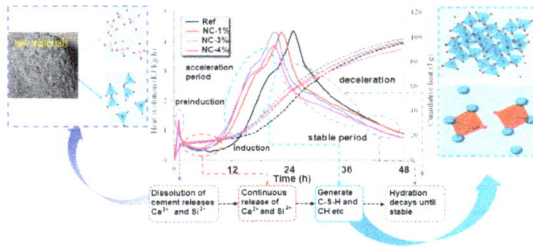

Figure 17. Hydration of cement in presence of NC [52]

It gradually reduces setting times and the workability. It increases the flexural, splitting, compressive, and bond strength and at the same time reduces chloride penetration, water sorptivity and acid attack [51].

5.4   Effect of nano Al_2O_3 (NA)

NA accelerate the hydration of silicate phases. It also accelerates the reactions of aluminate phases forming AFt and AFm. This is due to seeding effects of the NA. With increasing NA, paste's porosity is decreased. 4.0 wt% NA, the compressive strength is increased (Fig. 18) [53].

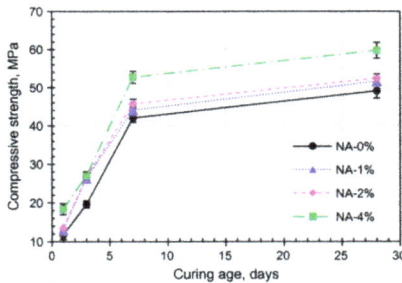

Figure 18. Compressive strengths of cement paste in presence of NA as a function of time

5.5   Effect of nano ZnO(NZ)

It has been shown that NZ retards cement hydration, decreasing early strength [54,55]. But NZ has little effect on hydration at the later age. Autogenous shrinkage in hardened pastes is reduced by NZ. Zn $(OH)_2$ layers are formed at the surface which later on disintegrates. NZ is considered to be an ideal inorganic retarder of cement based materials. The overall mechanism of OPC hydration can be shown in Fig.19 [55].

241

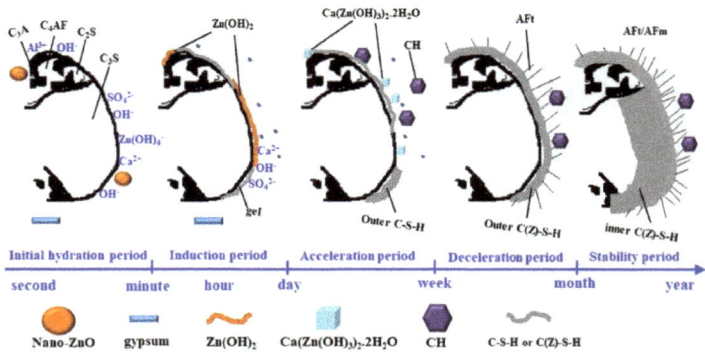

Figure 19. Mechanism of hydration of cement paste in presence of NZ[55]

5.6   Effect of nano TiO₂ (NT)

In general NT accelerates the hydration of cement, increases mechanical properties, reduces pore volume and pore size and modifies the microstructure [56]. It is also reported that NT increases the degree of hydration with rise of temperature. In presence of NT, increases of degree of hydration is due to increase of nucleation sites. Up to 2% addition of NT, both compressive and flexural strengths increased but after that it slowed down (Fig.17) [56]. The strength of mortar may be due to the uniform distribution NT and hydrated product (C-S-H). It appears that after 2% NT, clustering takes place and distribution of C-S-H gel is not uniform. SEM pictures (Fig.20) also support the increase of strength in presence of 2%NT.

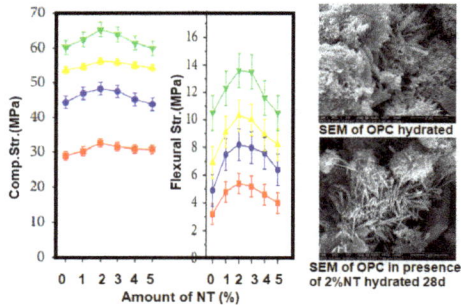

Figure 20. Strength and microstructure in presence and absence of NT

In addition to improving the physical properties of cement and concrete, NT is used as a cleaning agent in presence of sunlight. It is also used as antibacterial agent in hospitals.

TiO_2, particularly NT, is frequently used as a photocatalyst in structural materials because of its compatibility with building materials, such as cement. In construction materials, NT is usually incorporated in the concrete bulk or as a coating at the surface. As such it acts as a cleaning and antibacterial agent. Photocatalytic action of NT is given in Fig.21 [57]

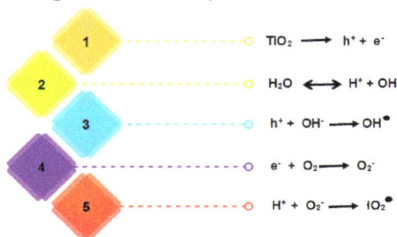

Figure 21. Photocatalytic action of NT

In the presence of light the overall cleaning action and antibacterial action can be shown by Fig.22 [58].

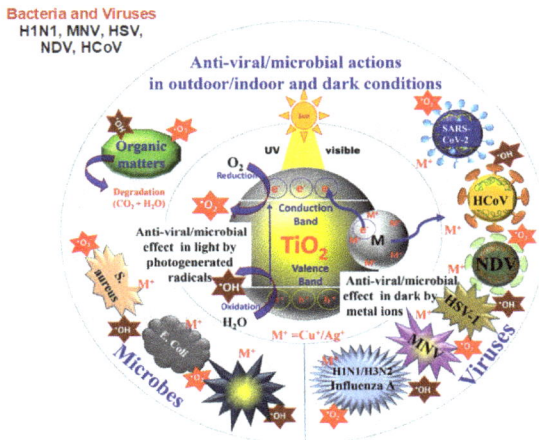

Figure 22. Photocatalytic activity of against several microbes and viruses (58)

One of the important applications of NT as cleaning agent is in Dives in Misericordia Church, Rome (Fig.23) [46,59].

Figure 23. NT as self-cleaning agent in Dives in Misericordia Church, Rome [46]

5.7   Effect of carbon nanotubes (CNT)

Basically CNTs are of two types: Single walled CNT and multiwalled CNT (Fig.24) [46] and have great potential in improving the properties of cement and concrete [60]. The basic problems with CNTs are their dispersion and interaction at the surfaces of CNTs. However, these problems can be minimized in cement and concretes by adopting suitable methods.

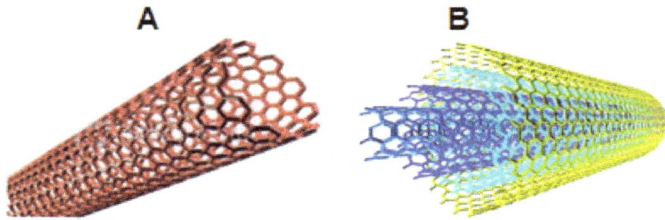

Figure 24. A-Single CNT, B-Multiwalled CNT [46]

Surfaces of CNTs have been modified by physical and chemical functionalization methods whereas the dispersion has been made by surfactant-assisted ultrasonication procedure.

Hydration of cement is accelerated in presence of CNT by providing additional nucleation sites and increase C-S-H formation. As a result of functionalization, the structure of hydration products also change through interfacial chemical bond formation. CNT also reduces porosity and may work as bridge during crack formation (Fig.25) [45]. Overall CNTs have positive effect on properties of cement and concrete.

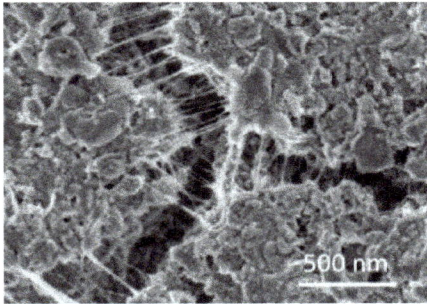

Figure 25. Bridge formation during crack in concrete in presence of CNT [46]

In the presence of CNTs in cementitious composites, electrical conductivities are improved and can detect the damage and can be used as self-sensing sensors, through the change in their electrical resistivity [60].

5.8   Effect of graphene based NMs

Basic structural unit for graphitic materials is graphene. It has different forms having extraordinary physical and chemical properties (Fig.26) [61].

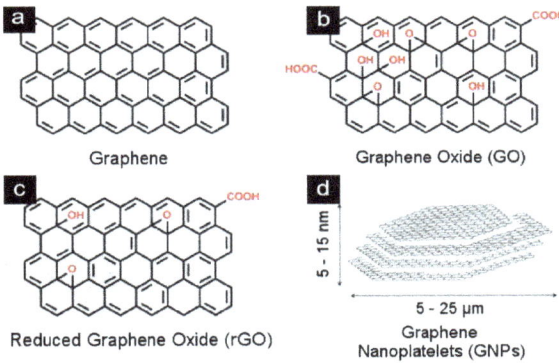

Figure 26. Modified forms of graphene [61]

Out of all, graphene oxide (GO) and reduced graphene oxide (rGO) are very important. The functional groups such as COOH, OH, and SO_3H on the surface of GO react

Materials Research Foundations 135 (2023) 226-264   https://doi.org/10.21741/9781644902172-10

preferentially with different phases of cement and form hydration products with flower like structures (Fig27) [62].

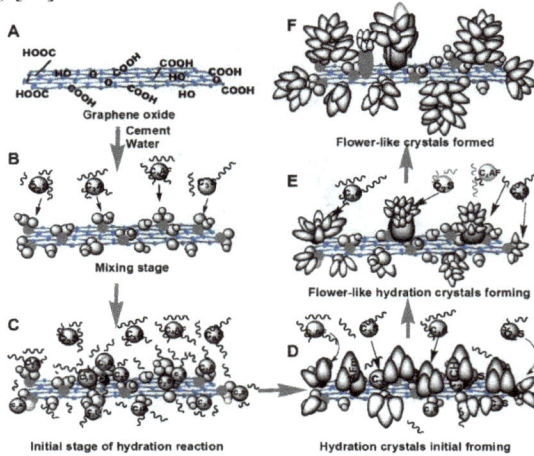

Figure 27. Role of GO on cement hydration crystals [62]

When cement is allowed to hydrate for different period of time in presence of 0.03% GO, the morphologies are changed (Fig.28) [62]

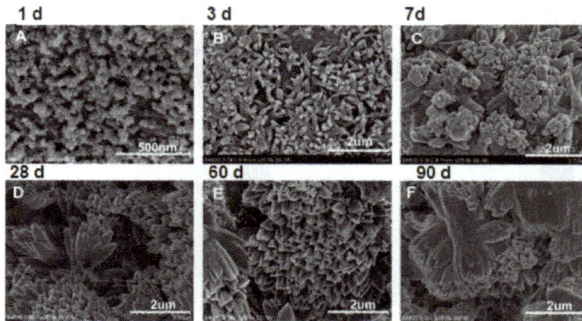

Figure 28. SEM images of cement hydration crystals with 0.03% GO with different hydration times [62]

5.9   Effect of nano-clay

Nano-clay has high pozzolanic activity and when added to cement and mortars, decreases setting time and workability. When added in appropriate amounts, increase hydration, mechanical properties and durability. Nano-clay has a filling effect, nucleation effect, pozzolanic effect, bridging effect and barrier effect. In presence of nano-clay, size of CH crystals is decreased and amounts of C-S-H and C-A-S-H are increased. In general, when appropriate amount of nano-clay is added to cement, hydration properties are increased [63].

It is reported that nano-metakaolin (NMK) accelerates the hydration of cement more than nano-SiO_2 and nano-$CaCO_3$. NMK due to seeding effect, promotes nucleation, reacts with available CH, and also forms secondary hydration products with needle shaped structures. It densifies the structure (Fig. 29) [64].

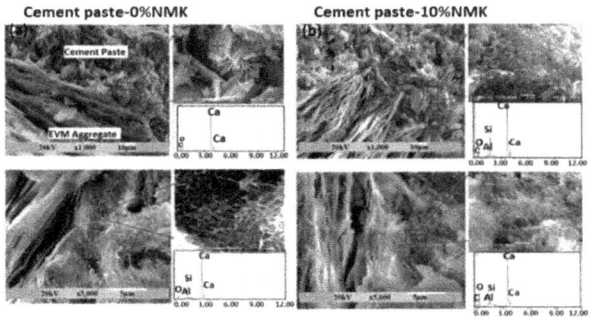

Figure 29. EDAX and SEM pictures of (a) 0% NMK and (b) 10% NMK [64]

5.10   Effect of nano-enhanced phase change materials

Materials holding and releasing thermal energy with phase change irreversibly in different temperature ranges are called phase change materials (PCMs). Therefore, PCM can be an important tool for thermal energy storage. PCMs can act as link or bridge between the energy supply and demand. There are variety of PCMs (Fig.30) [65]. During the daytime, building containing PCMs absorb heat due to melting and when there is a decrease in temperature in the night, PCMs release heat. Thus, PCMs during the daytime act as coolant and in the night as heater. The whole process can be shown as in Fig.31.

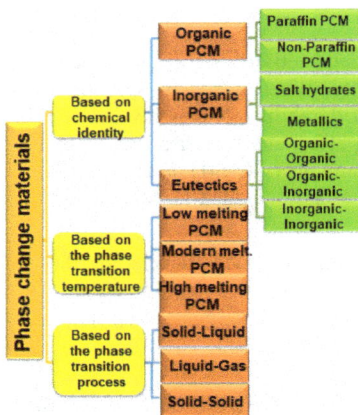

Figure 30. Phase change materials [65]

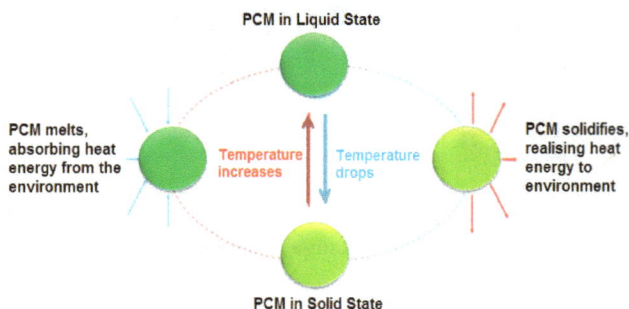

Figure 31. Functioning of PCM

Thermal performance of a PCM depends on melting energy storage density, temperature, and thermal conductivity. Many conventional PCMs have disadvantage of low thermal conductivity and hence do not fulfil the purpose. To overcome some of the problems, they are embedded with variety of nanoparticles (carbons, metals, metal oxides, metal carbides, metal nitrides, etc.) nanofibers, nanotubes and nanosheets. As a result of this the thermal conductivities of PCM are enhanced and due to this, PCM exhibits a number of other enhanced properties (Fig.32) [65].

Emerging Nanomaterials and Their Impact on Society in the 21st Century   Materials Research Forum LLC
Materials Research Foundations 135 (2023) 226-264   https://doi.org/10.21741/9781644902172-10

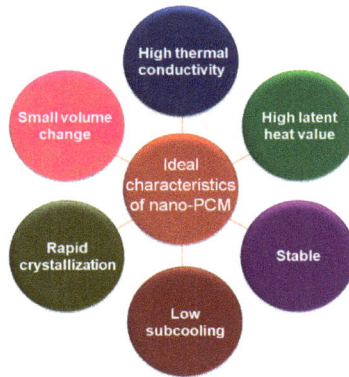

Figure 32. Characteristics of nano-PCM [65]

Nano-PCM can be used in different parts of the building (Fig.33) [66].

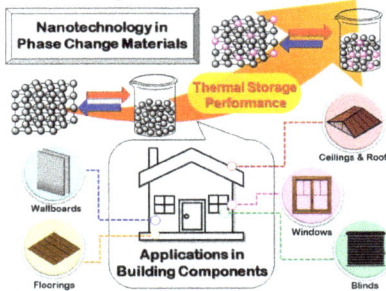

Figure 33. Use of nano-PCM in different parts of building [66]

6.   Nanomaterials and their impact on the properties of geopolymer cement concrete

Geopolymers are a new type of binders and can replace OPC composites. The production of geopolymer composites has a lower carbon footprint and utilises less energy as compared to OPC manufacture. In recent years, attempts have been made to add various kinds of nanoparticles (NPs) to geopolymer composites in order to improve the properties and performance [67]. Switching off one material from the macro scale to the nanoscale, resulted in significant changes in their chemical reactivity, mechanical properties, surface energy, shape, conductivity and optical properties. Numerous studies have been done using geopolymer concrete and NMs to examine the new composite's performance in terms of chemical durability, mechanical and structural performance, fresh and microstructural properties, physio-mechanical properties, fire resistance and permeability. The filler effect, which occurs when nanoparticles are added to cement composites, is the result of free water becoming immobile because the NMs fill the pores and spaces among cement particles. Additionally, the NMs contribute to the pozzolanic reaction's formation of fresh calcium silicate hydrate (C–S–H) gel, which improves the interfacial transition zone between the aggregates and binder pastes and, as a result, improves the bond strength properties of the mixture. Geopolymer composites have low compressive strength when cured in an ambient condition. In order to overcome this problem, NMs are mixed within geopolymer concrete composites. This is achieved by atomically modifying the geopolymer concrete's microstructure, which significantly improves the material's fresh and hardened state features as well as its structural properties. Numerous NMs such as nano-silica slurry (NS), nano-zinc oxide (NZ), waste glass nano powder (WGNP), colloidal nano-silica (CNS), nano-calcium carbonate (NC), nano-clay platelets (NCP), multi-wall carbon nanotubes (MWCNTs), nano-titanium oxide (NT), carbon nanotubes (CNTs), etc. have been used to enhance different properties of geopolymer composites. However, NS is the most used NM in geopolymer composites (Fig. 34) [67].

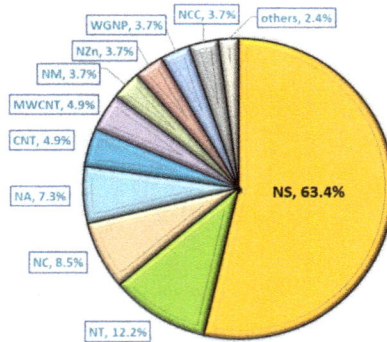

Figure 34. Uses of NPs in geopolymer composites [67].

In the presence of different NMs, polymerization, mechanical properties and resistance against aggressive atmosphere are increased and microstructures become dens.

6.1   Effect of nanosilica (NS)

With a primary diameter of 150 nm and a high specific surface area (15 to 25 m^2/g), NS generated by vaporising silica at temperatures between 1500 and 2000°C by reducing quartz (SiO_2) in an electric arc furnace has a higher fineness and when added to high calcium fly ash based geopolymer paste under ambient curing conditions, a noticeable

Emerging Nanomaterials and Their Impact on Society in the 21st Century   Materials Research Forum LLC
Materials Research Foundations 135 (2023) 226-264   https://doi.org/10.21741/9781644902172-10

increase in compressive and flexural strength is seen. The use of NS raises the water demand for concrete, mortar, and geopolymer paste. On the other hand, NS's spherical shape also contributes to the paste's improved workability. An increase in NS content from 1 to 3wt % in slag-fly ash-based geopolymer mixes increased water demand. The slump value decreased by a range of 18 to 46%. On the contrary, according to Rodrı́guez et al. [68], the hydroxide solution reduced the mixture's viscosity, making it easier to work with in presence of NS. Adak et al. [69] stated that the inclusion of colloidal nano-silica improved the workability of fly ash-based geopolymer mortar. NS particles having high surface area speeds up the geopolymerization process. A positive effect of NS on compressive strength of slag geopolymer concrete exposed to freezing-thawing cycles have been studied [70].

Due to the improved chemical reaction between the extra calcium present in high calcium fly ash and NS, additional geopolymeric gels, such as C-S-H or C-A-S-H and N-A-S-H, are formed, increasing the strength. Up to certain percentage of NS, the properties of geopolymers are improved considerably, whereas above optimum value, NS has some detrimental effect. Table 1 lists different geopolymers in presence of different amounts of NS.

Table 1 Different nanoparticles used to modify properties of geopolymer composites [67].

Nanomaterial	Properties of Nanoparticle			% Age of Nanoparticles	Binder	Geopolymer Composite
	Particle size/ Surface Area	Purity (%)	Density			
Nano silica	4-16 nm		2.37	0.0 and 6	Fly ash	Geopolymer composite
Nano silica	200±25 m^2/g/ 12nm	99.8		0, 0.5, 1, 1.5	Fly ash	Geopolymer composite
Nano aumina	20 nm	99.7		0,1,2,3	Ground granulated blast furnace slag	Geopolymer composite
Nano-TiO_2	20-10 μm			0,1,2,3,4,5	Fly ash	Geopolymer composite

Waste glass nano pozzolon	80 nm	69.14		0,5,10,15,20	Fly ash-Ground granulated blast furnace slag	Geopolymer mortar
Nano aumina	50 nm	99.9	0.4-0.5 g/cm^3	0,1,2,3	Fly ash	Geopolymer paste
Nano zinc oxide	30-50 nm	99.8		0, 2.5,5,10	Fly ash	Geopolymer paste

Fig. 35 shows SEM pictures of geopolymer pastes. The control paste had a less dense matrix with more fly ash particles embedded in a continuous matrix that had not fully or partially reacted (Fig. 35a). Less fly ash particles were seen, and the matrix appeared denser for the 1–2 percent NS (Fig. 35b and c) than for the control paste. High magnification (3,000) images with 1 % NS (Fig 35d) revealed denser matrices than the control paste. The addition of NS to high calcium fly ash geopolymers led the co-existence of CSH or CASH gels, which improved the compressive strength of the geopolymer as a whole. A substantial number of NS particles were noticed after the addition of 3 % NS (Fig. 35e), and the matrix seemed less dense than in the other combinations, indicating that this amount of NS has detrimental effect.

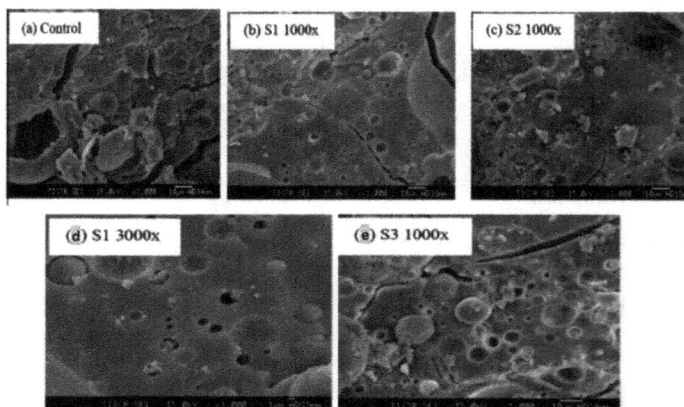

Figure 35 SEM images of geopolymer paste after 28 days [71].

Emerging Nanomaterials and Their Impact on Society in the 21st Century   Materials Research Forum LLC
Materials Research Foundations 135 (2023) 226-264   https://doi.org/10.21741/9781644902172-10

6.2   Effect of nano TiO$_2$ (NT)

One of the nanoparticles that is frequently employed in cement-based materials for various applications is NT. According to the studies, the addition of 5% NT to fluidized bed fly ash-based geopolymer paste causes the microstructure to become denser. The scientists also stated that the nucleation and nanopore filling effects of NT have positive effects on the properties. In fly ash and metakaolin-based geopolymers with NT, excellent mechanical bonding has been reported [72]. Two curing times, such as one week and four weeks, were used to test the specimens for the analysis of the compressive, split tensile, and flexural strengths. After one-week curing cycle, 2.0 % NT modified the compressive strength of geopolymer cement by a maximum of 37.04 % in comparison to the geopolymer cement without any nanoparticles.

Similar to this, following a one-week curing cycle, the 2.0 percent NT increased geopolymer cement's split tensile strength by 25.3% and after one week of curing, the flexural strength of the geopolymer cement increased by 37.60 %. With the addition of NT up to a 2.0 percent loading in geopolymer cement, the overall mechanical behaviour was expressly improved, and a further rise in NT content had a negative impact on the mechanical characteristics of geopolymer cement [73].

6.3   Effect of nanoclay

Similar to other nanomaterials, adding nanoclay reduces the flowability of geopolymer composites. In one sense, nanoclay makes the mixture more viscous and makes it harder to work with, but it also helps to prevent the segregation of the concrete. However, compared to the addition of NS or nano-alumina, nanoclay inclusion exhibits greater workability. However, to get more workability, researchers advise using superplasticizers. With the addition of an ideal concentration of nanoclay, the rate of geopolymerization increases. Higher concentrations, however, cause some of the nanoclay particles to remain dormant and cause the geopolymerization to stagnate. Shahrajabian and Behfarnia [70] published their experimental findings in examining the impact of nanoclay on slag-based geopolymer freeze and thaw resistance. They found that the specimen with the nanoclay admixture included more geopolymeric gel than usual. When added to the mixture, nanoclay particles produced micro-dispersion of particles in the geopolymer paste. Higher nanoclay content in the fly ash-based geopolymer did, however, result in some unreacted particles being present in the matrix, which led to excessive self-desiccation and fissures in the matrix, which became the source of the weakening. According to the investigations, nanoclay addition significantly affects the strength characteristics of geopolymer composites. The impact of adding nanoclay on the mechanical and durability characteristics of geopolymer composites, reinforced with flax fabric has also been examined. On adding nanoclay, adhesive force between the geopolymer matrix and the flax fibres is improved. The bond strength of the geopolymer composites was strengthened by this improved adhesion, which also raised their flexural strength. The ideal concentration of nanoclay reported is 2% by weight. Additionally, the geopolymer paste was densified by nanoclay by producing more geopolymer products. In order to improve the characteristics of geopolymer composites by speeding up the pace of geopolymerization and pore-filling action, 2wt % of nanoclay is the ideal content.

6.4   Effect of nanocarbons

The addition of carbon nanotubes (CNTs) is anticipated to raise the water requirement of the geopolymer mix and may reduce the paste's workability. Multiwalled carbon nanotubes (MWCNTs) were added to geopolymer paste in amounts of 0%, 0.50%, and 10% by weight to examine the microstructure and mechanical characteristics of the carbon nanotube-admixed metakaolin-based geopolymer. MWCNTs have an inner diameter in the range of 10–20 nm, a length of 10 m, and a specific surface area in the range of 250–280 m^2/g. Comparing 0.5wt% MWCNT to the reference mix resulted in a maximum (32%) increase in compressive strength, which is due to the fact that MWCNTs inhibit the development of microcracks. With 1 weight percent MWCNT, however, the strength is lower (by 9%) than it is with 0.5 weight percent MWCNT in the mixture. As explained earlier by Khater and Gawaad, a higher concentration of MWCNTs caused the agglomeration of MWCNT particles [74]. In the instance of geopolymer mortar, they reported that the specimens' reduced water absorption with increased curing age. With increasing amount of MWCNTs, the flow of fresh high calcium fly ash based geopolymer (HCFG) decreased. At 0.6 % MWCNTs, the flow of the control mix dropped to its lowest level of 32 percent. This is due to the carbon nanotube's tubular structure, which resembles short threads and leads to particle interlocking when used in high quantities. Additionally, the MWCNTs' high specific surface area decreased the amount of free water in the alkali activator solution and slowed the flow.

Electrical resistivity of HCFG with MWCNTs under compression load changed. As the MWCNT concentration or testing frequency increased, the resistivity decreased with increasing load levels. Electrical conductivity is enhanced and an electron flow path is created by the distribution of carbon nanotube in HCFG matrices [75].

6.5   Effect of nanoalumina (NA)

When NA is added to a geopolymer mixture, the geopolymerization process is improved and as a result, mechanical performance improves. In presence of NA, the thermal stability is changed [76]. The inclusion of NA improves the composite's durability by reducing water absorption and chloride ion penetration. A more crystalline form of Al_2O_3 can be added to the activator solution to lower the Si/Al ratio and increase the level of geopolymeric gel rearrangement, which in turn encourages geopolymerization. Alkali-activated slag concrete shows highest compressive strength at 2wt % NA compared to 1 and 3wt % NA [77].

Fig. 36 displays SEM images of the geopolymer paste. The control paste had a less dense matrix with more fly ash particles embedded in a continuous matrix that had not fully or partially reacted (Fig. 36a). Less fly ash particles were seen and the matrix appeared denser in the presence of 1–2 percent NA (Fig. 36b and c).

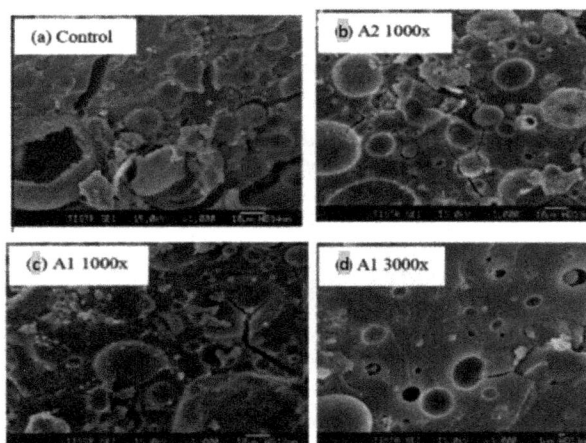

Figure 36. SEM images of geopolymer paste after 28 days [70].

High magnification (3,000) images of 1 percent NA (Figs. 40d) revealed denser matrices than the control paste. The addition of NA to high calcium fly ash geopolymers caused co-existence of CSH or CASH gels, which improved the compressive strength of the geopolymer as a whole. A substantial number of NA were noticed after the addition of 3 % NA, and the matrix seemed less dense than in the other combinations, indicating that the amount of nano-particles added at this level of 3 percent was excessive.

7.   LC^3 in presence of nano silica

In recent years, a group of researchers was working on the utility of LC^3 as it is economical and environment friendly. In LC^3, calcined clay is present, which is a pozzolanic material and forms C-(A)-S-H and AFm phases similar to that of OPC during hydration. In the presence of limestone, C$_3$A reacts to give monocarboaluminate (Mc) and hemicarboaluminate (Hc). In spite of a number of advantages of LC^3 in terms of durability, cost saving, energy savings and emission reduction, early strength of 1 day is lower. This affects construction schedule and increases labour cost. Due to pozzolanic nature and ability to provide large nucleation sites, NS improves the hydration properties and early compressive strength of concrete [78]. In the presence of NS, heat evolution is increased during 1-3 h of hydration [79]. Total heat evolved also increased and the values are higher in presence of 2% NS.

Compressive strengths of OPC and LC^3 mortars in the presence of NS at 1, 3, 7 and 28 days are shown in Fig.37. In presence of 2% NS, the compressive strengths increased [79].

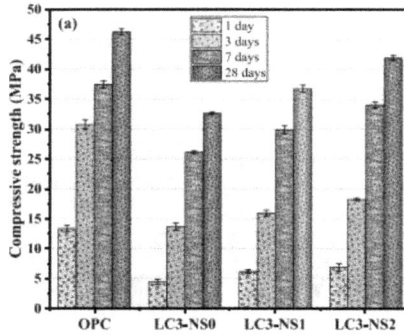

Figure 37. Compressive strength of LC³ in presence of NS

SEM pictures of LC³ in absence and presence of 2% NS after 1 d of hydration, are shown in Fig.38. All the pictures show needle-like ettringite crystals and honeycomb-like C-(A)-S-H but in the absence of NS, there are many pores and loose structure. However, in presence of 2% NS, the structure becomes denser and more compact due to pozzolanic reactions and filling of the pores [79].

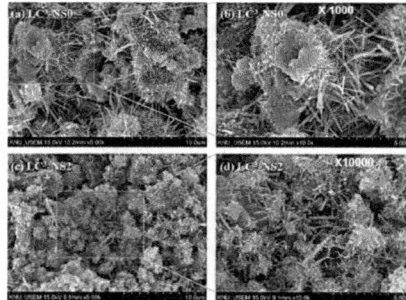

Figure 38 SEM images of LC³ in presence of NS at 1 day [79]

In general, NS in LC³, increases hydration rate, strength, refine pore structure, minimise carbonation and increase electrical conductivity [79].

Conclusions

In recent years, much attention is being paid on binding materials used in the construction industry. In order to make high tech binders, functionalized materials are being used. One of the most important such material is nanomaterial. Nanomaterials, when added to OPC, geopolymer cement and LC^3, bring changes in structure, strength, durability and in many other properties. However, addition of nanomaterials up to certain concentrations, has positive effect but beyond this optimum limit, has a negative impact. In addition to the above-mentioned properties, nanomaterials act as cleaning agent and helps the properties of PCM. Nanomaterials in general help in improving the performance of binders in the construction industry. The role of nanomaterials in the construction industry has not been fully explored and requires a detailed investigation.

References

[1] C.N. Waters, J. Zalasiewicz, Concrete: The Most Abundant Novel Rock Type of the Anthropocene, Elsevier Inc., 2018. doi: 10.1016/b978-0-12-809665- 9.09775-5. https://doi.org/10.1016/B978-0-12-809665-9.09775-5

[2] C. Shi, A.F. Jiménez, A. Palomo, New cements for the 21st century: The pursuit of an alternative to Portland cement, Cem. Concr. Res. 41 (7) (2011) 750-763, https://doi.org/10.1016/j.cemconres.2011.03.016. https://doi.org/10.1016/j.cemconres.2011.03.016

[3] F. Pacheco-Torgal, S. Miraldo, Y. Ding, J.A. Labrincha, Nanoparticles for high performance concrete (HPC), Nanotechnol. Eco-Efficient Constr. Mater. Process. Appl. (2013) 38-52, https://doi.org/10.1533/9780857098832.1.38. https://doi.org/10.1533/9780857098832.1.38

[4] M.I. Khan, Nanosilica/silica fume, Elsevier Ltd, 2018. doi: 10.1016/b978-0-08-102156-9.00014-6.

[5] G. Habert, E. Denarié, A. Šajna, P. Rossi, Lowering the global warming impact of bridge rehabilitations by using Ultra High Performance Fibre Reinforced Concretes, Cem. Concr. Compos. 38 (2013) 1-11, https://doi.org/10.1016/j. cemconcomp.2012.11.008. https://doi.org/10.1016/j.cemconcomp.2012.11.008

[6] D. Xuan, B. Zhan, C.S. Poon, Development of a new generation of eco-friendly concrete blocks by accelerated mineral carbonation, J. Cleaner Prod. 133 (2016) 1235-1241, https://doi.org/10.1016/j.jclepro.2016.06.062. https://doi.org/10.1016/j.jclepro.2016.06.062

[7] M.B. Ali, R. Saidur, M.S. Hossain, A review on emission analysis in cement industries, Renew. Sustain. Energy Rev. 15 (5) (2011) 2252-2261, https://doi. org/10.1016/j.rser.2011.02.014. https://doi.org/10.1016/j.rser.2011.02.014

[8] F. Pacheco-Torgal, S. Jalali, J.A. Labrincha, V.M. John, Eco-Efficient Concrete, 2013. doi: 10.1533/9780857098993. https://doi.org/10.1533/9780857098993

[9] R. Yu, P. Spiesz, H.J.H. Brouwers, Effect of nano-silica on the hydration and microstructure development of Ultra-High Performance Concrete (UHPC) with a low binder amount, Constr. Build. Mater. 65 (2014) 140-150, https://

doi.org/10.1016/j.conbuildmat.2014.04.063.
https://doi.org/10.1016/j.conbuildmat.2014.04.063

[10] Abhilash P. P. , Dheeresh Kumar Nayak , Bhaskar Sangoju , Rajesh Kumar, Veerendra Kumar, Effect of nano-silica in concrete; a review, Construction and Building Materials 278(2021)122347,
https://doi.org/10.1016/j.conbuildmat.2021.122347
https://doi.org/10.1016/j.conbuildmat.2021.122347

[11]. J. S. Belkowitz, W. B. Belkowitz, R. D. Moser, F. T. Fisher, and C. A. Weiss. Jr., "The influence of nano silica size and surface area on phase development, chemical chrinkage and compressive ctrength of cement composites," in Nanotechnology in Construction Proceedings of NICOM5, K. Sobolev and S. P. Shah, Eds., ed: Springer International Publishing, 2015, pp. 207-212. https://doi.org/10.1007/978-3-319-17088-6_26

[12] K. Sobolev, "Nanotechnology and nanoengineering of construction materials," in Nanotechnology in Construction Proceedings of NICOM5, K. Sobolev and S. P. Shah, Eds., ed: Springer International Publishing, 2015, pp. 3-13.

[13] G. Li, "Properties of high-volume fly ash concrete incorporating nano-SiO2," Cement and Concrete Research, vol. 34, pp. 1043-1049, 2004.
https://doi.org/10.1016/j.cemconres.2003.11.013

[14] D. Kong, Y. Su, X. Du, Y. Yang, S. Wei, and S. P. Shah, "Influence of nano-silica agglomeration on fresh properties of cement pastes," Construction and Building Materials, vol. 43, pp. 557-562, 2013.
https://doi.org/10.1016/j.conbuildmat.2013.02.066

[15] Q. Ye, Z. Zhang, D. Kong, and R. Chen, " Influence of nano-SiO2 addition on properties of hardened cement paste as compared with silica fume," Construction and building materials, vol. 21, pp. 539-545, 2007.
https://doi.org/10.1016/j.conbuildmat.2005.09.001

[16] W. Li, Z. Huang, F. Cao, Z. Sun, and S. P. Shah, "Effects of nano-silica and nano-limestone on flowability and mechanical properties of ultra-high-performance concrete matrix," Construction and Building Materials, vol. 95, pp. 366-374, 2015.
https://doi.org/10.1016/j.conbuildmat.2015.05.137

[17] P. Hou, S. Kawashima, K. Wang, D. J. Corr, J. Qian, and S. P. Shah, "Effects of colloidal nanosilica on rheological and mechanical properties of fly ash-cement mortar," Cement and Concrete Composites, vol. 35, pp. 12-22, 2013
https://doi.org/10.1016/j.cemconcomp.2012.08.027

[18] F. Sanchez and K. Sobolev, "Nanotechnology in concrete - A review," Construction and Building Materials, vol. 24, pp. 2060-2071, 2010
https://doi.org/10.1016/j.conbuildmat.2010.03.014

[19] Ali M. Onaizi , Ghasan Fahim Huseien , Nor Hasanah Abdul Shukor Lim , Mugahed Amran , Mostafa Samadi, Effect of nanomaterials inclusion on sustainability of cement-based concretes: A comprehensive review, Construction and Building Materials 306 (2021) 124850 https://doi.org/10.1016/j.conbuildmat.2021.124850

[20] Abbas Mohajerani, Lucas Burnett, John V. Smith, Halenur Kurmus, John Milas, Arul Arulrajah , Suksun Horpibulsuk and Aeslina Abdul Kadir, Nanoparticles in Construction Materials and Other Applications, and Implications of Nanoparticle Use, Materials 12(2019) , 3052; doi:10.3390/ma12193052 https://doi.org/10.3390/ma12193052

[21] J. Davidovits, Geopolymers, J. Therm. Anal. 37 (1991) 1633-1656. https://doi.org/10.1007/BF01912193

[22] P. Zhang, K.X. Wang, Q.F. Li, J. Wang, Y.F. Ling, Fabrication and engineering properties of concretes based on geopolymers/alkali-activated binders - A review, J. Clean. Prod. 258 (2020). 120896. https://doi.org/10.1016/j.jclepro.2020.120896

[23] H.Y. Zhang, V. Kodur, B. Wu, L. Cao, F. Wang, Thermal behavior and mechanical properties of geopolymer mortar after exposure to elevated temperatures, Constr. Build. Mater. 109 (4) (2016) 17-24. https://doi.org/10.1016/j.conbuildmat.2016.01.043

[24] N.B.Singh, Mukesh Kumar , Sarita Rai , Geopolymer cement and concrete: Properties, Materials Today: Proceedings 29(2020)743-748 https://doi.org/10.1016/j.matpr.2020.04.513

[25] B. Singh, G. Ishwarya, M. Gupta, S.K. Bhattacharyya, Geopolymer concrete: A review of some recent developments, Constr. Build. Mater. 85 (2015) 78-90. https://doi.org/10.1016/j.conbuildmat.2015.03.036

[26] M.M. Al-mashhadani, O. Canpolat, Y. Aygörmez, M. Uysal, S. Erdem, Mechanical and microstructural characterization of fiber reinforced fly ash based geopolymer composites, Constr. Build. Mater. 167 (2018) 505-513. https://doi.org/10.1016/j.conbuildmat.2018.02.061

[27] T.Y. Xie, P. Visintin, X.Y. Zhao, R. Gravina, Mix design and mechanical properties of geopolymer and alkali activated concrete: Review of the state of-the-art and the development of a new unified approach, Constr. Build. Mater. 256 (2020). 119380. https://doi.org/10.1016/j.conbuildmat.2020.119380

[28] V. Sreevidya, Investigations on the flexural behaviour of ferro geopolymer composite slabs, 2014.

[29] Karen Scrivenera, Fernando Martirena, Shashank Bishnoi, Soumen Maity, Calcined clay limestone cements (LC3) Cement and Concrete Research 114(2018)49-56 https://doi.org/10.1016/j.cemconres.2017.08.017

[30] R. Bawa, Drug delivery at the nanoscale: a guide for scientists, physicians and lawyers, in: S.A. Mousa, R. Bawa, G.F. Audette (Eds.), The Road from Nanomedicine

to Precision Medicine, Jenny Stanford Publishing, New York, 2019, pp. 1-134. https://doi.org/10.1201/9780429295010-1

[31] R.P. Feynman, There's plenty of room at the bottom, Eng. Sci. 23 (1960) 22-36.

[32] N. Taniguchi, C. Arakawa, T. Kobayashi, On the basic concept of nano-technology, in: Proceedings of the International Conference on Production Engineering, 1974.

[33] S. Bayda, M. Adeel, T. Tuccinardi, M. Cordani, F. Rizzolio, The history of nanoscience and nanotechnology: from chemical-physical applications to nanomedicine, Molecules 25 (2019) 112 https://doi.org/10.3390/molecules25010112

[34] (https://www.nano.gov/nanotech-101/what)

[35] Ali M. Onaizi , Ghasan Fahim Huseien , Nor Hasanah Abdul Shukor Lim , Mugahed Amran, Mostafa Samadi, Effect of nanomaterials inclusion on sustainability of cement-based concretes: A comprehensive review, Construction and Building Materials 306 (2021) 124850 https://doi.org/10.1016/j.conbuildmat.2021.124850

[36] Tawfik A. Saleh, Nanomaterials: Classification, properties, and environmental toxicities, Environmental Technology & Innovation 20 (2020) 101067 https://doi.org/10.1016/j.eti.2020.101067

[37] Mousumi Das and Saptarshi Chatterjee, Green synthesis of metal/metal oxidenanoparticles toward biomedical applications: Boon or bane, Green Synthesis, Characterization and Applications of Nanoparticles, (2019) 265-301 https://doi.org/10.1016/B978-0-08-102579-6.00011-3

[38]Mahmoud Nasrollahzadeh, S. Mohammad Sajadi, Mohaddeseh Sajjadi and Zahra Issaabadi, Applications of Nanotechnology in Daily Life -Chapter 4, Interface Science and Technology, 28, (2019) 113-143 https://doi.org/10.1016/B978-0-12-813586-0.00004-3

[39]Dimitra Papadaki, George Kiriakidis, Theocharis Tsoutsos, Applications of nanotechnology in construction industry-Chapter-11, Fundamentals of Nanoparticles (2018)343-370 https://doi.org/10.1016/B978-0-323-51255-8.00011-2

[40] F. Pacheco-Torgal, S. Jalali, Nanotechnology: advantages and drawbacks in the field of construction and building materials, Construction and Building Materials (2011) 582-590. https://doi.org/10.1016/j.conbuildmat.2010.07.009

[41] A.K. Rana, et al. Significance of nanotechnology in construction engineering, Int. J. Recent Trends Eng. 1 (4) (2009) 6-8.

[42]Harald R. Tschiche, Frank S. Bierkandt, Otto Creutzenberg , Valerie Fessard, Roland Franz, Bernd Giese, Ralf Greiner, Karl Heinz Haas, Andrea Haase, Andrea Hartwig, Kerstin Hund Rinke, Pauline Iden, Charlotte Kromer, Katrin Loeschner, Diana Mutz, Anastasia Rakow, Kirsten Rasmussen, Hubert Rauscher, Hannes Richter, Janosch Schoon, Otmar Schmid, Claudia Som, Günter E. M.Tovar, Paul Westerhoff, Wendel Wohlleben , Andreas Luch, Peter Laux, Environmental considerations and current

status of grouping and regulation of engineered nanomaterials, Environmental Nanotechnology, Monitoring & Management 18 (2022) 100707 https://doi.org/10.1016/j.enmm.2022.100707

[43] S. Shah, P. Hou, M. Konsta-Gdoutos, Nano-modification of cementitious material: toward a stronger and durable concrete, J. Sustain. Cem. Based. Mater., 5 (2016)1-22 https://doi.org/10.1080/21650373.2015.1086286

[44] Konstantin Sobolev,Modern developments related to nanotechnology and nanoengineering of concrete, Frontiers of Structural and Civil Engineering 10(2016)131-141 https://doi.org/10.1007/s11709-016-0343-0

[45] Natt Makul, Modern sustainable cement and concrete composites: Review of current status, challenges and guidelines, Sustainable Materials and Technologies 25 (2020) e00155 https://doi.org/10.1016/j.susmat.2020.e00155

[46] N.B.Singh, Properties of cement and concrete in presence of nanomaterials (Chapter-2) in Smart Nanoconcretes and Cement-Based Materials, 2020,9-39 https://doi.org/10.1016/B978-0-12-817854-6.00002-7

[47] M.A. Kewalramani, Z.I. Syed, Application of nanomaterials to enhance microstructure and mechanical properties of concrete, Int. J. Integr. Eng. 10 (2) (2018). https://doi.org/10.30880/ijie.2018.10.02.019

[48] M.A.A. Rajak, Z.A. Majid, M. Ismail, Morphological characteristics of hardened cement pastes incorporating nano-palm oil fuel ash, Procedia Manuf. 2 (2015) 512-518. https://doi.org/10.1016/j.promfg.2015.07.088

[49] Qiang Fu, Xu Zhao, Zhaorui Zhang, Wenrui Xu, Ditao Niu, Effects of nanosilica on microstructure and durability of cement-based materials, Powder Technology 404 (2022) 117447 https://doi.org/10.1016/j.powtec.2022.117447

[50] Xiaoyan Liu, Li Liu , Kai Lyu , Tianyu Li, Pingzhong Zhao, Ruidan Liu , Junqing Zuo, Feng Fu, Surendra P. Shah, Enhanced early hydration and mechanical properties of cement-based materials with recycled concrete powder modified by nano-silica, Journal of Building Engineering 50 (2022) 104175 https://doi.org/10.1016/j.jobe.2022.104175

[51] Jamal A. Abdalla, Blessen Skariah Thomas, Rami A. Hawileh , Jian Yang , Bharat Bhushan Jindal , Erandi Ariyachandra, Influence of nano-TiO2, nano-Fe2O3, nanoclay and nano-CaCO3 on the properties of cement/geopolymer concrete 4(2022)100061 https://doi.org/10.1016/j.clema.2022.100061

[52] Qiang Fu, Zhaorui Zhang, Xu Zhao, Wenrui Xu, Ditao Niu, Effect of nano calcium carbonate on hydration characteristics and microstructure of cement-based materials: A review, Journal of Building Engineering 50(2022)104220 https://doi.org/10.1016/j.jobe.2022.104220

[53] Z.B. Jian, X.D. Xing, P.C. Sun, The effect of nanoalumina on early hydration and mechanical properties of cement pastes. Construction and Building Materials 202(2019) 169-176. https://doi.org/10.1016/j.conbuildmat.2019.01.022

[54] Raju Goyal, Vinay K. Verma , N.B. Singh, Effect of nano TiO2 & ZnO on the hydration properties of Portland cement, Materials Today: Proceedings (2022) (in press) https://doi.org/10.1016/j.matpr.2022.05.206

[55] Xiaoying Li, Jun Li , Zhongyuan Lu, Jiakun Chen, Properties and hydration mechanism of cement pastes in presence of nano-ZnO, Construction and Building Materials 289(2021)123080 https://doi.org/10.1016/j.conbuildmat.2021.123080

[56] Li Wang, Hongliang Zhang , and Yang Gao, Effect of TiO2 Nanoparticles on Physical and Mechanical Properties of Cement at Low Temperatures, Advances in Materials Science and Engineering , Volume 2018, Article ID 8934689, 12 pages https://doi.org/10.1155/2018/8934689 https://doi.org/10.1155/2018/8934689

[57] Fatemeh Hamidi and Farhad Aslani, TiO2-based Photocatalytic Cementitious Composites: Materials, Properties, Influential Parameters, and Assessment Techniques, Nanomaterials 9(2019) 1444; doi:10.3390/nano9101444 https://doi.org/10.3390/nano9101444

[58] Jai Prakash, Junghyun Cho, Yogendra Kumar Mishra, Photocatalytic TiO2 nanomaterials as potential antimicrobial and antiviral agents: Scope against blocking the SARS-COV-2 spread,14(2022)100100 https://doi.org/10.1016/j.mne.2021.100100

[59] https://www.archdaily.com/20105/church-of-2000-richard-meier.

[60]Mahyar Ramezani , Ayoub Dehghani, Muhammad M. Sherif, Carbon nanotube reinforced cementitious composites: A comprehensive review , Construction and Building Materials 315(2022)125100 https://doi.org/10.1016/j.conbuildmat.2021.125100

[61] Ezzatollah Shamsaei, Felipe Basquiroto de Souza , Xupei Yao , Emad Benhelal , Abozar Akbari , Wenhui Duan, Graphene-based nanosheets for stronger and more durable concrete: A review, Construction and Building Materials 183(2018)642-660 https://doi.org/10.1016/j.conbuildmat.2018.06.201

[62]Shenghua Lv , Yujuan Ma, Chaochao Qiu, Ting Sun, Jingjing Liu, Qingfang Zhou, Effect of graphene oxide nanosheets of microstructure and mechanical properties of cement composites , Construction and Building Materials 49(2013)121-127 https://doi.org/10.1016/j.conbuildmat.2013.08.022

[63] Xu-Jing Niu , Qing-Bin Li , Yu Hu , Yao-Sheng Tan , Chun-Feng Liu, Properties of cement-based materials incorporating nano-clay and calcined nano-clay: A review, Construction and Building Materials 284(2021)122820 https://doi.org/10.1016/j.conbuildmat.2021.122820

[64] N.B. Singh, Clays and Clay Minerals in the Construction Industry, Minerals, 12 (2022)301 https://doi.org/10.3390/min12030301

[65] Muhammad Aamer Hayat , Yong Chen , Mose Bevilacqua, Liang Li, Yongzhen Yang, Characteristics and potential applications of nano-enhanced phase change materials: A critical review on recent developments, Sustainable Energy Technologies and Assessments 50 (2022) 101799 https://doi.org/10.1016/j.seta.2021.101799

[66] Kwok Wei Shah, Pin Jin Ong, Ming Hui Chua, Sheng Heng Gerald Toh, Johnathan Joo Cheng Lee, Xiang Yun Debbie Soo, Zhuang Mao Png, Rong Ji, Jianwei Xu, Qiang Zhu, Application of phase change materials in building components and the use of nanotechnology for its improvement, Energy and Buildings 262(2022) 112018 https://doi.org/10.1016/j.enbuild.2022.112018

[67] Hemn Unis Ahmed , Azad A. Mohammed, Ahmed S. Mohammed, The role of nanomaterials in geopolymer concrete composites: A state-of-the-art review, Journal of Building Engineering 49 (2022) 104062. https://doi.org/10.1016/j.jobe.2022.104062

[68] E.D. Rodrı'guez, S.A. Bernal, J.L. Provis, J. Paya, J.M. Monzo, M.V. Borrachero, Effect of nanosilica-based activators on the performance of an alkali-activated fly ash binder. Cement and Concrete Composites 35 (2013) 1-11. https://doi.org/10.1016/j.cemconcomp.2012.08.025

[69] D. Adak, M. Sarkar, S. Mandal, Effect of nano-silica on strength and durability of fly ash based geopolymer mortar. Construction Building Materials 70 (2014) 453-459. https://doi.org/10.1016/j.conbuildmat.2014.07.093

[70] F. Shahrajabian, K. Behfarnia, The effects of nano particles on freeze and thaw resistance of alkali activated slag concrete, Construction and building materials, 176 (2018) 172-178. https://doi.org/10.1016/j.conbuildmat.2018.05.033

[71] Tanakorn Phoo-ngernkham, Prinya Chindaprasirt , Vanchai Sata , Sakonwan Hanjitsuwan , Shigemitsu Hatanaka, The effect of adding nano-SiO2 and nano-Al2O3 on properties of high calcium fly ash geopolymer cured at ambient temperature, Materials and Design 55 (2014) 58-65. https://doi.org/10.1016/j.matdes.2013.09.049

[72] D. Syamsidar, Nurfadilla, Subaer, The Properties of Nano TiO2-Geopolymer Composite as a Material for Functional Surface Application, in: MATEC Web Conf., EDP Sciences, 97 (2017) 1013. https://doi.org/10.1051/matecconf/20179701013

[73] Sudhir Singh Bhadauria, Ashita Singh, Investigating mechanical characteristics of geopolymer concrete under the influence of Nano-fillers, Materials Today: Proceedings xxx (xxxx) xxx.

[74] H.M. Khater, H.A. Abd el Gawaad, H.A.A. Gawaad, H.A. Abd el Gawaad, Characterization of alkali activated geopolymer mortar doped with MWCNT. Construction Building Materials 102 (2016) 329-337. https://doi.org/10.1016/j.conbuildmat.2015.10.121

[75] Buchit Maho, Piti Sukontasukkul, Gritsada Sua-Iam, Manote Sappakittipakorn, Darrakorn Intarabut, Cherdsak Suksiripattanapong, Prinya Chindaprasirt, Suchart Limkatanyu, Mechanical properties and electrical resistivity of multiwall carbon nanotubes incorporated into high calcium fly ash geopolymer, Case Studies in Construction Materials 15 (2021) e00785. https://doi.org/10.1016/j.cscm.2021.e00785

[76] C.S.B. Xavier, A. Rahim, Nano aluminium oxide geopolymer concrete: An experimental study, Nano aluminium oxide geopolymer concrete: An experimental study, 56 (2022) 1643-1647. https://doi.org/10.1016/j.matpr.2021.10.070

[77] Tanakorn Phoo-ngernkham, Prinya Chindaprasirt , Vanchai Sata , Sakonwan Hanjitsuwan , Shigemitsu Hatanaka, The effect of adding nano-SiO2 and nano-Al2O3 on properties of high calcium fly ash geopolymer cured at ambient temperature, Materials and Design 55 (2014) 58-65. https://doi.org/10.1016/j.matdes.2013.09.049

[78] P.P. Abhilash, D.K. Nayak, B. Sangoju, R. Kumar, V. Kumar, Effect of nano-silica in concrete; a review, Constr Build Mater 278 (2021), 122347 https://doi.org/10.1016/j.conbuildmat.2021.122347

[79] Run-Sheng Lin Seokhoon Oh, Wei Du, Xiao-Yong Wang, Strengthening the performance of limestone-calcined clay cement (LC3) using nano silica, Construction and Building Materials 340(2022) 127723 https://doi.org/10.1016/j.conbuildmat.2022.127723

Emerging Nanomaterials and Their Impact on Society in the 21st Century   Materials Research Forum LLC
Materials Research Foundations 135 (2023) 265-283   https://doi.org/10.21741/9781644902172-11

Chapter 11

Nanoparticles Incorporated Soy Protein Isolate for Emerging Applications in Medical and Biomedical Sectors

Priya Rani[1], Priyaragini Singh[1], Abhay Pandit[1], Rakesh Kumar[1]*

[1]Central University of South Bihar

*krrakesh72@gmail.com; rakeshkr@cusb.ac.in

Abstract

Proteins are deemed as ideal material for nanoparticles preparation and incorporation. Soy protein isolate (SPI) is chosen as matrix for fabrication of bionanocomposite materials because of versatile functional groups offered by amino acids side chain responsible for inter and intramolecular interactions. Complexion of SPI with nanoparticle makes it feasible for drug delivery as it offers excellent loading capacity. Encapsulation of drugs inside protein matrix not only protects drug from harsh environment but also helps in targeted delivery. Various antibiotics such as gentamicin and ciprofloxacin can be incorporated into the SPI matrix for targeted controlled release and antibacterial effects.

Keywords

Soy Protein Isolate, Biomedical Application, Targeted Drug Delivery, Wound Dressing, Cytotoxicity, Bone Fillers

Contents

Nanoparticles Incorporated Soy Protein Isolate for Emerging
Applications in Medical and Biomedical Sectors...265

1.   Introduction...266

2.   Soy protein based nanocomposites..267

3.   Techniques for preparation of nanocomposites based soy protein
materials...268

    3.1   Wet process...268

    3.2   Dry process 269

4.   Different types of SPI-nanocomposites 269

4.1   Soy protein/inorganic filler nanocomposites..................................269

4.2   SPI – Clay nanocomposites 270

4.3   SPI – Metal and metal oxide nanocomposites.................................270

4.4   Soy protein/organic nanocomposites...270

5.   **Applications of soy protein based nanocomposites in
biomedical sector**...**271**

5.1   Drug delivery 271

5.2   Antibacterial properties 274

5.3   Wound healing and cytotoxicity..275

5.4   Tissue regeneration and artificial bone fillers 276

Conclusions...**278**

References...**278**

1.   Introduction

Glycine max, commonly known as soybean is of captivating interest for the design of biomaterials in the health and medicine sector. Soy protein is an abundantly available biopolymer derived from renewable sources that constitute about 30-45% mass of the total soybean [1]. The major constituents of soy protein are aspartic acid and glutamic acid and their amide that is asparagine and glutamine, nonpolar amino acids comprising leucine, alanine, valine, glycine as well as basic amino acids such as lysine and arginine and relatively small percentage of cysteine [1]. Commercially, three forms of soy protein are available based on different protein content i.e., soy flour (SF), soy protein concentrate (SPC) and soy protein isolate (SPI). SPI is rich in protein (92%) while SF contains least protein content (50%). SF is prepared by grinding soy bean seeds into fine powder with very low protein content around 50%. SPC is basically defatted soy flour with protein content around 70% which is free form carbohydrates. Defatted soy beans are processed further for removal of water soluble carbohydrates to obtain SPC. SPI is the most refined form of soy protein that contains 90-92% of protein content. It is obtained from SF by further processing to remove its carbohydrates, fats and other non-protein components. Soy protein is an impure globular protein that consists of different fractions of 2S, 7S,11S and 15S fractions [2]. About 90-95% fractions feature two subunits, namely 52% of glycinin (11S) and 35% of conglycinin (7S) constitutes the major fraction of soy protein [3]. In aqueous medium soy protein fractions are present as globular structure with hydrophilic interior and hydrophobic kernel studded with some water soluble aggregates. Formation of various structures such as microspheres, hydrogels and polymer blends can be regulated through addition of dissolvent or cross linking agents [4].

In recent years, substantial efforts have been invested for the development of alternative biodegradable material for biomedical applications. However, the range of biopolymers that are available for clinical research is rather relatively smaller in size and they also present several limitations for their commercial application. The polymers obtained by microbial systems such as poly lactic acid (PLA), polyhydroxy butyrate (PHB) can offer better physical and mechanical properties compared to plant derived protein polymers [5,6]. However, the production cost of such polymers too high, and challenges associated with their applications are mainly degradation at low pH inside human body which leads to inflammatory response [7]. Carbohydrate derived biopolymers like starch has been also tested for their application in bio-medicine but found to be extremely water sensitive that leads to immediate disintegration in aqueous environment and poor storage stability [8]. Overcoming such barriers, soy protein is seen as a better alternative in bio-medicine owing to its cost effective nature, improved storage stability and significant hydrophobic behaviour [9]. Additionally, soy protein is seen as a biocompatible material of choice because of its similarity with tissue composition. Moreover, it is also advised to explore different approaches for biomaterials fabrication that do not require extensive thermal treatment as soy protein is susceptible to higher temperature compared to other proteins.

2. Soy protein based nanocomposites

Materials with particle size in nano scale that is less than 100 nanometre are classified as nanoparticles. The last few years has witnessed significant change in the biomedical sector where nanoparticles are exploited for application in bioimaging, cancer therapy, targeted delivery for drugs and genes [10]. Indeed, at some circumstances nanoparticles are the only reliable approach for therapy which can simply not be performed without their use. Moreover, nanoparticles come with their associated environmental and societal challenges particularly they cause environmental toxicity. There are various properties associated with nanoparticles which differentiate themselves from bulk materials such as their size, chemical reactivity, energy absorption and biological mobility [11].

Nanoparticles are widely explored in medical sector especially in targeted drug delivery systems (DDS). To enhance therapeutic effects and lessen the side effects, nanoparticles are seen as vehicle for drugs, antibiotics and bioactive molecules [12]. Such DDS are performed by using various nano structure materials like liposomes, polymers, dendrimers, silicon and silver nanoparticles [13]. Among these silver nanoparticles has taken over the rest owing to its low toxicity, biodegradability and biocompatible nature. The incorporation of silver nanoparticles in organic and inorganic materials lead to increase in transparency, mechanical toughness, conductance, swelling/deswelling rates and bactericidal effects [14].

Based on the method of preparation nanoparticles, nanospheres or nano capsules can be obtained. Nano capsules are designed in such a way that the drug is inserted inside a cavity surrounded by a polymeric membrane while nanospheres are matrix in which drug is uniformly distributed [15].

Different kinds of matrix materials such as proteins, polysaccharides and synthetic polymers are time to time tested for their suitability in formation of nanoparticles [16]. The selection of matrix is influenced by factor such as size of nanoparticles desired, intrinsic nature of drug like its solubility and stability inside human system, charge and permeability of nanoparticles, biodegradability, biocompatibility, toxicity, drug release profile and antigenicity of the final conjugated product. Generally, they are prepared by following three methods [17] given as

a) dispersion of preformed polymers,

b) polymerization of monomers,

c) ionic gelation of hydrophilic polymers.

Soy protein nanoparticles are expanding in the biomedical sector due to its high abundance and low cost. Further, its biodegradable nature and low immunogenicity adds to its desirable characteristics. Hydrophobic drugs are packaged with much ease inside soy protein nanoparticles because of its unique amino acid composition that facilitates drug and matrix interaction [18]. They can be used successfully in oral drug delivery owing to its water soluble nature unlike its counterpart such as zein protein. Moreover, SPI based nanoparticles also used for coatings along with other materials for protection and surface modification [19].

3.   Techniques for preparation of nanocomposites based soy protein materials

Preparation of SPI films involves the denaturation of the native state of the protein to form unique configuration through new linkages within the protein molecule. The process of denaturation can be induced by changes in pH, electrical force, mechanical force and heat [20]. The reported isoelectric point for SPI is 4.2-4.6 hence, changing the pH away from the given isoelectric point leads to protein unfolding and thus increases its solubility. In solution casting, SPI is solubilized under heating condition that is casted on a plain glass surface and forms a film upon drying. Apart from pH, forces like pressure and shear are also reported to break bonds and initiates intermolecular rearrangements. SPI generally denatures at a temperature ranging from 65 to 70°C. The process of denaturation exposes the sulphide and hydrophobic groups on the surface and several disulphide bonds are reformed [21]. Generally, two common processes are employed for formation of SPI protein films that are discussed below:

3.1   Wet process

This process is popularly known as solution casting and based on solubilization of proteins in a suitable solvent medium using alkaline pH to unfold the native proteins before pouring the solution on a flat mould surface. Under appropriate relative humidity condition such films are peeled off from the flat surface of the mould. Several parameters that need to be considered while using the solvent casting process for film formation are temperature, surface properties of the mold used for casting, pH and drying time [22].

Emerging Nanomaterials and Their Impact on Society in the 21st Century   Materials Research Forum LLC
Materials Research Foundations 135 (2023) 265-283   https://doi.org/10.21741/9781644902172-11

3.2   Dry process

Such processes are combination of heat and mechanical process that could be employed to disrupt inter and intramolecular interactions of bio polymers by extrusion or mechanical mixing. Such protein materials combined with plasticizer can be extruded, injection or compression moulded under which the protein side chains make interaction through disulphide and amide linkages offered by carboxylic and amino group of the protein.

In both processes, it is clearly understood that a protein films could not be formed at pH values near its isoelectric point as protein precipitates at this point. However, proteins can be easily denatured and precipitated at pH values at different isoelectric point which unravel its functional groups and thus leads to increased intermolecular interactions. As a thumb rule, most of the protein based films demonstrate improved physical properties when casted under alkaline condition compared to those processed under acidic pH [23].

If we describe a generalised procedure for formulation of SPI based films by the solution casting method it involves preparation of SPI solution by slowly dissolving the protein under constant stirring condition. Generally, SPI concentration of 5% to 7% w/w is used within the pH range of 9-10 [24]. Since neat SPI is not easily processible hence to increase its processibility plasticizers are used in concentration of 30% to 55% w/w with respect to SPI. Sometimes cross linking agents such as glyoxal are also recommended. Then, SPI solution is heated at 65 to 70°C for 30 min under constant stirring and further nanoparticles in different concentration could be added to the solution followed by additional 30 min heating. After complete removal of air bubbles from the solution it is casted onto a glass mold and dried at ambient temperature and humidity [24].

4.   Different types of SPI-nanocomposites

A combination of nanoparticles and SPI is considered a promising way to upgrade the physical properties of neat soy protein. The amalgamation of filler materials to increase polymer barrier characteristics is a well-known approach in the food packaging sector [25]. Considering the wide range of fillers available, SPI- nanocomposites can be categorised into various types based on their chemical makeup. Organic fillers include carbon atoms and are typically generated directly from biomass, whereas inorganic fillers can be of natural origin but, mostly, do not contain any carbon atoms.

For increasing the physical characteristics of soy protein composites, combining nanometric particles with soy protein as a base material could be a potential alternative. The qualities of these nanocomposites materials are determined by their shape, interfacial features, and their separate parents' attributes. The common approach to generate nanoparticles in a protein matrix is an *in-situ* synthesis of nanoparticles within the protein matrix using a precursor. This method leads to enhanced interaction between fillers and base materials.

4.1   Soy protein/inorganic filler nanocomposites

Soy-protein-based polymers greatest drawbacks are their poor mechanical characteristics and hydrophilicity. The incorporation of inorganic fillers significantly improves physical and mechanical properties. This remarkable change in the mechanical properties depends on the size of nanoparticles (fillers), the compatibility of nanoparticles and polymers, the surface area of fillers, etc.

4.2   SPI – Clay nanocomposites

The incorporation of clay as fillers with SPI minimises flammability and permeability to liquids and gases. This combination can reinforce a variety of plastics. Clay-based composites have also been approved for use as meals, medications, drinks, and biomedical devices by the US Food and Drug Administration [26]. Hydroxyapatite (HA) is an example of new reinforcing material for biomedical applications such as tissue scaffolds and bone regeneration. Recently, hydroxyapatite nanoparticles were effectively produced from eggshell waste using wet chemical precipitation and then mixed into a soy-based resin with regular dispersion. HA worked as excellent filler for SPI-biopolymer. The incorporation of evenly disseminated HA in SPI sheets resulted in a significant increase in tensile modulus and strength [27].

4.3   SPI – Metal and metal oxide nanocomposites

The incorporation of metal nanoparticles in SPI-based polymer tends to induce antibacterial or antimicrobial properties to the polymer. Titanium dioxide (TiO_2), silver, and other metals might be incorporated into soy protein films to extend their shelf-life and can be used in food packaging and medicinal industries. It is well reported that silver nanoparticles possess excellent antibacterial activity against various bacteria, fungi, and viruses. In-situ synthesis of silver nanoparticles (AgNPs)/SPI films was performed using SPI/ $AgNO_3$ solution, by utilizing the benefits of tyrosine reduction in SPI [28]. The antimicrobial test revealed that the films exhibit exceptional antibacterial characteristics, effectively killing both Gram-positive and Gram-negative bacteria.

4.4   Soy protein/organic nanocomposites

Organic fillers used for reinforcement of SPI-based polymers are mostly biodegradable in nature. Organic fillers include starch, cellulose, chitin, protein, etc. Owing to their natural origin and eco-compatible nature, organic fillers are preferred over inorganic fillers in composite preparation. Polysaccharide nanocrystals are emerging as a good option of organic fillers because of their inexpensive cost, low density, high mechanical properties and low toxicity [29]. The nanocrystals of polysaccharides like starch, chitin and cellulose are among the most studied inorganic filler options. Nanocrystals of cellulose and chitin have a rod-like appearance. As a result of their morphology, they are commonly referred to as whiskers. In contrast, the morphology of nanocrystals of starch is platelet like. These nanocrystals can be blended into the soy protein base to synthesise nanocomposites [30]. Whiskers of chitin and cellulose and crystals of starch can be obtained from natural sources like natural sources wheat straw, wood, bacterial cellulose etc.

Uniform distribution of fillers and strong chemical bonds (intramolecular hydrogen bonds) between fillers and matrix leads to enhanced mechanical properties and improved composite performance [31]. An example of soy protein isolate composite reinforced with electrospun cellulose nanofibrous mats (CNM) has been dealt in detail [32]. The ultrafine diameter and high length-to-diameter ratio of nanofibers help them to interact with SPI matrix in such a way that leads to a 13- and 6-fold increase in mechanical strength and Young's modulus over neat SPI film, respectively. Few natural polymer-based nanoparticles like citric acid blended with starch nanoparticles and phthalic anhydride blended with soy protein were also employed as fillers in SPI based nanocomposites [33-34]. These nanocomposites showed significantly enhanced mechanical properties as these fillers were found to be highly compatible. These nanocomposites can find their market in the packaging sector.

5.   Applications of soy protein based nanocomposites in biomedical sector

The biocompatible nature of naturally occurring polymeric material makes them a suitable choice for formulation of several biomedical devices. Plant based proteins also provide additional benefit to their animal counterparts as the latter could cause immunogenic reaction and offers possible routes for disease transmission. Such plant derived natural polymers share structural similarity with extracellular matrix of cells that facilitates the process of cell adhesion [35, 36]. Proteins are composed of polar as well as non-polar groups, polar groups are reported to facilitate cell matrix attachment. Generally, protein based polymeric materials are chosen because of versatile functional groups offered by side chain of amino acids that can initiate different kinds of reaction and modifications. Additionally, proteins are also susceptible to slight changes in pH that may result in protonation and deprotonation of -COOH and $-NH_2$ groups that further adds to their cross linking tendency [37]. Thanks to these benefits of plant derived protein that makes them a promising candidate for wound healing, tissue engineering and drug delivery [38].

5.1   Drug delivery

Protein, as a material for drug delivery is garnering interest now-a-days because of their capacity to retain degradative neutraceuticals and bioactive compounds without compromising their functionality. By encapsulating the drug inside a protein matrix not only protects the drug from the harsh environment but also helps in its targeted delivery. Soy protein is also tested for its ability to withhold bioactive compound delivery because of their stiffer gel forming ability and degradative ability. A study by Jafari et al. has reported that glycinin agglomeration favours formation of stiffer hydrogels that can be an excellent tool for delivery, capturing and protecting various bioactive compound in food and pharmaceutical industry [39]. The SPI matrix coated with PLA favours intimate interaction with protein that imparts better mechanical properties and make it suitable in ocular treatment. Igarzabal et al., has shown that such SPI-PLA combination can decrease the intraocular pressure significantly [40].

The journey of a drug inside the human body is not a smooth rather a bumpy one with too many unwanted barriers. Orally administered drugs must travel through circulatory system, passes several layers of tissue and cell layers in order to reach its destination. Such long route travel not only reduces the efficacy of drugs but also makes healthy cells susceptible towards it and hence can cause side effects. Targeted drug delivery here comes as a saviour which provides opportunity to biomedical engineers for design of many polymer based formulation. Such systems are capable in regulating the rate of drug release and its destination for delivery. Moreover, drug delivery system also ensures minimum side effects while maximizing its benefits.

To enhance the therapeutic effect of drugs and bioactive compounds they are generally complexed with nanoparticles. Several nanostructure materials like liposomes, silicon, polymers and silver nanoparticles. Owing to the low toxicity, biocompatibility and biodegradability silver nanoparticles are seen as promising candidates for drug delivery [14]. Plant derived soy protein increases the storage time, stability and imparts water resistance.

Ciprofloxacin (Cfx) is a broad spectrum antibiotic effective against both gram positive as well as gram negative bacteria specifically used to treat diverticulitis and various bacteria borne infection [41]. Cfx is partially soluble in aqueous medium, and presence of π-π interactions due to aromatic ring structure facilitates intermolecular bonding. Generally, when drugs are taken orally, they must travel a long journey in which they come across where it interacts with healthy tissue types and may cause side effects on it. To avoid such unwanted effects of the drug on healthy cells they are delivered by targeted delivery system that also maximize its benefit along with increase in release rate, like if we see the case of ciprofloxacin that could lead to gastro intestinal disorders in human if it is orally administered [42]. In case of Cfx, it is reported that the mucoadhesive modification of hydrogels not only increases the GI retention time of hydrogels but also improves the bioavailability and efficiency for eliminating microbial infection. A study by Basu and group has reported a green synthesis process where they reported the release rate of IPN hydrogels to be around 95% after 350 min for delivery of Cfx [43]. Similar results have been reported by Garcia et al. where they synthesized an ionic complex of ciprofloxacin for targeted delivery of this antibiotic with maximum release date of 47% within 24 h [44]. Another group of Aminabhavi et al., has surfaced out that acrylamide-grafted-guar gum incorporated with chitosan has the capability to penetrate polymeric mesh of hydrogel for targeted delivery of Cfx with highest release rate of 60% within 12 h of administration [45].

Authors have reported pH dependent release of ciprofloxacin drugs from prepared nano silver embedded soy protein polyacrylamide (Pam-SP-Ag) hydrogel (Figure 1). The drug release rate is faster at basic pH compared to acidic pH in case of both hydrogels and nano composite hydrogel. The release of drug is lower in pH range of 1.2- 6.5, but it is significantly higher in pH range of 7.5 in case of both hydrogels [46]. Generally, the release rate of a drug is dependent on mesh size as well as the cross linking intensity of the matrix, the higher the mesh size the greater will be the release rate.

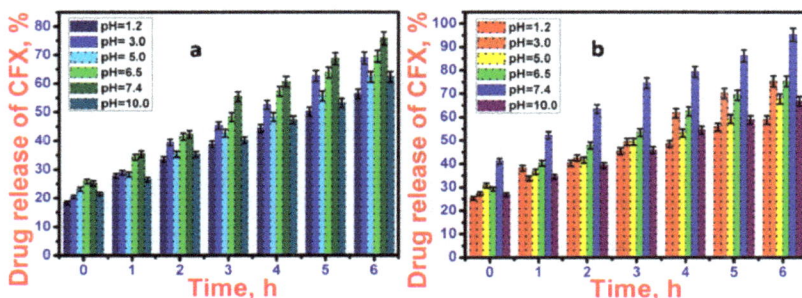

Figure 1: Ciprofloxacin drug release profile from (a) PAM-SP hydrogel and (b) PAM-SP-Ag nanocomposite hydrogel under different pH conditions

The rate of release of ciprofloxacin increases from 75.4% to 95.27% with increasing time from 0 h to 6 h (Figure 2). PAM-SP-Ag nano composite showed noticeable increase in drug release rate of around 95.27% as compared to PAM-SP with release rate of 75%. It is evident that the structural complexity of hydrogel imparts significant resistance in the release of drug. The *in vitro* release rate of drug is higher for PAM-SP-Ag as compared to PAM-SP. The higher release rate of ciprofloxacin can be understood by the hydrophilic behaviour of cross linkers that increases the diffusion of drug and subsequently increased drug release rate [46].

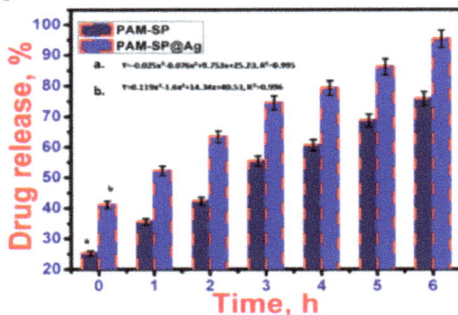

Figure 2. Ciprofloxacin drug release profile In vitro condition from PAM-SP hydrogel and PAM-SP-Ag nanocomposite hydrogel at 32ºC [46]

Study by Nayak and group has reported rapid release of ofloxacin from SPI/MMT composites at pH 7.4 compared to 3.4. The drug release kinetics is influenced by nature of matrix and pH of intracellular milieu [47]. The basic pH medium facilitates the breakage of electrostatic interaction between matrix and drugs hence rapid release is observed at alkaline pH range. Such blends can be easily employed for drug release in alkaline medium of large intestine and colon. The release half times, t_{50} (time required for releasing 50 wt% of drug) for 10%, 20%, 30%, 40%, 50% drug loadings are 2.8, 1.8 and 1.7 h at pH 7.4, and 6.0, 5.0 and 4.4 h at pH 3.4, respectively. Approximately, 80 wt% of loxacin is released at pH 7.4 while only 44 wt% release was observed at pH 3.3 [47]. Another study has reported the release of gentamicin from SPI-glyoxal film prepared by thermal treatment [24]. Films were loaded with 1% w/w and 3% w/w gentamicin, and the drug release kinetics was studied for 2 months which showed three main stages of release profiles that are:

a) Immediate rapid release during the first six hours,

b) Followed by exponentially decreasing release rate,

c) Ultimately a constant release rate was observed in last stage during second week.

It is also seen that heating/thermal treatment enhances the drug release by several fold in case of gentamicin for both samples containing 1% and 3%. The non-treated films had approximately 40% higher release rate than that of the films prepared by thermal treatment [24].

5.2   Antibacterial properties

Interestingly, silver nanoparticles are powerful antibacterial agent against several bacteria that are mostly resistant towards many antibiotics [48]. Prusty et al., have observed that incorporation of silver nanoparticles leads to increase in antibacterial activity against *Escherichia coli* (EC), *Shigella flexneri* (SF), *Bacillus cereus* (BC), and *Listeria inuaba* (LI). The PAM-SP and nanoparticles embedded PAM-SP-Ag hydrogels lead to increase in antibacterial efficacy with gradual increase in the silver nanoparticles concentration (Figure 3). Incorporation of bacteria inside PAM-SP-Ag nano composite hydrogels restrict growth of bacteria due to extensive cross linking present between silver nanoparticles. The plausible reason behind this could be presence of hydrogen bond amide group of acrylamide chain and silver nanoparticles. Additionally, the surface area to volume ratio of nanocomposite hydrogel is significantly increased by silver nanoparticles that increase bacterial resistivity. Hence PAM-SP-Ag nano composites could be viewed as promising alternatives in DDS [46].

Emerging Nanomaterials and Their Impact on Society in the 21st Century   Materials Research Forum LLC
Materials Research Foundations 135 (2023) 265-283   https://doi.org/10.21741/9781644902172-11

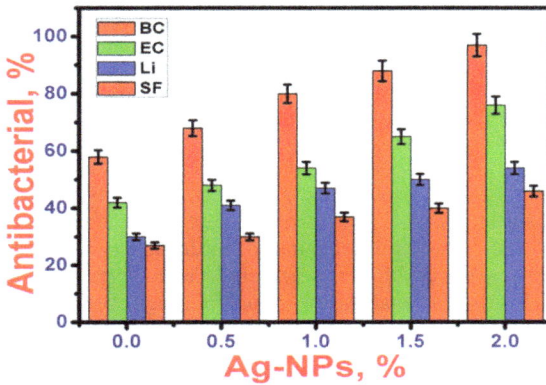

Figure 3: Antibacterial activity of prepared PAM-SP-Ag nanocomposite hydrogel at different concentration against Escherichia coli (EC), Shigella flexneri (SF), Bacillus cereus (BC), and Listeria inuaba (LI)((Reproduced with permission from Ref. [46]

Gentamicin loaded SPI films are tested for their antibacterial activity against *S. aureus*, *S. albus* and *P. aeruginosa*. The bacterial strains were effectively inhibited till 14th day of culture where a prominent zone of inhibition was observed. However, in case of *P. aeruginosa*, initially for first three days it was inhibited later it showed resistance towards it. Thermal treatment has a crucial effect on the antibacterial activity as it is evident from the zone of inhibition where a constant zone of inhibition was observed for thermally treated films after burst effect. However, the zone of inhibition of non-thermal treated films showed decreasing radius overtime. As expected 3% gentamicin showed a larger zone of inhibition as compared to 1% gentamicin [24].

5.3   Wound healing and cytotoxicity

Being the largest organ in the human body, skin acts as a guard against several wear and tears including mechanical, chemical, and biological agents and even protects the underlying cells from harmful damaging UV spectrum. It further prevents excessive loss of water, facilitates hydration, and thereby regulates the temperature of the body. Whenever skin is encountered with any cut, puncture, or torn, it leads to the progression of condition termed as wound. Normally healing of wound or wound healing is the normal biological event that includes cascade of reactions: hemostasis, inflammation, cell-proliferation, re-epithelialization and restoration [48]. Based on wound recovery period, it has been classified into two types namely: acute and chronic ones. Acute wounds take very short period and heal through the regular repair pathway whereas chronic wound consumes longer period for the recovery. It is well established that the delay in the healing of acute wound takes the aggressive form and changes into the chronic wound. An open wound often facilitates the proliferation and colonization of aggressive microbial species and leads to the progression into a chronic wound. Subsequent contamination of the wound by pathogens leads to the condition of wound infection.

To encounter the risk associated with wound infection, wound dressing materials have been developed. The wound dressing materials are functionalized with active agents to prevent the contamination of the wound for faster healing. The usual dressing materials are made up of cotton and wool, representing the traditional mode of wound dressing. These traditional dressings have proven beneficial and facilitate the suitable environment required for wound healing. In these directions numbers of natural and synthetic combinations have been developed with numerous different preparations including sponges, hydrogels, films, hydrocolloids and hydro fiber mats specific for the types of wounds [49]. The natural abundance, attractive water solubility, significant biocompatibility, biodegradability of soy protein with non-immunogenic and anti-cancerous properties has aided it for use in bioactive compounds drug delivery. SPI derived matrix shows profound stability with longer storage time due to which it becomes a capitative base material for the manufacture of biomedical devices [50].

Study by Vaz et al., has reported the biomedical application of soy protein-derived thermoplastic reinforced with tricalcium phosphate for use in orthopedic [51]. *In-vitro* analysis of these thermoplastic derived of soy proteins were reported to be nontoxic, rather it was found to encourage the cell proliferation rate [52]. The report also claims the potential application of these materials in wound healing, due to its integration into blood clot thereby stimulating the deposition of collagen. The deposition of collagen favors the formation of new tissues thereby eliminating the need of expensive growth factors for same activities [53]. In an effort to develop wound dressing material based on soy protein film, Peles et al., incorporated gentamicin and studied its release profile from the soy based film. It also demonstrated significant antibacterial activity against *S. aureus* and *S. albus* for at least 2 weeks and *P. aeruginosa* for 3 days. *In-vitro* cytotoxicity of these films reported no sign of cell cytotoxicity, rather it shows high biocompatibility enabling its application as wound dressings [24].

While selecting any matrix material for delivery of drug it is very important to determine its cytotoxicity. The prepared matrix will enter inside the human body where it could lead damage to healthy cells so toxicity behaviour should be investigated well. In case of PAM-SP-Ag, authors claimed that the prepared hydrogel is nontoxic and biocompatible. Silver nanoparticles embedded hydrogel demonstrated a low level of cell viability that is around 35.67% hence, it clearly indicates that such nano composite hydrogels are free from any kind unreacted molecules [46].

5.4   Tissue regeneration and artificial bone fillers

The success of regenerative medicine is largely dependent on the development of suitable biomaterials that hold the ability to regulate *in vitro* and *in vivo* tissue regeneration. Whatever products that are currently available in the markets are not able to combine the entire desired characteristic for biodegradable biomaterials that includes adjustable degradation rate, minimum inflammation, least toxicity and initiation of tissue regeneration. Additionally, the material must be prepared by cost effective processes and provides ease while handling surgery. In trauma patients where tissues lose the capacity for self and rapid regeneration, their biodegradable biomaterials are a suitable choice to strengthen healing and recovery. However, the clinical performances of such biomaterials are often limited to either protective or scaffolding role. They are also accompanied by adverse reactions that could impair the tissue regeneration potential and leads to intense inflammation [54]. To fulfil the desired role in physiological condition such materials need to possess five crucial properties which are tuneable biodegradation rate, homeostatic properties, controlled inflammation potential, lack of toxicity and stimulation of tissue cell activity [53].

Formulations of soybean based biomaterials are largely focused on use of its protein and oil components. In general, silicon oils are used in breast implants that could be successfully replaced by soy oil [55]. Use of soy oil components is also extended to bone fillers where its lipid fractions are generally used. Commercially, such bone fillers are available as various demineralized bone matrix formulation and in combination with hydroxyapatite granules. Report suggests that defeated soybean floor can be formulated into Gamma sterilized membrane, granules, gels, paste and bio glue under different physical and chemical environment [38]. The biocompatibility of such biomaterials is largely evaluated where it has shown no cytotoxicity and least degradation product in *in vitro* studies. Study have shown that no significant lactose hydrolase activity has been seen in fibroblast when incubated with hydrogels [56]. Recent studies have also reported the induction of osteoblast formation along with involvement of macrophages and osteoclast activity. Additionally, soybean-based materials possess the capacity to induce collagen synthesis by both fibroblast and osteoblast cells responsible for formation of mineralised bone nodule in osteoblast cultures [38]. A pilot study performed for bone regeneration shows that tissue regeneration in presence of such materials initiated with formation of dense trabecular mess that gradually surrounded the biomaterial granules [57]. The tissue regeneration pattern for soy-based biomaterials is like other commercially available bone fillers for periodontal application. Study by Lin et al. has reported that SPI with addition of 0.05% poly ethylene oxide holds the capacity to induce skin regeneration. Soy derived scaffolds supported the addition and proliferation of culture primary human dermal fibroblasts. They are reported to up regulate several genes involved in skin regeneration such as MMP-10, MMP-1, collagen VII, integrin-α2 and laminin-β3 [57].

Several cases where bone injury leads to removal of diseased bone do create a gap which needs to be filled either by organic or synthetic bone fillers. Synthetic bone filler such as hydroxyapatite, tricalcium phosphate and bioactive glasses are generally employed. However, the substances are brittle and hence create hindrance while embedding them inside the defect site. Additionally, residual components of such synthetic polymers are also reported to induce inflammatory response, thus natural polymers such as SPI are generally used. The release of isoflavones specifically daidzein and genistein are reported to decrease the level of cytokines and osteoclast activity upon interaction with receptors tyrosine kinase present on the cell membrane [58].

Wu et al., have reported that montmorillonite-modified nanocomposites were successfully developed using a combination of cross linkers, blending and freeze-drying processes [59]. The prepared nano composite exhibited improved physical, chemical and biological properties within the tissue regeneration range owing to its strong interface interaction between nano clay and SPI chains. The addition of montmorillonite led to enhancement in cell adhesion properties, proliferation, spreading and differentiation during bone regeneration. The biocompatibility assay showed uniform cell distribution *in vitro* along with up regulation of osteogenesis related jeans such as Runx2, Col-1, OPN, and OCN [59]. Within 12 week of treatment the defects were completely healed as evident from formation of neo vascular tissue throughout the defect site. Hence the bioactive MMT nano sheet combination could be a better substitute for the design of bones substitute in bone repair [59].

Conclusions

The development of materials in the biomedical sector has switched towards naturally derived polymeric products. Natural systems contain substantial amounts of protein that could be targeted for fabrication of biomaterials. SPI is obtained as a byproduct from soybean oil industry. The limitations like poor mechanical strength, hydrophilicity, poor processibility associated with commercial application of SPI could be addressed through application of plasticizer, nanoparticles reinforcement and blending with organic and inorganic fillers. The properties of the end products are influenced by the inherent nature of the matrix material, formulation methods, interaction between matrix material and fillers and dispersion effects. Taking into account the world-wide research investments and future advancements, it is anticipated that soy protein-based biomaterials would garner much research attention and may capture the market as a potential substitute for synthetic biomaterials in health and medical sector. The application in the biomedical sector comprises like bone fillers, hydrogels, adhesives and other biomedical materials.

References

[1] R. A. Grant, Industrial Soya Protein Technology, in: Applied protein chemistry, Applied Science Publishers Ltd., London, 1980, pp. 87-112.

[2] Z. Teng, C. Liu, X. Yang, L. Li, C. Tang, Y. Jiang, Fractionation of soybean globulins using Ca(2+) and Mg(2+): a comparative analysis, J. Am. Oil Chem.' Soc. 86 (2009) 409-417. https://doi.org/10.1007/s11746-009-1367-6

[3] J.E. Kinsella, Functional properties of soy proteins, J. Am. Oil Chem. Soc. 56 (1979) 242-258. https://doi.org/10.1007/BF02671468

[4] D. Liu, H. Tian, J. Zeng, P. R. Chang, Core-shell nanoblends from soy protein/polystyrene by emulsion polymerization, Macromol. Mater. Eng. 293 (2008) 714-721. https://doi.org/10.1002/mame.200800119

[5] A. U. Daniels, K. P. Andriano, W.P. Smutz, M.K. Chang, J. Heller, Evaluation of absorbable poly (ortho esters) for use in surgical implants, J. Appl. Biomater. 5 (1994) 51-64. https://doi.org/10.1002/jab.770050108

[6] K.P. Andriano, T. Pohjonen, P. Törmälä, Processing and characterization of absorbable polylactide polymers for use in surgical implants, J. Appl. Biomater. 5 (1994) 133-140. https://doi.org/10.1002/jab.770050206

[7] G. Chen, T. Ushida, T. Tateishi, Hybrid biomaterials for tissue engineering: a preparative method for PLA or PLGA-collagen hybrid sponges, Adv. Mater. 12 (2000) 455-457. https://doi.org/10.1002/(SICI)1521-4095(200003)12:6<455::AID-ADMA455>3.0.CO;2-C

[8] J. J. van Soest, P. M. Kortleve, The influence of maltodextrins on the structure and properties of compression-molded starch plastic sheets, J. Appl. Polym. Sci. 74 (1999) 2207-2219. https://doi.org/10.1002/(SICI)1097-4628(19991128)74:9<2207::AID-APP10>3.0.CO;2-3

[9] I. Paetau, C. Z. Chen, J. L. Jane, Biodegradable plastic made from soybean products. 1. Effect of preparation and processing on mechanical properties and water absorption, Ind. Eng. Chem. Res. 33 (1994) 1821-1827. https://doi.org/10.1021/ie00031a023

[10] Chouke, P.B., Potbhare, A.K., Meshram, N.P., Rai, M.M., Dadure, K.M., Chaudhary, K., Rai, A.R., Desimone, M.F., Chaudhary, R.G., Masram, D.T. Bioinspired NiO nanospheres: Exploring in-vitro toxicity using Bm-17 and L. rohita liver cells, DNA degradation, docking and proposed vacuolization mechanism, ACS Omega, 7 (2022) 6869−6884. https://doi.org/10.1021/acsomega.1c06544

[11] Sonkusare, V., Chaudhary, R.G., Bhusari, G., Juneja, H.D. Microwave-mediated synthesis, photocatalytic degradation and antibacterial activity of α-Bi2O3 Microflower/γ-Bi2O3 Microspindle. Nano-Structures & Nano-Objects, 13 (2018) 121-131. https://doi.org/10.1016/j.nanoso.2018.01.002

[12] X. Qi, W. Wei, J. Shen, W. Dong, Salecan polysaccharide-based hydrogels and their applications: a review, J. Mater. Chem. 7 (2019) 2577-2587. https://doi.org/10.1039/C8TB03312A

[13] A. Latorre, P. Couleaud, A. Aires, A. L. Cortajarena, A. Somoza, Multifunctionalization of magnetic nanoparticles for controlled drug release: A general approach, Eur. J. Med. Chem. 82 (2014) 355-362. https://doi.org/10.1016/j.ejmech.2014.05.078

[14] L. Wei, J. Lu, H. Xu, A. Patel, Z. S. Chen, G. Chen, Silver nanoparticles: synthesis, properties, and therapeutic applications, Drug Disc. Tod. 20 (2015) 595-601. https://doi.org/10.1016/j.drudis.2014.11.014

[15] V. J. Mohanraj, Y. Chen, Nanoparticles-a review, Trop. J. Pharm. Res. 5(2006) 561-573. https://doi.org/10.4314/tjpr.v5i1.14634

[16] Catalano, P.N. Chaudhary, R.G. Martin F. Desimone, P.L. Santo, A Survey on Analytical methods for green synthesized nanomaterials. Current

[17] J. P. Rolland, B. W. Maynor, L. E. Euliss, A. E. Exner, G. M. Denison, J. M. DeSimone, Direct fabrication and harvesting of monodisperse, shape-specific nano biomaterials, J. Am. Chem. Soc. 127 (2005) 10096-10100. https://doi.org/10.1021/ja051977c

[18] Z. Teng, Y. Luo, Q. Wang, Nanoparticles synthesized from soy protein: preparation, characterization, and application for nutraceutical encapsulation, J. Agric. Food Chem. 60 (2012) 2712-2720. https://doi.org/10.1021/jf205238x

[19] J. Hadzieva, K. Mladenovska, C. M. Simonoska, M. Glavaš Dodov, S. Dimchevska, N. Geškovski, K. Goracinova, Lactobacillus casei encapsulated in soy protein isolate and alginate microparticles prepared by spray drying, Food Tech. Biotech. 55 (2017) 173-186. https://doi.org/10.17113/ftb.55.02.17.4991

[20] C. J. Verbeek, L. E. van den Berg, Extrusion processing and properties of protein-based thermoplastics, Macromol. Mater. eng. 295 (2010) 10-21. https://doi.org/10.1002/mame.200900167

[21] S. Y. Cho, J. W. Park, H. P. Batt, R. L. Thomas, Edible films made from membrane processed soy protein concentrates, LWT-Food Sci. Technol. 40 (2007) 418-423. https://doi.org/10.1016/j.lwt.2006.02.003

[22] R. R. Koshy, S. K. Mary, S. Thomas, L. A. Pothan, Environment friendly green composites based on soy protein isolate-A review, Food Hydrocoll. 50 (2015) 174-192. https://doi.org/10.1016/j.foodhyd.2015.04.023

[23] J. T. Kim, A. N. Netravali, Development of aligned-hemp yarn-reinforced green composites with soy protein resin: Effect of pH on mechanical and interfacial properties, Compos. Sci. Technol. 71 (2011) 541-547. https://doi.org/10.1016/j.compscitech.2011.01.004

[24] Z. Peles, I. Binderman, I. Berdicevsky, M. Zilberman, Soy protein films for wound-healing applications: antibiotic release, bacterial inhibition and cellular response, J. Tissue Eng. Regen. Med. 7(2013) 401-412. https://doi.org/10.1002/term.536

[25] O. Ozcalik, F. Tihminlioglu, Barrier properties of corn zein nanocomposite coated polypropylene films for food packaging applications, J. Food Eng. 114 (2013) 505-513. https://doi.org/10.1016/j.jfoodeng.2012.09.005

[26] M. Alexandre, P. Dubois, Polymer-layered silicate nanocomposites: preparation, properties and uses of a new class of materials, Mater. Sci. Eng. R Rep. 28 (2000) 1-63. https://doi.org/10.1016/S0927-796X(00)00012-7

[27] M. M. Rahman, A. N. Netravali, B. J. Tiimob, V. Apalangya, V. K. Rangari, Bio-inspired "green" nanocomposite using hydroxyapatite synthesized from eggshell waste and soy protein, J. Appl. Polym. Sci. 133 (2016). https://doi.org/10.1002/app.43477

[28] S. Zhao, J. Yao, X. Fei, X. Chen, An antimicrobial film by embedding in situ synthesized silver nanoparticles in soy protein isolate, Mater. Lett. 95 (2013) 142-144. https://doi.org/10.1016/j.matlet.2012.12.106

[29] M. N. Angles, A. Dufresne, Plasticized starch/tunicin whiskers nanocomposites, Macromol. 33(2000), 8344-8353. https://doi.org/10.1021/ma0008701

[30] H. Zheng, F. Ai, P. R. Chang, J. Huang, A. Dufresne, Structure and properties of starch nanocrystal-reinforced soy protein plastics, Polym. Compos. 30 (2009) 474-480. https://doi.org/10.1002/pc.20612

[31] S. Zhang, C. Xia, Y. Dong, Y. Yan, J. Li, S. Q. Shi, L. Cai, Soy protein isolate-based films reinforced by surface modified cellulose nanocrystal, Ind. Crops. Prod. 80 (2016) 207-213. https://doi.org/10.1016/j.indcrop.2015.11.070

[32] Y. Chen, L. Zhang, Blend membranes prepared from cellulose and soy protein isolate in NaOH/thiourea aqueous solution, J. Appl. Polym. Sci. 94 (2004) 748-757. https://doi.org/10.1002/app.20956

[33] H. Tian, G. Xu, Processing and characterization of glycerol-plasticized soy protein plastics reinforced with citric acid-modified starch nanoparticles, J. Polym. Environ. 19 (2011) 582-588. https://doi.org/10.1007/s10924-011-0304-6

[34] G. Guo, C. Zhang, Z. Du, W. Zou, A. Xiang, H. Li, Processing and properties of phthalic anhydride modified soy protein/glycerol plasticized soy protein composite films, J. Appl. Polym. Sci. 132 (2015). https://doi.org/10.1002/app.42221

[35] N, Reddy, Y. Yang, Soy protein fibers with high strength and water stability for potential medical applications, Biotechnol. Prog. 25 (2009) 1796-1802. https://doi.org/10.1002/btpr.244

[36] R.G. Chaudhary, M.S. Umekar, A.K. Potbhare, G.S. Bhusari, M.F. Desimone. Bioinspired reduced graphene oxide based nanohybrids for photocatalysis and antibacterial applications, Current Pharmaceutical Biotechnology, 22 (2021) 1759 - 1781. https://doi.org/10.2174/1389201022666201231115826

[37] N. Reddy, Y. Yang, Potential of plant proteins for medical applications, Trends Biotechnol. 29 (2011) 490-498. https://doi.org/10.1016/j.tibtech.2011.05.003

[38] M. Santin, C. Morris, G. Standen, L. Nicolais, L. Ambrosio, A new class of bioactive and biodegradable soybean-based bone fillers, Biomacromol. 8 (2007) 2706-2711. https://doi.org/10.1021/bm0703362

[39] A. Abaee, M. Mohammadian, S. M. Jafari, Whey and soy protein-based hydrogels and nano-hydrogels as bioactive delivery systems, Trends Food Sci. Technol. 70 (2017) 69-81. https://doi.org/10.1016/j.tifs.2017.10.011

[40] A. González, L. I. Tártara, S. D. Palma, C. I. A. Igarzabal, Crosslinked soy protein films and their application as ophthalmic drug delivery system, Mater. Sci. Eng. C. 51 (2015) 73-79. https://doi.org/10.1016/j.msec.2015.02.048

[41] P. Lavanya, N. Vijayakumari, Fabrication of Poly (d, l-Alanine)/minerals substituted hydroxyapatite bio-composite for bone tissue applications, Mater. Discov. 11 (2018) 14-18. https://doi.org/10.1016/j.md.2018.07.001

[42] T. Kelesidis, J. Fleisher, S. Tsiodras, Anaphylactoid reaction considered ciprofloxacin related: a case report and literature review, Clin. Ther. 32 (2010) 515-526. https://doi.org/10.1016/j.clinthera.2010.03.002

[43] S. Basu, H. S. Samanta, J. Ganguly, Green synthesis and swelling behavior of Ag-nanocomposite semi-IPN hydrogels and their drug delivery using Dolichos biflorus Linn, Soft Mater. 16 (2018) 7-19. https://doi.org/10.1080/1539445X.2017.1368559

[44] M. C. García, J. C. Cuggino, C. I. Rosset, P. L. Páez, M. C. Strumia, R. H. Manzo, A. F. Jimenez-Kairuz, A novel gel based on an ionic complex from a dendronized polymer and ciprofloxacin: Evaluation of its use for controlled topical drug release, Mater. Sci. Eng. 69 (2016) 236-246. https://doi.org/10.1016/j.msec.2016.06.071

[45] P. B. Kajjari, L. S. Manjeshwar, T. M. Aminabhavi, Novel interpenetrating polymer network hydrogel microspheres of chitosan and poly (acrylamide)-grafted-guar gum for controlled release of ciprofloxacin, Ind. Eng. Chem. Res. 50 (2011) 13280-13287. https://doi.org/10.1021/ie2012856

[46] K. Prusty, A. Biswal, S. B. Biswal, S. K. Swain, Synthesis of soy protein/polyacrylamide nanocomposite hydrogels for delivery of ciprofloxacin drug, Mater. Chem. Phy. 234 (2019) 378-389. https://doi.org/10.1016/j.matchemphys.2019.05.038

[47] P. Nayak, S. K. Sahoo, A. Behera, P. K. Nanda, P. L. Nayak, B. C. Guru, Synthesis and characterization of soy protein isolate/MMT nanocomposite film for the control release of the drug ofloxacin, World J. Nanosci. Eng. 1(2011) 27. https://doi.org/10.4236/wjnse.2011.12005

[48] A.K. Potbhare, P.B. Chouke, A. Mondal, R.U. Thakare, S. Mondal, Rhizoctonia solani assisted biosynthesis of silver nanoparticles for antibacterial assay, Materials Today: Proceedings, 29 (2020) 939-945. https://doi.org/10.1016/j.matpr.2020.05.419

[49] J. Koehler, F. P. Brandl, A. M. Goepferich, Hydrogel wound dressings for bioactive treatment of acute and chronic wounds, Eur. Polym. J. 100 (2018) 1-11. https://doi.org/10.1016/j.eurpolymj.2017.12.046

[50] C. M. Vaz, P. F. Van Doeveren, G. Yilmaz, L. A. De Graaf, R. L. Reis, A. M. Cunha, Processing and characterization of biodegradable soy plastics: Effects of crosslinking with glyoxal and thermal treatment, J. Appl. Polym. Sci. 97(2005) 604-610. https://doi.org/10.1002/app.20609

[51] C. M. Vaz, J. F. Mano, M. Fossen, R. F. Van Tuil, L. A. De Graaf, R. L. Reis, A. M. Cunha, Mechanical, dynamic-mechanical, and thermal properties of soy protein-based thermoplastics with potential biomedical applications, J. Macromol. Sci. Phys. 41 (2002) 33-46. https://doi.org/10.1081/MB-120002344

[52] G. A. Silva, C. M. Vaz, O. P. Coutinho, A. M. Cunha, R. L. Reis, In vitro degradation and cytocompatibility evaluation of novel soy and sodium caseinate-based membrane biomaterials, J. Mater. Sci.: Mater. Med. 14 (2003) 1055-1066. https://doi.org/10.1023/B:JMSM.0000004002.11278.30

[53] M. Santin, L. Ambrosio, Soybean-based biomaterials: preparation, properties and tissue regeneration potential, Expert Rev. Med. Devices. 5 (2008) 349-358. https://doi.org/10.1586/17434440.5.3.349

[54] J. M. Anderson, Biological responses to materials, Annu. Rev. Mater. Res. 31 (2001) 81-110. https://doi.org/10.1146/annurev.matsci.31.1.81

[55] Medicines and Health Care products Regulatory Agency. MDA/2004/047: Trilucent (soy bean oil filled). (Breast implants 2004)

[56] G. Standen, J. Salvage, C. Morris, Manufacture process control and cytotoxicity evaluation of soybean based biomaterials. Proceedings of the 19th European Society for Biomaterials conference. Sorrento, Italy 11-15 september, 2005

[57] L. Lin, A. Perets, Y. E. Har-el, D. Varma, M. Li, P. Lazarovici, D. L. Woerdeman, P. I. Lelkes, Alimentary 'green' proteins as electrospun scaffolds for skin regenerative engineering, J. Tissue Eng. Regen. Med. 7 (2013) 994-1008. https://doi.org/10.1002/term.1493

[58] S. Tansaz, A. R. Boccaccini, Biomedical applications of soy protein: A brief overview, J. Biomed. Mater. Res. 104 (2016) 553-569. https://doi.org/10.1002/jbm.a.35569

[59] M. Wu, F. Chen, P. Wu, Z. Yang, S. Zhang, L. Xiao, L. Cai, Nanoclay mineral-reinforced macroporous nanocomposite scaffolds for in situ bone regeneration: In vitro and in vivo studies, Mater. Des. 205 (2021) 109734. https://doi.org/10.1016/j.matdes.2021.109734

Materials Research Foundations 135 (2023) 284-303   https://doi.org/10.21741/9781644902172-12

Chapter 12

Emerging Nanomaterials in Healthcare

Ephraim Felix Marondedze[1]*, Lukman O. Olasunkanmi[1], Atheesha Singh[2], Penny Poomani Govender[1,3]

[1]Department of Chemical Sciences, University of Johannesburg, P. O. Box 17011, Doornfontein Campus, 2028, Johannesburg, South Africa

[2]Water and Health Research Centre, University of Johannesburg, Doornfontein Campus

[3]Research Capacity Development, Postgraduate School: Research & Innovation, University of Johannesburg, Kingsway Campus, South Africa

*ephraimm18@gmail.com

Abstract

Applications of nanomaterials in the field of medicine and healthcare have been on a rapid rise in recent years. Modifiable physical, optical and electronic structures of nanomaterials enable them to be fabricated for various uses. Examples of nanoparticles widely used in the healthcare sector include, but are not limited to silica (Si), aluminium oxide (Al_2O_3), silver (Ag), copper (Cu), titanium (Ti), gold (Au) and zinc (Zn). Application of nanomaterials in healthcare range from, bioimaging, sensing, diagnosis, targeted drug delivery, prosthetics, cancer therapy and antibiotics. Although mechanisms behind sensing and other functions are well known, mechanisms behind the antibiotic properties need more scientific validation. In this book chapter, we focus on current uses of nanomaterials in healthcare and give a brief insight on future perspectives on nanomaterials in medicine and healthcare.

Keywords

Nanomaterials, Healthcare, Drug-Delivery, Metal Oxide, Anti-Microbial Resistance

Contents

Emerging Nanomaterials in Healthcare 284

1.   Introduction..285

2.   Applications of graphene and other 2d materials in healthcare.......286

3.   Metal and metal oxides nanomaterials in healthcare.........................288

**4.   Selected specific applications of nanomaterials in
healthcare systems**..**290**

    4.1   Bioimaging 290

    4.2   Disease detection, sensing and diagnosis 290

    4.3   Prosthetic applications..290

    4.4   Nanomaterials in the fight against anti-microbial resistance 291

    4.5   Nanomaterials and drug delivery..291

    4.6   Nanomaterials in cancer therapy 292

5.   Future benefits and potential risks **293**

Conclusions..**294**

References **295**

1.   Introduction

Nanomaterials are popularly known to have nanoscale dimensions in the range of less than 100 nm and below. These materials exhibit special characteristics that have made them to be widely researched. They are known to possess unique molecular and electronic structures. Their fascinating properties have earned them numerous applications. Nanomaterials based researches have expanded from the investigations of their physical and chemical properties by nano-chemists, through applications in modern technology by nanotechnologists to disease diagnostic and therapeutic applications in nanomedicine. Nanomaterials could be purely inorganic, organic, or a hybrid. Their physicochemical properties and applications vary with the nature and/or size of material. In certain instances, properties of nanomaterials have been enhanced by composite formulations of two or more materials. The superior properties of nanomaterials compared to traditional materials are well established. This has made nanomaterials to gain tremendous applications in the health care sector.

Nanomaterials have been applied in diverse areas of modern medicine. These include biological imaging in which organic nanomaterials predominate. Inorganic nanomaterials have been widely used in the fabrication of biosensors. Hybrid and composite nanomaterials often display enhanced charge transport properties for more efficient biosensing devices.

This chapter presents comprehensive literature on applications of nanomaterials in health care. This is presented under a range of topics, which include graphene nanoparticles, metal oxide nanoparticles, biocompatible nanomaterials, nanomaterials in health care products etc.

2.   Applications of graphene and other 2d materials in healthcare

First, it must be mentioned that a 2018 review article by Jayakumar *et al.* [1] presents a view on a similar topic. Graphene is a sheet or network of sp2 hybridized carbon atoms with planar configuration. Graphite, carbon nanotube (CNT) and fullerene are all composed of graphene as basic structural unit [2]. The bonding system in graphene earns it stability, large surface area and electron conduction characteristics. This is because, graphene unit comprises C-C σ-bonds as strongest bonds and network of π-bonds for electron transportation. The π-bonds in graphene also allows weak interactions within graphene layers and between graphene and biomolecular substrate. The large surface area of graphene as compared to graphite and CNT is associated with the special electron configuration that earns it expanded layers [2, 3]. Some fascinating characteristics of graphene in comparison to other materials are listed in Table 1.

Graphene has found applications in the development of biosensors because of its unique properties, which include large electrochemical potential window, fast electron transportation (ET), large surface to volume ratio, biocompatibility, and excellent thermal and electrical conductivity [4-6]. Graphene is reported to have some advantages over metal nanoparticles and carbon nanotube (CNT). Discrepant amplification of electrochemical signal by metal nanoparticles and presence of toxic impurities that could change electrochemistry of CNT material make them to be less advantageous compared to graphene [5, 7]. These demerits of metal nanoparticles and CNT, which are overcome by graphene had been highlighted to promote the use of graphene-based materials in electrochemical biosensors fabrication.

A review article by Kuila *et al.* [7] gave an account of advances in the use of graphene-based materials in electrochemical biosensors. The review also compared the sensing efficiency of graphene-based and CNT sensors. Superior performances of graphene-based biosensors had been attributed to the ability of graphene to act as electron-transporting agent between electrochemically active biomolecules such as enzyme or protein and electrode surface [7]. Efficiency of graphene-based biosensors could be tuned by specially functionalizing the graphene and varying the degree of immobilization of biomaterial on its surface [7, 8].

Wu *et al.* [9] reported highly sensitive surface plasmon resonance (SPR) based graphene biosensor, which comprised graphene sheet coated on gold (Au) thin film, designated as graphene-on-Au(III) electrode system. In their study, Wu et al. used graphene as a biomolecular recognition element (BRE) with which Au was functionalized to enhance adsorption of biomolecules on Au surface.

Other graphene-based 2D materials that are famously used in biosensors are graphene oxide (GO) and reduced graphene oxide (rGO). GO is oxygenated form of graphene that contains oxygen-based functional groups such as alcohols, epoxides, hydroxyls, carboxyls etc. It is also in the form of layered sheets. The oxygen functional groups are on the edge and surface of the sheets and the carbon to oxygen ratio is about 3:1 [2, 10]. Oxygen functional groups in the GO might enhance aqueous solubility, biocompatibility and charge transport property compared to graphene [2, 11]. Special characteristics are also conferred on rGO, which is the reduced form of GO with structural defects and high thermal conductivity [2, 12]. The oxygen functional groups in GO and rGO enrich them in negative charges, making them to adsorb positively charged biomolecules easily [13]. A list of some 2D graphene-based and non-graphene materials that have shown potentials in health care applications is contained in Table 2. A more elaborate list and broad classification could be found in the work of Application of graphene-based 2D materials in health care has been reported extensively in the area of biosensors fabrication. A 2020 review article by Jiang *et al.* [13] gave an extensive overview of applications of graphene-based materials in the development of biosensors for bacterial and viral pathogens. The review also serves as a good literature on preparation, functionalization and properties of graphene. Jiang *et al.* presented graphene biosensors for pathogens in three categories based on functionalization materials. This was because coupling of biomolecules such as nucleic acids, lectins, peptides etc. with nanomaterials afford interesting interactions with pathogens [13], leading to a more sensitive and efficient sensing and diagnostic device [14]. A review on graphene-based "wearable sensors for healthcare monitoring" was produced by Zhang *et al.* [15]. Few specific cases of applications of graphene-based 2D materials in sensing devices as well as for other healthcare purposes are hereby highlighted.

- Rather *et al.* [16] reported biosensing of hydroxylated polychlorobipheyls (OH-PCBs) with a fabricated platform comprising composite materials - tyrosinase (Tyr) and zinc oxide quantum dots (ZnO QDs), immobilized on graphene oxide (GO).

- Drug delivery assistance functionality and therapeutic characteristics of graphene has been documented. Gelatin conjugated graphene was reported by Liu *et al.* [17]to exhibit good efficiency as a drug career. In a similar study, GO material functionalized with polyethylene glycol (PEG), termed PEGylated nano-GO was found to be a good nanocarrier for water-insoluble cancer drugs. Functionalization of GO with PEG convers high aqueous solubility on GO [18].

- Health-related unfavourable characteristics of GO such as hemolysis and thrombotoxicity had been reportedly suppressed by amine modification of GO. Amine-modified GO (GO-NH$_2$) has been reported to find good applications in intravenous injections. Injections based on GO-NH$_2$ would not arouse platelets or cause thromboembolism [19].

- In a case where biocompatibility is sought in healthcare applications of GO, polymer-modified GO could be suitable. This biocompatibility of GO as healthcare material could be tuned via polymer modification such as functionalization with polyacrylic acid (PAA), polyacryl amide (PAM) and PEG [20].

Table 1: Selected properties of graphene in comparison to some other materials.

Selected Properties of Graphene	Value	Comparison with Other Materials	Ref.
Surface Area	$2630 \ m^2 \ g^{-1}$	About 2 times that of CNT. About 263 times that of graphite	[2, 21]
Tensile Strength	130 gigapascal	About 200 times that of steel	[2, 21]
Electrical Conductivity: Graphene Graphene Particles	$10^8 \ mS \ cm^{-1}$ $64 \ mS \ cm^{-1}$	- About 60 times that of single-walled CNT	[21, 22]

Table 2: Some 2D graphene-based and non-graphene materials used in health care [1].

Classification	Example
Carbon-based Materials	Graphene
	Graphene oxide (GO)
	Reduced graphene oxide (rGO)
	Graphane
	Fluorinated graphene
	Graphyne
	Graphdiyne
2D Clay Materials	Hydrotalcite
	Kaolinite
	Laponite
	Layered double hydroxide (LDH)
Transition metal dichalcogenides (TMDs)	Molybdenum disulphide (MoS_2)
	Tungsten disulphide (WS_2) etc.
Transition metal oxides (TMOs)	NiO, TiO, ZnO etc.
Black phosphorous (BP)	-
Graphitic carbon nitride (g-C_3N_4)	-
Hexagonal boron nitride (hBN)	-
Silicene and germanene	-
Arsenene and antimonene	-

3.   Metal and metal oxides nanomaterials in healthcare

Metal and metal oxides are usually included in formulations for healthcare products. Some of the inorganic materials often added in healthcare products formulations include silver, zinc, titanium, iron, gold etc. and their respective oxides. Metal oxides or materials that contain them often display special characteristics such as fascinating catalytic, magnetic, electrical, optical, light-absorption and emission properties. They are also known to be thermally stable and cost-effective [23]. These features enhanced their applications in many areas, including biosensor fabrications, drug formulations and drug delivery. Biomolecules such as DNA could be anchored on metal oxide surface by using small molecules such as dopamine as a linking bridge or immobilizer. This is because, metal oxide nanomaterials can bind small molecules with high affinity and these molecules in turn serve as a bridge between oxide surface and biomolecules [23-25]. This is an essential mechanism associated with the functions of metal oxides in biosensors as transduction and/or target recognition enhancer. Interactions between metal oxide nanomaterials and biomolecules mainly form the basis of applications of metal oxide nanomaterials in healthcare. Some of these interactions and associated mechanism have been discussed in literature [23, 26]. Stability and tuneable dissolution of metal oxides in biological media are other reasons for their famous applications in healthcare. Zhang *et al.* [27] presented a comparative assessment of dissolution of metal oxide nanoparticles in water and biological media and showed that ZnO, CuO and WO_3 have high dissolution rate in cell culture.

Catalytic properties of metal oxides nanomaterials also make them suitable for biosensor development. Their catalytic activities have been greatly discussed in literature. Gao *et al.* [28] explored the "intrinsic peroxidase-like activity of ferromagnetic nanoparticles" and reported that nanoparticles of Fe_3O_4 could be modified with antibody for the development of immunoassay with potential applications in medicine. Enzyme-mimic activity of Fe_2O_3 nanoparticles and its potential applications in biosensor development was reported by Chaudhari *et al.* [29]. Similarly, peroxidase-like activities of V_2O_5 [30], Co_3O_4 [31, 32], and CuO nanoparticles have also been reported [33]. Few cases of applications of metal oxides nanomaterials in biosensor fabrication are presented in Table 3.

Table 3: Application of metal oxide nanomaterials

Metal/Metal oxide	Biological target	References
Nanoceria (CeO_2)	Ampliflu	[34]
Fe_3O_4-chitosan	Glucose	[35]
[a]nFe_2O_3@PEDOT:PSS	carcinoembryonic antigen	[36]
Thiolated AuNP	Glucose	[37]
MnO_2-Co_3O_4	Chloride ions	[38]
GO/CCNTs/AuNPs@MnO2	L-cysteine	[39]
Mn_3O_4@SiO_2	Rifampicin	[40]
TiO_2	dissolved CO	[41]
Pd@TiO_2–SiC	hydroquinone and bisphenol	[42]
[b]MIP-TiO_2	4-nonylphenol	[43]

[a]poly(3,4-ethylenedioxythiophene):poly(styrenesulfonate) and nanostructured iron oxide
[b]Molecularly imprinted

4.   Selected specific applications of nanomaterials in healthcare systems

4.1   Bioimaging

Nanomaterials have been proven by various researchers to be applicable in healthcare systems for diagnosis and monitoring purposes particularly because of their modifiable optical and electronic properties in addition to biological compatibility [44]. Among other methods, imaging, simply referred to as bioimaging is one of the many uses of nanomaterials. Several nanoprobes have been fabricated for imaging biological tissues for the purposes diagnosis and treatment of diseases such as Alzheimer's disease or cancer [45]. Bioimaging is also important for in-vivo monitoring of various responses during disease treatment more especially when using targeted therapy approach [46]. Currently, imaging techniques widely used include X-ray imaging [47], magnetic resonance [48], fluorescence [49], luminescence [50], photoacoustics [51], radionuclides [52], and mass spectrometry [53].

4.2   Disease detection, sensing and diagnosis

Perhaps one of the most widely known uses of nanomaterials in healthcare applications. Metal-based nanomaterials such as metal platinum, gold, silver, and nickel, for example, together with graphene composites often are central in fabricating nanomaterials for biosensing and disease diagnostic purposes. Qin et al [54] employed graphene oxide nanosheets in conjunction with gold nanowire arrays to fabricate sensors with high selectivity and sensitivity towards glucose. The findings of this study can be used for diagnosing and or monitoring blood glucose levels. Another example is by Jin et al [55]. Their research aimed to develop a nanomaterial-based electrochemical sensor using graphene and gold nanoparticles. Their work yelded a biosensor with limit of detection in the nanomolar range paving way to early and accurate diagnosis of cancer. Several other review articles [56-60] have dissected in detail various nanoparticles and their use in biosensors for disease detection further demonstrating the importance of nanomaterials in healthcare.

4.3   Prosthetic applications

In recent years, many people have become increasingly depended on prosthetics. To be in synchrony with various functions of the human body, several prosthetics require biosensing abilities. As a result, nanomaterials are now widely used in various prosthetics because of their biosensing properties. Examples include, but are not limited to, electronic skin (e-skin), e-skin's neurotactility sensing and related biosensors [61]. Yao and co-workers employed a facile approach to fabricate graphene arrays on a latex film to develop an e-skin. Their research exploited not only morphological properties, but electronic and optical properties of nanomaterials used to fabricate a highly flexible e-skin with multifunctional sensing abilities comparable to those found on human skin [62]. In other studies [63, 64], piezoresistive sensors exhibiting high-sensitivity were fabricated and embedded in a prosthetics and intelligent robotics.

4.4   Nanomaterials in the fight against anti-microbial resistance

The continual rise of antimicrobial resistance (AMR) including resistance to antibiotics is a global burden on with negative social and economic impacts worldwide. Annually, billions of dollars are invested towards the fight against emergence of new, drug resistant microbial strains. If improperly managed, AMR poses serious threats on human, animal, and aquatic lives. As a result, recent years have seen increased research efforts aimed at novel ways to combat AMR. Among these, the use of nanomaterials has been and still is being investigated by various researchers to solve the AMR problem. Examples of such nanoparticles include, but are not limited to silica (Si), aluminium oxide (Al_2O_3), silver (Ag), copper (Cu), titanium (Ti), gold (Au) and zinc (Zn). Wang et al [65] presented research that utilized nanowires fabricated with Ag nanoparticles for antibiotic purposes. Their findings demonstrated improved potency of various antibiotics and reduced rate of antibiotic resistance build-up. In a separate study, Besinis et al investigated antibiotic properties of Ag nanoparticles against *streptococcus mutans*. Chlorhexidine, commonly used as an oral disinfectant, was used as a control. Their research findings indicated that silver-based nanoparticles had better antibiotic activity in comparison to the control, chlorhexidine. Rate of bacterial growth in with Ag nanoparticles was reduced 25-fold when compared to that of the control, chlorhexidine [66].

Nanomaterials possess properties that enable them to neutralize the growth of pathogens such as fungi, protozoa, viruses, and bacteria. Biomechanisms behind the mode of actions of nanoparticles still require more scientific investigations. However, mechanisms such as metal ion release, oxidative stress induction, and non-oxidative mechanisms have been hypothesized [67]. The research cited research by Wang et al [65] hypothesized that the nanoparticles presented therein function via a combination of physical and chemical pathways to induce antibiotic effects. Chemical pathways suggested includes, reactive oxygen species (ROS), adenosine triphosphate (ATP) consumption, biofilm destruction and membrane depolarization.

Another study by Nagy et al focused on antibacterial properties of silver nanoparticles embedded within a zeolite membrane (AgNP-ZM) [68]. Their findings proved that AgNP-ZM induce antibiotic properties through the gradual release of Ag^+ ions. Similar observations of metal ion release were noted in the research presented by Kaushik and colleagues [69]. A dual mechanism involving both meta ion (Zn^{2+}) release and oxidative stress was reported bt Yang et al [70]. Their work focused on the antibiotic properties for zinc oxide (ZnO) nanoparticles against *Escherichia coli (E. coli)* and *Staphylococcus aureus (S. aureus)* [70].

4.5   Nanomaterials and drug delivery

Although several nanoparticles possess antibiotic properties, they can also be used as carrier mediums of several therapeutic drugs – more especially in targeted drug delivery [67]. Examples nanoparticles often used in drug delivery systems include, but are not limited to silver carbon-based (for example graphene-based nanomaterials), (Ag), copper

Emerging Nanomaterials and Their Impact on Society in the 21st Century   Materials Research Forum LLC
Materials Research Foundations 135 (2023) 284-303   https://doi.org/10.21741/9781644902172-12

(Cu), gold (Au), zinc (Zn), and polymer-based nanomaterials (for example chitosan and its derivatives) [71]. Nanomaterials aim to achieve targeted drug delivery and controlled (rapid or extended) drug release. Delayed/extended release is important in minimizing interaction of antibiotics with unintended sites/microorganisms thereby decreasing the chances of the pathogen developing antimicrobial resistance. [71]. Currently, various methods have been developed to enhance nanomaterials-based drug delivery systems. These strategies include, but are not limited to, coating, adsorption, impregnation, conjugation/bioconjugation and loading [71]. Mehrabi eta al [72] fabricated Ag-coated nanoparticles for use as drug delivery agents for trimethoprim and sulfamethoxazole, antibiotics commonly used against *E. Coli* and *S. aureus.* Their results demonstrated enhanced minimum inhibitory concentration (MIC) values against the selected bacteria species. In a separate report, Paudel et al [73] impregnated enrofloxacin, an antibiotic potent against gram-negative bacteria, in poly(lactic-co-glycolic acid) nanoparticles. Controlled/delayed release was noted to take up to 4.2 days. In terms of bioactivity, high efficacy against *E. coli* and *S. aureus* [73]. Both these studies by [72] and [73] demonstrate that the use of the characteristics for use nanomaterials-based drug delivery systems on antibiotics leads to improved controlled release and therefore smaller doses compared to their traditional forms. Morphologically, various nanomaterials possess is drug delivery systems. Examples of such nanomaterials include, carbon nanomaterials, quantum dots, liposomal nanomaterials and polymer-based nanomaterials [67].

4.6   Nanomaterials in cancer therapy

Besides AMR stated above, cancer as a disease is one of the biggest concerns and leading cause of death globally. Although chemotherapy is often the method of choice for cancer treatment, the method lacks target specificity and drugs used exhibit high toxicity levels in various organs and body tissues. Furthermore, drug resistance is one of the drawbacks associated with chemotherapy [74]. Novel methods that exhibit both high target specificity and low toxicity are required. Among current efforts to identify these methods, researchers have investigated the use of nanomaterials to target tumour cells so as to overcome toxicity and lack of specificity, in addition to enhancing drug capacity as well as bioavailability [75].

Lin et al [76] incorporated folic acid into nanomaterials with the aim for controlled release and targeted distribution of metformin towards lung cancer cells. Findings indicated that, the drug metformin was released quicker at pH 6.4% than pH 7.4% demonstrating that the nanomaterials enabled controlled release of the drug around lung cancer cells and restricted release of the drug around non-cancerous cells. Another research involving a pH responsive nanomaterial was presented by Li et al [77]. Both these research [76, 77] demonstrate that nanomaterials can be biocompatible and achieve targeted therapy and controlled release of drugs. Several other nanomaterials were discussed in review articles such as [78-80]

5.   Future benefits and potential risks

Nanotechnology in medicine has seen many setbacks and advances through the past decade with the most promising developments being improving the characteristics and the processing of nano-medical agents and devices for the effective treatment of cardiovascular disease including cancer and tumour therapies. This manipulative molecular level technology has given rise to the development of innovative specific drug targeting, improved biocompatibility, and security without compromising its therapeutic effect [81]. Nano-medical tools (genetic engineering, particle imaging and computational chemistry) has improved the treatment of human disease by architecting better delivery systems to accelerate treatment outcomes [82]. As one of the most inventive nodes in healthcare technology, the application of nanomedicine holds the potential to endorse radical changes in medical communication and automation in human health. Nanomedicine has offered great treatment potential in the COVID-19 pandemic spanning from *viral detection, neutralization,* and *vaccine developments* [83, 84]. Thousands of laboratories around the world are currently working on the development of COVID-19 preventive measures and treatments [85]. In neutralization studies by Zhang and colleagues [86], cellular nanoparticles were created similar to ACE-2 and CD147 receptors that mediate viral infection of host cells, and they showed that once the SARS- CoV-2 virus is bonded to the nanoparticle surface receptor, it is no longer able to infect its normal cell target. Another benefit of nanomedicine is the targeting of viral RNA using RNA interference (RNAi) for inhibition and antiviral activity [83]. Lipid nanoparticles and viral-like nanoparticles have been used for vital detection and production of protective equipment. These technologies although promising are still in the pre-clinical phase and requires further validation before its true impact is realised and can be applied to clinical trials. "Smart drug delivery systems" for cancer therapy that have active targeting and stimuli sensitivity is a positive strategy that can eradicate cancer tissues while shielding regular tissue [87]. The precise release of targeted drugs, assessment technology of the drug's effect in diseased tissues as well as theoretical models of predication have not yet been achieved. The most beneficial feature of nanomedicine is that it provides a general platform that can easily adapt to the application required.

Despite the benefit accompanied with this new era medicine, the risk profile and hazard susceptibility still exist and differ for each nanotechnological application. The robustness of nanomedicines has been questioned by restrictive agencies, due to the lack of precise procedures or guidelines to characterise their physico-chemical properties. This shortfall impacts the accountability of the failure of clinical tests and subsequent potential for clinical development. Collaboration amongst the various stakeholders, as well as trade and restrictive agencies remains. The United States Food and Drug Administration approval for nanomedicines are based on three aspects of nano-preparation: chemical effects (medicine), mechanical effects (devices), biological sources (biological), whilst some preparations cover all aspects (blended product) [88]. Additionally, the complex formulations, production costs and approval processes prolong the health benefits to users. Although nanomedicine can be more effective than conventional medicine, the complex design, manufacturing and testing requirements is more costly and clinical transformation may never be realized.

Nanotoxicology, the toxic consequence of nanomedicine on human health, is a "double-edged sword" and poses a challenge for future medicine due to personalised genetic susceptibility, its overuse, and abuse [89, 90]. Toxicity of the complex nanomaterials, triggered by cellular interactions has been shown by Next generation technologies, including proteomics. This toxicity may be due to size, shape, movement, aggregation, and the amplified surface area increases chemical reactivity, causing instability of these particles [91]. The diverse physico-chemical properties from these micron-sized elements, can cause changes in body distribution, blood-brain barrier passage, and blood coagulation pathways [81]. Occupational exposure either through manufacturing or direct application can also result in residues of the nanomedicine being released into the environment, with greatest concern being the inhalation and absorption of residues on skin [91]. Clear validation to control the related risks, requires further medical trials involving nanomedicine, and these should be implemented together with multidisciplinary investigative research in the fields of nanotechnology, biomedical engineering, and medicine.

The future of nanomedicine is beneficial, with the industrialised technologies refining and enhancing treatments and diagnostics, and machine learning functions augmenting to save time and resources. Situational approaches should be used to assess the advantages and disadvantages of using nanoparticles for medical diagnosis and treatment.

Conclusions

This chapter highlights the importance on nanomaterials in healthcare. Nanomaterials often exhibit a wide range of uses because of their easiness to fabricate and modifiable physical, optical and electronic properties. As a result of these properties nanomaterials have been applied in diverse areas of modern medicine. These include biological imaging in which organic nanomaterials predominate. Sensing and diagnosis and treatment of diseases all exploit properties of nanomaterials. In recent times, antimicrobial resistance has become a global concern and new methods to combat this are required. Among these, the use of nanomaterials has been and still is being investigated by various researchers to solve the AMR problem. Examples of nanoparticles include, but are not limited to silica (Si), aluminium oxide (Al_2O_3), silver (Ag), copper (Cu), titanium (Ti), gold (Au) and zinc (Zn). Furthermore, nanomaterials are being used in drug delivery systems. Current results on drug delivery demonstrate that the use of nanomaterials-based drug delivery systems on antibiotics leads to improved controlled release and therefore smaller doses compared to their traditional forms. Lastly, nanomaterials to target tumour cells are key to overcome toxicity and lack of specificity, in addition to enhancing drug capacity as well as bioavailability

Materials Research Foundations 135 (2023) 284-303   https://doi.org/10.21741/9781644902172-12

References

[1] A. Jayakumar, A. Surendranath, P. Mohanan, 2D materials for next generation healthcare applications, International journal of pharmaceutics 551 (2018) 309-321. https://doi.org/10.1016/j.ijpharm.2018.09.041

[2] C.I. Justino, A.R. Gomes, A.C. Freitas, A.C. Duarte, T.A. Rocha-Santos, Graphene based sensors and biosensors, TrAC Trends in Analytical Chemistry 91 (2017) 53-66. https://doi.org/10.1016/j.trac.2017.04.003

[3] X. Huang, Z. Yin, S. Wu, X. Qi, Q. He, Q. Zhang, Q. Yan, F. Boey, H. Zhang, Graphene-based materials: synthesis, characterization, properties, and applications, small 7 (2011) 1876-1902. https://doi.org/10.1002/smll.201002009

[4] T. Terse-Thakoor, S. Badhulika, A. Mulchandani, Graphene based biosensors for healthcare, Journal of Materials Research 32 (2017) 2905-2929. https://doi.org/10.1557/jmr.2017.175

[5] M. Pumera, Graphene-based nanomaterials and their electrochemistry, Chemical Society Reviews 39 (2010) 4146-4157. https://doi.org/10.1039/c002690p

[6] M.J. Allen, V.C. Tung, R.B. Kaner, Honeycomb carbon: a review of graphene, Chemical reviews 110 (2010) 132-145. https://doi.org/10.1021/cr900070d

[7] T. Kuila, S. Bose, P. Khanra, A.K. Mishra, N.H. Kim, J.H. Lee, Recent advances in graphene-based biosensors, Biosensors and bioelectronics 26 (2011) 4637-4648. https://doi.org/10.1016/j.bios.2011.05.039

[8] Y. Shao, J. Wang, H. Wu, J. Liu, I.A. Aksay, Y. Lin, Graphene based electrochemical sensors and biosensors: a review, Electroanalysis: An International Journal Devoted to Fundamental and Practical Aspects of Electroanalysis 22 (2010) 1027-1036. https://doi.org/10.1002/elan.200900571

[9] L. Wu, H.-S. Chu, W.S. Koh, E.-P. Li, Highly sensitive graphene biosensors based on surface plasmon resonance, Optics express 18 (2010) 14395-14400. https://doi.org/10.1364/OE.18.014395

[10] J. Chang, G. Zhou, E.R. Christensen, R. Heideman, J. Chen, Graphene-based sensors for detection of heavy metals in water: a review, Analytical and bioanalytical chemistry 406 (2014) 3957-3975. https://doi.org/10.1007/s00216-014-7804-x

[11] X. Deng, H. Tang, J. Jiang, Recent progress in graphene-material-based optical sensors, Analytical and bioanalytical chemistry 406 (2014) 6903-6916. https://doi.org/10.1007/s00216-014-7895-4

[12] Q. Bao, K.P. Loh, Graphene photonics, plasmonics, and broadband optoelectronic devices, ACS nano 6 (2012) 3677-3694. https://doi.org/10.1021/nn300989g

[13] Z. Jiang, B. Feng, J. Xu, T. Qing, P. Zhang, Z. Qing, Graphene biosensors for bacterial and viral pathogens, Biosensors and Bioelectronics 166 (2020) 112471. https://doi.org/10.1016/j.bios.2020.112471

[14] C. Zhu, D. Du, Y. Lin, Graphene and graphene-like 2D materials for optical biosensing and bioimaging: A review, 2D Materials 2 (2015) 032004. https://doi.org/10.1088/2053-1583/2/3/032004

[15] H. Zhang, R. He, Y. Niu, F. Han, J. Li, X. Zhang, F. Xu, Graphene-enabled wearable sensors for healthcare monitoring, Biosensors and Bioelectronics 197 (2022) 113777. https://doi.org/10.1016/j.bios.2021.113777

[16] J.A. Rather, S. Pilehvar, K. De Wael, A graphene oxide amplification platform tagged with tyrosinase-zinc oxide quantum dot hybrids for the electrochemical sensing of hydroxylated polychlorobiphenyls, Sensors and Actuators B: Chemical 190 (2014) 612-620. https://doi.org/10.1016/j.snb.2013.09.018

[17] K. Liu, J.-J. Zhang, F.-F. Cheng, T.-T. Zheng, C. Wang, J.-J. Zhu, Green and facile synthesis of highly biocompatible graphene nanosheets and its application for cellular imaging and drug delivery, Journal of Materials Chemistry 21 (2011) 12034-12040. https://doi.org/10.1039/c1jm10749f

[18] Z. Liu, J.T. Robinson, X. Sun, H. Dai, PEGylated nanographene oxide for delivery of water-insoluble cancer drugs, Journal of the American Chemical Society 130 (2008) 10876-10877. https://doi.org/10.1021/ja803688x

[19] S.K. Singh, M.K. Singh, P.P. Kulkarni, V.K. Sonkar, J.J. Grácio, D. Dash, Amine-modified graphene: thrombo-protective safer alternative to graphene oxide for biomedical applications, ACS nano 6 (2012) 2731-2740. https://doi.org/10.1021/nn300172t

[20] M. Xu, J. Zhu, F. Wang, Y. Xiong, Y. Wu, Q. Wang, J. Weng, Z. Zhang, W. Chen, S. Liu, Improved in vitro and in vivo biocompatibility of graphene oxide through surface modification: poly (acrylic acid)-functionalization is superior to PEGylation, Acs Nano 10 (2016) 3267-3281. https://doi.org/10.1021/acsnano.6b00539

[21] E.B. Bahadır, M.K. Sezgintürk, Applications of graphene in electrochemical sensing and biosensing, TrAC Trends in Analytical Chemistry 76 (2016) 1-14. https://doi.org/10.1016/j.trac.2015.07.008

[22] S. Alwarappan, A. Erdem, C. Liu, C.-Z. Li, Probing the electrochemical properties of graphene nanosheets for biosensing applications, The Journal of Physical Chemistry C 113 (2009) 8853-8857. https://doi.org/10.1021/jp9010313

[23] B. Liu, J. Liu, Sensors and biosensors based on metal oxide nanomaterials, TrAC Trends in Analytical Chemistry 121 (2019) 115690. https://doi.org/10.1016/j.trac.2019.115690

[24] T. Paunesku, T. Rajh, G. Wiederrecht, J. Maser, S. Vogt, N. Stojićević, M. Protić, B. Lai, J. Oryhon, M. Thurnauer, Biology of TiO2-oligonucleotide nanocomposites, Nature materials 2 (2003) 343-346. https://doi.org/10.1038/nmat875

[25] C. Xu, K. Xu, H. Gu, R. Zheng, H. Liu, X. Zhang, Z. Guo, B. Xu, Dopamine as a robust anchor to immobilize functional molecules on the iron oxide shell of magnetic nanoparticles, Journal of the American Chemical Society 126 (2004) 9938-9939. https://doi.org/10.1021/ja0464802

[26] M.J. Limo, A. Sola-Rabada, E. Boix, V. Thota, Z.C. Westcott, V. Puddu, C.C. Perry, Interactions between metal oxides and biomolecules: from fundamental understanding to applications, Chemical reviews 118 (2018) 11118-11193. https://doi.org/10.1021/acs.chemrev.7b00660

[27] H. Zhang, Z. Ji, T. Xia, H. Meng, C. Low-Kam, R. Liu, S. Pokhrel, S. Lin, X. Wang, Y.-P. Liao, Use of metal oxide nanoparticle band gap to develop a predictive paradigm for oxidative stress and acute pulmonary inflammation, ACS nano 6 (2012) 4349-4368. https://doi.org/10.1021/nn3010087

[28] L. Gao, J. Zhuang, L. Nie, J. Zhang, Y. Zhang, N. Gu, T. Wang, J. Feng, D. Yang, S. Perrett, Intrinsic peroxidase-like activity of ferromagnetic nanoparticles, Nature nanotechnology 2 (2007) 577-583. https://doi.org/10.1038/nnano.2007.260

[29] K.N. Chaudhari, N.K. Chaudhari, J.-S. Yu, Peroxidase mimic activity of hematite iron oxides (α-Fe 2 O 3) with different nanostructures, Catalysis Science & Technology 2 (2012) 119-124. https://doi.org/10.1039/C1CY00124H

[30] R. André, F. Natálio, M. Humanes, J. Leppin, K. Heinze, R. Wever, H.C. Schröder, W.E. Müller, W. Tremel, V2O5 nanowires with an intrinsic peroxidase-like activity, Advanced Functional Materials 21 (2011) 501-509. https://doi.org/10.1002/adfm.201001302

[31] J. Mu, Y. Wang, M. Zhao, L. Zhang, Intrinsic peroxidase-like activity and catalase-like activity of Co 3 O 4 nanoparticles, Chemical Communications 48 (2012) 2540-2542. https://doi.org/10.1039/c2cc17013b

[32] J. Dong, L. Song, J.-J. Yin, W. He, Y. Wu, N. Gu, Y. Zhang, Co3O4 nanoparticles with multi-enzyme activities and their application in immunohistochemical assay, ACS Applied Materials & Interfaces 6 (2014) 1959-1970. https://doi.org/10.1021/am405009f

[33] W. Chen, J. Chen, A.L. Liu, L.M. Wang, G.W. Li, X.H. Lin, Peroxidase-like activity of cupric oxide nanoparticle, ChemCatChem 3 (2011) 1151-1154. https://doi.org/10.1002/cctc.201100064

[34] A. Asati, C. Kaittanis, S. Santra, J.M. Perez, pH-tunable oxidase-like activity of cerium oxide nanoparticles achieving sensitive fluorigenic detection of cancer biomarkers at neutral pH, Analytical chemistry 83 (2011) 2547-2553. https://doi.org/10.1021/ac102826k

[35] A. Kaushik, R. Khan, P.R. Solanki, P. Pandey, J. Alam, S. Ahmad, B. Malhotra, Iron oxide nanoparticles-chitosan composite based glucose biosensor, Biosensors and bioelectronics 24 (2008) 676-683. https://doi.org/10.1016/j.bios.2008.06.032

[36] S. Kumar, M. Umar, A. Saifi, S. Kumar, S. Augustine, S. Srivastava, B.D. Malhotra, Electrochemical paper based cancer biosensor using iron oxide nanoparticles decorated PEDOT: PSS, Analytica Chimica Acta 1056 (2019) 135-145. https://doi.org/10.1016/j.aca.2018.12.053

[37] P. Pandey, S.P. Singh, S.K. Arya, V. Gupta, M. Datta, S. Singh, B.D. Malhotra, Application of thiolated gold nanoparticles for the enhancement of glucose oxidase activity, Langmuir 23 (2007) 3333-3337. https://doi.org/10.1021/la062901c

[38] M.M. Rahman, S.B. Khan, G. Gruner, M.S. Al-Ghamdi, M.A. Daous, A.M. Asiri, Chloride ion sensors based on low-dimensional α-MnO2-Co3O4 nanoparticles fabricated glassy carbon electrodes by simple I-V technique, Electrochimica Acta 103 (2013) 143-150. https://doi.org/10.1016/j.electacta.2013.04.067

[39] X. Wang, C. Luo, L. Li, H. Duan, Highly selective and sensitive electrochemical sensor for L-cysteine detection based on graphene oxide/multiwalled carbon nanotube/manganese dioxide/gold nanoparticles composite, Journal of Electroanalytical Chemistry 757 (2015) 100-106. https://doi.org/10.1016/j.jelechem.2015.09.023

[40] T. Gan, Z. Shi, K. Wang, J. Sun, Z. Lv, Y. Liu, Rifampicin determination in human serum and urine based on a disposable carbon paste microelectrode modified with a hollow manganese oxide@ mesoporous silica oxide core-shell nanohybrid, Canadian Journal of Chemistry 93 (2015) 1061-1068. https://doi.org/10.1139/cjc-2015-0017

[41] E. Topoglidis, T. Lutz, R.L. Willis, C.J. Barnett, A.E. Cass, J.R. Durrant, Protein adsorption on nanoporous TiO 2 films: a novel approach to studying photoinduced protein/electrode transfer reactions, Faraday Discussions 116 (2000) 35-46. https://doi.org/10.1039/b003313h

[42] L. Yang, H. Zhao, S. Fan, B. Li, C.-P. Li, A highly sensitive electrochemical sensor for simultaneous determination of hydroquinone and bisphenol A based on the ultrafine Pd nanoparticle@ TiO2 functionalized SiC, Analytica Chimica Acta 852 (2014) 28-36. https://doi.org/10.1016/j.aca.2014.08.037

[43] J. Huang, X. Zhang, S. Liu, Q. Lin, X. He, X. Xing, W. Lian, D. Tang, Development of molecularly imprinted electrochemical sensor with titanium oxide and gold nanomaterials enhanced technique for determination of 4-nonylphenol, Sensors and Actuators B: Chemical 152 (2011) 292-298. https://doi.org/10.1016/j.snb.2010.12.022

[44] Y.Y. Hui, H.-C. Chang, H. Dong, X. Zhang, Carbon nanomaterials for bioimaging, bioanalysis, and therapy, John Wiley & Sons2019. https://doi.org/10.1002/9781119373476

[45] Y. Yang, L. Wang, B. Wan, Y. Gu, X. Li, Optically Active Nanomaterials for Bioimaging and Targeted Therapy, Frontiers in Bioengineering and Biotechnology 7 (2019). https://doi.org/10.3389/fbioe.2019.00320

[46] W. Fan, B. Yung, P. Huang, X. Chen, Nanotechnology for Multimodal Synergistic Cancer Therapy, Chemical Reviews 117 (2017) 13566-13638. https://doi.org/10.1021/acs.chemrev.7b00258

[47] L.E. Cole, R.D. Ross, J.M. Tilley, T. Vargo-Gogola, R.K. Roeder, Gold nanoparticles as contrast agents in x-ray imaging and computed tomography, Nanomedicine 10 (2015) 321-341. https://doi.org/10.2217/nnm.14.171

[48] C. Rümenapp, B. Gleich, A. Haase, Magnetic nanoparticles in magnetic resonance imaging and diagnostics, Pharmaceutical research 29 (2012) 1165-1179. https://doi.org/10.1007/s11095-012-0711-y

[49] N. Arias-Ramos, L.E. Ibarra, M. Serrano-Torres, B. Yagüe, M.D. Caverzán, C.A. Chesta, R.E. Palacios, P. López-Larrubia, Iron oxide incorporated conjugated polymer nanoparticles for simultaneous use in magnetic resonance and fluorescent imaging of brain tumors, Pharmaceutics 13 (2021) 1258. https://doi.org/10.3390/pharmaceutics13081258

[50] U. Kostiv, M.M. Natile, D. Jirák, D. Pulpanova, K. Jiráková, M. Vosmanská, D. Horák, PEG-Neridronate-Modified NaYF4: Gd3+, Yb3+, Tm3+/NaGdF4 Core-Shell Upconverting Nanoparticles for Bimodal Magnetic Resonance/Optical Luminescence Imaging, ACS omega 6 (2021) 14420-14429. https://doi.org/10.1021/acsomega.1c01313

[51] H. Zhang, T. Wu, Y. Chen, Q. Zhang, Z. Chen, Y. Ling, Y. Jia, Y. Yang, X. Liu, Y. Zhou, Hollow carbon nanospheres dotted with Gd-Fe nanoparticles for magnetic resonance and photoacoustic imaging, Nanoscale 13 (2021) 10943-10952. https://doi.org/10.1039/D1NR02914B

[52] R. Zou, Y. Gao, Y. Zhang, J. Jiao, K.-L. Wong, J. Wang, 68Ga-labeled magnetic-nir persistent luminescent hybrid mesoporous nanoparticles for multimodal imaging-guided chemotherapy and photodynamic therapy, ACS applied materials & interfaces 13 (2021) 9667-9680. https://doi.org/10.1021/acsami.0c21623

[53] F. Moradnia, S.T. Fardood, A. Ramazani, B.-k. Min, S.W. Joo, R.S. Varma, Magnetic Mg0. 5Zn0. 5FeMnO4 nanoparticles: green sol-gel synthesis, characterization, and photocatalytic applications, Journal of Cleaner Production 288 (2021) 125632. https://doi.org/10.1016/j.jclepro.2020.125632

[54] Y. Qin, J. Cui, Y. Zhang, Y. Wang, X. Zhang, H. Zheng, X. Shu, B. Fu, Y. Wu, Integration of microfluidic injection analysis with carbon nanomaterials/gold nanowire arrays-based biosensors for glucose detection, Science Bulletin 61 (2016) 473-480. https://doi.org/10.1007/s11434-016-1013-2

[55] B. Jin, P. Wang, H. Mao, B. Hu, H. Zhang, Z. Cheng, Z. Wu, X. Bian, C. Jia, F. Jing, Q. Jin, J. Zhao, Multi-nanomaterial electrochemical biosensor based on label-free graphene for detecting cancer biomarkers, Biosensors and Bioelectronics 55 (2014) 464-469. https://doi.org/10.1016/j.bios.2013.12.025

[56] T.-T. Wang, X.-F. Huang, H. Huang, P. Luo, L.-S. Qing, Nanomaterial-based optical- and electrochemical-biosensors for urine glucose detection: A comprehensive review, Advanced Sensor and Energy Materials 1 (2022) 100016. https://doi.org/10.1016/j.asems.2022.100016

[57] C. Padmakumari Kurup, S. Abdullah Lim, M.U. Ahmed, Nanomaterials as signal amplification elements in aptamer-based electrochemiluminescent biosensors, Bioelectrochemistry 147 (2022) 108170. https://doi.org/10.1016/j.bioelechem.2022.108170

[58] Y. Yang, Q. Huang, Z. Xiao, M. Liu, Y. Zhu, Q. Chen, Y. Li, K. Ai, Nanomaterial-based biosensor developing as a route toward in vitro diagnosis of early ovarian cancer, Materials Today Bio 13 (2022) 100218. https://doi.org/10.1016/j.mtbio.2022.100218

[59] R. Eivazzadeh-Keihan, E. Bahojb Noruzi, E. Chidar, M. Jafari, F. Davoodi, A. Kashtiaray, M. Ghafori Gorab, S. Masoud Hashemi, S. Javanshir, R. Ahangari Cohan, A. Maleki, M. Mahdavi, Applications of carbon-based conductive nanomaterials in biosensors, Chemical Engineering Journal 442 (2022) 136183. https://doi.org/10.1016/j.cej.2022.136183

[60] B. Pérez-Fernández, A. de la Escosura-Muñiz, Electrochemical biosensors based on nanomaterials for aflatoxins detection: A review (2015-2021), Analytica Chimica Acta 1212 (2022) 339658. https://doi.org/10.1016/j.aca.2022.339658

[61] Q. Tan, C. Wu, L. Li, W. Shao, M. Luo, Nanomaterial-Based Prosthetic Limbs for Disability Mobility Assistance: A Review of Recent Advances, Journal of Nanomaterials 2022 (2022). https://doi.org/10.1155/2022/3425297

[62] D. Yao, L. Wu, S. A, M. Zhang, H. Fang, D. Li, Y. Sun, X. Gao, C. Lu, Stretchable vertical graphene arrays for electronic skin with multifunctional sensing capabilities, Chemical Engineering Journal 431 (2022) 134038. https://doi.org/10.1016/j.cej.2021.134038

[63] G. Selleri, M.E. Gino, T.M. Brugo, R. D'anniballe, J. Tabucol, M.L. Focarete, R. Carloni, D. Fabiani, A. Zucchelli, Self-sensing composite material based on piezoelectric nanofibers, Materials & Design (2022) 110787. https://doi.org/10.1016/j.matdes.2022.110787

[64] X. Tang, W. Yang, S. Yin, G. Tai, M. Su, J. Yang, H. Shi, D. Wei, J. Yang, Controllable graphene wrinkle for a high-performance flexible pressure sensor, ACS Applied Materials & Interfaces 13 (2021) 20448-20458. https://doi.org/10.1021/acsami.0c22784

[65] Z. Wang, C. Yin, Y. Gao, Z. Liao, Y. Li, W. Wang, D. Sun, Novel functionalized selenium nanowires as antibiotic adjuvants in multiple ways to overcome drug resistance of multidrug-resistant bacteria, Biomaterials Advances 137 (2022) 212815. https://doi.org/10.1016/j.bioadv.2022.212815

[66] A. Besinis, T. De Peralta, R.D. Handy, The antibacterial effects of silver, titanium dioxide and silica dioxide nanoparticles compared to the dental disinfectant chlorhexidine on Streptococcus mutans using a suite of bioassays, Nanotoxicology 8 (2014) 1-16. https://doi.org/10.3109/17435390.2012.742935

[67] L. Wang, C. Hu, L. Shao, The antimicrobial activity of nanoparticles: present situation and prospects for the future, Int J Nanomedicine 12 (2017) 1227-1249. https://doi.org/10.2147/IJN.S121956

[68] A. Nagy, A. Harrison, S. Sabbani, R.S. Munson, Jr., P.K. Dutta, W.J. Waldman, Silver nanoparticles embedded in zeolite membranes: release of silver ions and mechanism of antibacterial action, Int J Nanomedicine 6 (2011) 1833-1852. https://doi.org/10.2147/IJN.S24019

[69] M. Kaushik, R. Niranjan, R. Thangam, B. Madhan, V. Pandiyarasan, C. Ramachandran, D.-H. Oh, G.D. Venkatasubbu, Investigations on the antimicrobial activity and wound healing potential of ZnO nanoparticles, Applied Surface Science 479 (2019) 1169-1177. https://doi.org/10.1016/j.apsusc.2019.02.189

[70] T. Yang, S. Oliver, Y. Chen, C. Boyer, R. Chandrawati, Tuning crystallization and morphology of zinc oxide with polyvinylpyrrolidone: Formation mechanisms and antimicrobial activity, Journal of colloid and interface science 546 (2019) 43-52. https://doi.org/10.1016/j.jcis.2019.03.051

[71] G. Herrera, J. Peña-Bahamonde, S. Paudel, D.F. Rodrigues, The role of nanomaterials and antibiotics in microbial resistance and environmental impact: an overview, Current Opinion in Chemical Engineering 33 (2021) 100707. https://doi.org/10.1016/j.coche.2021.100707

[72] F. Mehrabi, T. Shamspur, H. Sheibani, A. Mostafavi, M. Mohamadi, H. Hakimi, R. Bahramabadi, E. Salari, Silver-coated magnetic nanoparticles as an efficient delivery system for the antibiotics trimethoprim and sulfamethoxazole against E. Coli and S. aureus: release kinetics and antimicrobial activity, BioMetals 34 (2021) 1237-1246. https://doi.org/10.1007/s10534-021-00338-5

[73] S. Paudel, C. Cerbu, C.E. Astete, S.M. Louie, C. Sabliov, D.F. Rodrigues, Enrofloxacin-Impregnated PLGA Nanocarriers for Efficient Therapeutics and Diminished Generation of Reactive Oxygen Species, ACS Applied Nano Materials 2 (2019) 5035-5043. https://doi.org/10.1021/acsanm.9b00970

[74] J.H. Kim, M.J. Moon, D.Y. Kim, S.H. Heo, Y.Y. Jeong, Hyaluronic acid-based nanomaterials for cancer therapy, Polymers 10 (2018) 1133. https://doi.org/10.3390/polym10101133

[75] Z. Cheng, M. Li, R. Dey, Y. Chen, Nanomaterials for cancer therapy: Current progress and perspectives, Journal of Hematology & Oncology 14 (2021) 1-27. https://doi.org/10.1186/s13045-020-01025-7

[76] X. Lin, Y. Bai, Q. Jiang, Precise fabrication of folic acid-targeted therapy on metformin encapsulated β-cyclodextrin nanomaterials for treatment and care of lung cancer, Process Biochemistry 118 (2022) 74-83. https://doi.org/10.1016/j.procbio.2022.04.003

[77] J. Li, S. Wu, X. Tian, X. Li, Fabrication of a multifunctional nanomaterial from a mussel-derived peptide for multimodal synergistic cancer therapy, Chemical Engineering Journal 446 (2022) 136837. https://doi.org/10.1016/j.cej.2022.136837

[78] Z. Jing, Q. Du, X. Zhang, Y. Zhang, Nanomedicines and nanomaterials for cancer therapy: Progress, challenge and perspectives, Chemical Engineering Journal 446 (2022) 137147. https://doi.org/10.1016/j.cej.2022.137147

[79] X. Zhang, S. Wang, G. Cheng, P. Yu, J. Chang, Light-Responsive Nanomaterials for Cancer Therapy, Engineering (2021). https://doi.org/10.1016/j.eng.2021.07.023

[80] T. Liu, K. Yang, Z. Liu, Recent advances in functional nanomaterials for X-ray triggered cancer therapy, Progress in Natural Science: Materials International 30 (2020) 567-576. https://doi.org/10.1016/j.pnsc.2020.09.009

[81] M. Jena, S. Mishra, S. Jena, S.S. Mishra, Nanotechnology-future prospect in recent medicine: A review, (2013). https://doi.org/10.5455/2319-2003.ijbcp20130802

[82] M.A. Dobrovolskaia, Pre-clinical immunotoxicity studies of nanotechnology-formulated drugs: Challenges, considerations and strategy, Journal of Controlled Release 220 (2015) 571-583. https://doi.org/10.1016/j.jconrel.2015.08.056

[83] Z.S. Al-Ahmady, H. Ali-Boucetta, Nanomedicine & nanotoxicology future could be reshaped post-COVID-19 pandemic, Frontiers in Nanotechnology 2 (2020) 610465. https://doi.org/10.3389/fnano.2020.610465

[84] M.A. Mujawar, H. Gohel, S.K. Bhardwaj, S. Srinivasan, N. Hickman, A. Kaushik, Nano-enabled biosensing systems for intelligent healthcare: towards COVID-19 management, Materials Today Chemistry 17 (2020) 100306. https://doi.org/10.1016/j.mtchem.2020.100306

[85] M.-F. Xiao, C. Zeng, S.-H. Li, F.-L. Yuan, Applications of nanomaterials in COVID-19 pandemic, Rare Metals (2021) 1-13. https://doi.org/10.1007/s12598-021-01789-y

[86] Q. Zhang, R. Xiang, S. Huo, Y. Zhou, S. Jiang, Q. Wang, F. Yu, Molecular mechanism of interaction between SARS-CoV-2 and host cells and interventional therapy, Signal transduction and targeted therapy 6 (2021) 1-19. https://doi.org/10.1038/s41392-020-00451-w

[87] J.K. Patra, G. Das, L.F. Fraceto, E.V.R. Campos, M.d.P. Rodriguez-Torres, L.S. Acosta-Torres, L.A. Diaz-Torres, R. Grillo, M.K. Swamy, S. Sharma, Nano based drug delivery systems: recent developments and future prospects, Journal of nanobiotechnology 16 (2018) 1-33. https://doi.org/10.1186/s12951-018-0392-8

[88] J.A. Kemp, Y.J. Kwon, Cancer nanotechnology: current status and perspectives, Nano convergence 8 (2021) 1-38. https://doi.org/10.1186/s40580-021-00282-7

[89] R. Akçan, H.C. Aydogan, M.Ş. Yildirim, B. Taştekin, N. Sağlam, Nanotoxicity: A challenge for future medicine, Turkish journal of medical sciences 50 (2020) 1180-1196. https://doi.org/10.3906/sag-1912-209

[90] D.K. Scoville, D. Botta, K. Galdanes, S.C. Schmuck, C.C. White, P.L. Stapleton, T.K. Bammler, J.W. MacDonald, W.A. Altemeier, M. Hernandez, Genetic determinants of susceptibility to silver nanoparticle-induced acute lung inflammation in mice, The FASEB Journal 31 (2017) 4600-4611. https://doi.org/10.1096/fj.201700187R

[91] T. Sahu, Y.K. Ratre, S. Chauhan, L. Bhaskar, M.P. Nair, H.K. Verma, Nanotechnology based drug delivery system: Current strategies and emerging therapeutic potential for medical science, Journal of Drug Delivery Science and Technology 63 (2021) 102487. https://doi.org/10.1016/j.jddst.2021.102487

Emerging Nanomaterials and Their Impact on Society in the 21st Century   Materials Research Forum LLC
Materials Research Foundations 135 (2023) 304-340   https://doi.org/10.21741/9781644902172-13

Chapter 13

Toxicological Effects of Nanomaterials on the Aquatic Biota

Usman Lawal Usman*[1, 2], Sushmita Banerjee[1], Nakshatra Bahadur Singh[3] and Neksumi Musa[1]

[1] Department of Environmental Sciences, Sharda University, Greater Noida, India

[2] Department of Biology, Umaru Musa Yar'adua University, Katsina- Nigeria

[3]Department of Chemistry and Biochemistry, Sharda University, Greater Noida, India

* usman.usman@umyu.edu.ng

Abstract

Globally nanotechnology has transformed many sectors of human endeavour due to the innovative introduction of different unique materials and products in various areas. Presently large quantities of nanomaterials are synthesized and are essential for several industrial applications. This facilitates and enhances research in biophysics, biochemistry, wastewater treatment, etc. Nanotechnology and other science-related discipline have been combined to synthesize novel nanomaterials through modification of their shapes, composition, and dimension which can be used for various applications in medicines, agriculture, environmental remediation, etc. Despite their peculiar advantages and wide application in several industrials and domestic sectors, fabricated materials of nano-size scale have been raising concerns of unsafety for both human and aquatic biota, due to their incessant release into the aquatic ecosystem. These nanomaterials are ingested by the aquatic biota, from which they bioaccumulate in the organism's body and hence pass to humans via the food chain. Within the organism's body, nanomaterials serve as foreign material with different physicochemical characteristics because of their peculiar small sizes. Thus, the harmful effect of nanomaterials on aquatic biota has been generating great concern among the scientific communities. Thus, they can interfere with the normal mechanisms of physiological activities. The present chapter focuses on the sources of the nanomaterials in the aquatic ecosystem, the route of exposure, their process of transformation and interactions, their possible toxicity to the aquatic biota, challenges posed by the nanomaterials, and prospects.

Keywords

Nanomaterials, Characteristics, Aquatic Ecosystem, Toxicity, Exposure

Contents

Toxicological Effects of Nanomaterials on the Aquatic Biota **304**

1.   **Introduction**..**305**

2.   **Nature of nanomaterials** **308**

3.   **Toxicity of nanomaterials** **310**

4.   **Transformation of nanomaterial in water** **311**

5.   **Aquatic toxicity fate of nanomaterials**......................................**311**

6.   **Nanomaterials pathways to aquatic ecosystem**...........................**312**

7.   **Nanomaterials toxicity in aquatic ecosystem** **313**

        7.1   Factors affecting nanomaterials toxicity 315

        7.2   Nanomaterials toxicity on algae and aquatic plants 316

        7.3   Plant and nanomaterials interactions 317

        7.4   Bacteria and aquatic microbe nanomaterials toxicity.......318

        7.5   Semiaquatic nanomaterial toxicity 319

        7.6   Nanomaterial toxicity on aquatic invertebrates...............320

        7.7   Nanomaterial Toxicity on the Aquatic Vertebrates........322

8.   **Challenges and regulations of nanomaterials** **325**

Conclusion and future outlooks..**325**

References **326**

1.   Introduction

Nanomaterials are substances that are of nanometer scale that is within the range of 1-100 nm. They differ from other materials due to their peculiar physical and chemical characteristics which comprise their dimensions, phases, composition, porosity, uniformity, shape, and sizes [1,2]. Nanomaterials are immensely beneficial starting with their capability to eliminate persistent pollutants in the wastewater, improve the agricultural yield, and diagnose and treatment of ailments in medicine, electric devices, cosmetics, pharmaceuticals, sport wears, clothing, coating, and food additives are some of the substances that can be produced using nanomaterials [3,4]. The carbon nanotube and silver (Ag) nanoparticles are among the most widely used nanomaterials [5]. Some of these nanomaterials are known to have antibacterial characteristic in which it denatures and prevent the rapid development of the microbes [6]. Furthermore, nanomaterials are also used in other various applications such as photography and jewelry fabrication [7]. A major percentage of the nanomaterials also cooperate in some products such as cement and rubber to enhance and strengthen their quality [8]. However, despite the wide and potential application of these nanomaterials, they pose some threat to humans, aquatic biota, and the environment at large. Special consideration is given to the aquatic ecosystem because the aquatic is considered the main site for the entrance of nanoparticles and distribution to other resources, thus, serving as a link [9]. The nanoparticles concentration is the water surface varies in ranges starting from nanogram per litre (ng/L) to micro gram per litre (µg/L), depending on the nanomaterial type. The report of European Union (EU) depicts that TiO2 nanoparticles recorded the highest water surface concentration with a 2.2 µg/L value, followed by 1.5 ng/L value of Ag nanoparticles [10,11]. These results were assessed based on the models of probabilistic which take the specific nanoparticles variation over rate of input, mechanisms fate, time, stocks used and increase in the volume of productions [12].

However, the majority of these nanomaterials find their way to the aquatic environment, thereby endangering the aquatic biota. An investigation conducted on sewage treatment in Europe and USA has shown different nanomaterials such as TiO_2, ZnO, and Ag nanoparticles may be noxious to the aquatic biotas [13,14]. Nanomaterials have been found in the aquatic ecosystem from where they can be ingested by the aquatic biota and tend to bioaccumulate before their removal using the immune system or other processes. Nanomaterials are known to be foreign material within the body of an organism, due to their peculiar characteristics they can change the normal mechanisms of physiological activities in an organism thereby affecting the embryogenesis and growth of the animal. The nanomaterials at the embryo stage can disrupt the growth of the organism leading to deformities which can lead to the death of the animal, this can be due to the certain chemical reaction alongside the nanoscale size of the material itself which can interfere with the organism's biological, chemical and physical activities [15]. Thus, because of their nano-scale size, the nanomaterial can simply pass across the cell membrane, circumventing the body's defense system. Afterward, the material penetrates the organism's cell and passes into some organelles like mitochondria, thereby modifying the entire cell-metabolism and facilitating the death of the cell. Similarly, when the material size couldn't infiltrate the organism cell, they tend to interfere with the membranes of the cell, by disrupting the functions of the membrane such as signal transduction or ion transport [16].

Nanomaterials can also be cytotoxic due to their physical characteristics and chemical composition. The positive electrical charges of the nanomaterials can lethal the bilayer of the lipid membrane and the coating surface of the nanomaterial can as well disrupt the cell structure [17]. Furthermore, nanomaterials adsorption efficiency can be affected using other materials from the pollutants from which the adsorbed substances can also have detrimental effects on the animal that ingested the material [16]. several investigations have been carried out in recent decades to understand the toxicological effect of the nanomaterials, especially on the aquatic biota, however, their full mechanisms of action

Emerging Nanomaterials and Their Impact on Society in the 21st Century   Materials Research Forum LLC
Materials Research Foundations 135 (2023) 304-340   https://doi.org/10.21741/9781644902172-13

are yet to be ascertained. Thus, because of the cosmopolitan nature of nanomaterials, it is imperative to comprehend their possible direct and or indirect toxic effects on the biota. Studies have shown that nanomaterials can be noxious to algae, bacteria, invertebrates as well as vertebrates [18]. Nowadays, several biotas are used as models to ascertain the influence of the nanomaterials on them, for example, studies on some vertebrates such as fishes and mouse revealed the toxic impact of nanomaterial on the reproduction process which involve the development and growth of their embryos [19]. Another investigation involving an invitro and in vivo cultured organisms revealed that silver (Ag) nanoparticles resulted in oxidative stress in the animals which is characterized by many reactive molecules comprising of free radicals of oxygen, that is reactive oxygen species (ROS), DNA breakage due to apoptosis of a cell or genotoxicity [20]. Similarly, activation of the neutron has been adopted to investigate the impacts of cobalt (Co) nanoparticles on the *Eisenia fetida* (earthworm). Autoradiography and scintillation counting has also been used to detect cobalt nanoparticles of 4 nm size in the blood, cocoons, and spermatogenic cells [21]. However, investigation concerning the impacts of the nanomaterials on the environment is hindered due to the non-availability of equipment to confine and measure them in the soils, water, organisms, and sediment [16]. Some nanomaterials such as ZnO, CuO, and Ag nanoparticles are often used to control bacteria (bactericides). Nevertheless, after usage, these nanoparticles end up in the environment thereby causing a toxic effect on the nontarget biota. These adverse effects were reported in some cell cultures, fishes, crustaceans, protozoa and nematodes [22]. However, the effects differ from inhaling those nanomaterials [23,24]. But, currently, the harmful impacts of nanomaterials remain ambiguous. Thus, various ways of systematizing the study on the impact of nanomaterials were proposed by Kahru and Dubourguier [25], this includes; identifying the highest detrimental impacts of the nanomaterials on the most delicate living organisms, and collecting ecotoxicological data that can ascertain the risks associated with the types of the nanomaterials, for example; using some important biotas such as mammals, amphibians, fishes, crustaceans, earthworms, nematodes, protozoa, yeast, algae and bacteria in conducting experiments with nanomaterials like Ag, CuO, NiO2, ZnO nanoparticles, C60 fullerene, multiwall nanotubes (MWCNTs), single-walled carbon nanotubes (SWCNT-NTs) and or single wall nanotubes (SWNTs) [25]. The nanomaterials sample preparation, nature of the cytotoxic impacts and choices of the target cell are some of the conditions highlighted by Kong et al. [26] for the investigation of the nanomaterials influence on the organism. The experimental study on the toxicity influence of nanomaterials on the animals in all the sphere of ecosystem is of paramount important, these ecosystems comprise of terrestrial environment, both marine and fresh water and air by inhalation. In order to examine the noxious influence of the nanomaterials, investigation have been ongoing on several kinds of animals, examples in different vertebrates and invertebrates, protozoa, embryos alongside with adults' organisms. In view of that, the present review compile current studies on some toxicological impacts of the nanomaterials on the aquatic biota such as amphibians, crustaceans, fishes, etc. and also challenges posed by nanomaterials and future prospects.

2.   Nature of nanomaterials

Various kinds of nanomaterials exist based on their origin, these include, (i) natural; which occur during mechanical or bio-geochemical phenomena (such as forest fires, ocean spray volcanic ash, etc.) (ii) incidental; these nanomaterials are produced inadvertently through process either influence by a human directly or indirectly (examples include industrial or mechanical processes, like welding gases, exhaust gases from the vehicle, heating solid fuel and cooking combustion, (iii) engineered and bioinspired nanomaterials, these are materials design with specific desire functions by modifying their composition, shape porosity and dimension to mimic the natural materials [27] (Table 1). The engineered nanomaterials are of great concern lately, since they are being manufactured by industries and are used in medicines, agriculture, cosmetology, electronics, water remediation, etc. [28].

Table 1: Classification of nanomaterials based on their origin

Nanomaterials Origin	Types	Sources	References
Naturally occurring nanomaterials	Plant nanostructure and nanoparticles from nano-organisms	Comprises of insects, yeast, algae, fungi, viruses, nanobacteria, and nanostructures from the organisms.	[28]
Incidental nanomaterials	Cosmic Storms of dust	Stardust consists of many compositions of organic materials, silicate, nitride, oxide, carbon, and carbide. Method of their formation includes collision, waves shock, temperature, gradient pressure, electric and magnetic radiation	[29,28,30]
	Eruptions from volcanoes	Ashes released from a single eruption of volcanoes can yield 30×10^6 tons of nanomaterials in the atmosphere.	[31]
	Forest [conflagration] Fires	Smokes and ashes caused by the forest conflagration contain nanosized and micro particles.	[32]
	Evaporation of Sea and Ocean water	Aerosols [nanoparticles] made up of sea salt are produced from the droplet of water caused by evaporation or water waves	[33]

Engineered synthesized nanoparticles	Exhaust from automobile	Gasoline and diesel engines emit nanomaterials of various sizes ranging from 20 to 60 nm and 20 to 130 nm respectively. Carbon nanoparticles are also emitted from the combustion of gases and diesel	[34,35]
	Smokes from cigarette	Smokes from cigarettes comprise hundreds of thousands of nanomaterials which range between 10 to700 nm	[36]
	Healthcare and biomedical products	Antireflectants and antioxidant products that are used in personal care products	[37]
	Destruction of buildings	During the destruction of buildings, nanomaterials of less than 10 μm are emitted into the atmosphere from the debris of the buildings and other related toxic household particles	[38]

However, engineered nanomaterials exert some negative consequences because of their possible toxicity [39]. According to their chemical structure and composition, nanomaterials; specially engineered nanoparticles can be categorized into different groups namely; inorganic engineered nanoparticles, organic engineered nanoparticles, polymeric engineered nanoparticles, and miscellaneous engineered nanoparticles. Tables 2 depicted their categories, examples, and the potential application of the engineered nanoparticle products.

Table 2: Categories of the engineered nanoparticles and their possible applications

Group	Characteristics	Application	Nanomaterials	References
Inorganic engineered nanoparticles	Optoelectrical, antifungal, antimicrobial, and Photocatalytic	Drug carriers, dyes, plastics, paper, textiles, varnishes, paints, cosmetics, and Sunscreens	CeO_2, ZnO, TiO_2, Zero-valent metals; Au, Ag, and Fe	[40,41,42]
Organic engineered nanoparticles	Strength high material, optical, chemical, and Electrical conductivity	Aerospace, optics, and electronics.	Carbon nanotubes, Graphene and Fullerenes	[43,44,45]

| Polymeric engineered nanoparticles | Biocompatibility and magnetic | Chemical sensors, agents of transfecting DNA, and Drug carriers | Dendrimers [polymeric branched molecules] | [46,47] |
| Miscellaneous engineered nanoparticles | Chelation, electrical and optical | Solar cells and medical imaging | Quantum Dots and Nano foils | [48,49] |

3.   Toxicity of nanomaterials

Nanotechnology has portrayed propitious results in various fields of endeavor, such as medicine, agriculture, wastewater remediation, etc. However, despite their substantial application, there are some drawbacks associated with nanomaterials that stand as a barrier in the field of nanotechnology. Among these shortfalls of nanotechnology are; toxicity of the nanomaterial, social acceptability, and economic value of the nanomaterials e.g., gold nanoparticles. The hazards associated with nanotechnology comprise nanomaterials transformation, water pollution, and engineered nanomaterials which are associated with ecotoxicity [23]. Size and component of the molecular structures are the main properties that control the cellular uptake and endpoint toxicity of the nanomaterials respectively, because of the nanoscale size, nanomaterials can penetrate via the epithelial and also endothelial layers and the lymph node and blood or haemolymph and pass into various tissues and organs, including the liver, nervous system, bone marrow, spleen, kidney, heart, and brain [9]. Shape and size toxicity dependent on nanomaterials such as CNTs, Ag nanoparticles, and several other metal nanoparticles are reported. The nanomaterial sizes which range between 1 to 100 nm are similar to the globules of protein which range between 2 to 10 nm, cell membrane thickness which is 10 nm, and helix of DNA which is also 2 nm, all these allow the passage of nanomaterials easily within cells and between cells to organelles [50]. For example, gold nanoparticles having less than 6-nanometer size were reported by Zuber et al. [51] to have efficiently penetrated a nucleus, meanwhile, those with sizes ranging between 10 to 16 nm were observed to pass via the membranes of a cell and deposited in the cytoplasm. The TiO_2 nanoparticles were reported to impede polymerization by changing the configuration of tubulin, hence disorganizing the migration of cells, intracellular transportation, and cell division [52].

The toxicity of nanomaterials can be influenced by certain properties which comprise the composition of its materials, purity, surface area, dimension, agglomeration, shape, size, structure, etc. Thus, it is imperative to examine, analyze and envisage the toxicity of any novel nano-scale material be it natural, incidental, or engineered nanomaterial to suggest limited approaches (Table 1). The toxicity of nanomaterials can be evaluated through the cell lining, culturing conditions and time of incubation, however, a comparison of sample data does not accurately ascertain whether the relevant cytotoxicity impact observed is physiological. Numerous experimental organisms are used as models for the in vivo toxicity test, this includes rats, mice, zebrafish (*Danio rerio),* and cultured cells for the in vivo. Particulate matter of various sizes, as well as nanoscale ranges inclusive, are produced via different sources of combustion example motor vehicles. Most studies have been focused on evaluating the nanomaterial's toxicity impacts, which leads to increasingly stringent policies. Consequently, experimental results suggest that particles fine-sized have more adverse effects [53].

There are various shapes of nanomaterials including; rods, cubes, sheets, cylinders, ellipsoids, and spheres. The spherical nanoparticles are known to be more predisposed to endocytosis compared to the nanofibers and nanotubes. However, blocking the channels of calcium was efficiently found in the SWCNTs as compared to the spherical nanoparticles [54]. The toxicity of various shapes of hydroxyapatite nanoparticles such as spherical, needle-like, rod-like, and plate-like shapes were examined and the result reveal that death of some major parts of the lung non-tumorigenic epithelial cell (BEAS-2B cells) was caused by needle-like and plate-like nanoparticles as compared to the rod-like and spherical shapes nanoparticles.

4.   Transformation of nanomaterial in water

Nanomaterials' fate in the aquatic environment solely depends on their interaction mode with biotic and abiotic factors, their solubility, and dispersibility. Nanomaterials usually settle down gradually as compared to their bulk particles. However, due to their larger surface area to volume ratio, they tend to adsorb much more particles from the soil and sediment. Nanomaterials are non-soluble in water examples include fullerenes, CNTs, etc., thus, they can be removed easily within the column of water. The engineered nanoparticles are often utilized for some special activity, for instance, pristine nanomaterials can be converted into different forms of materials such as weathered products, modified products, and transferred environmentally engineered nanomaterials, from which the oxidation-reduction and conversion can be observed using photochemical process. These conversions tend to change the interactions between the environment and the nanomaterials, which finally control the water contaminants' adsorption and process of desorption [46].

5.   Aquatic toxicity fate of nanomaterials

The disturbances of the aquatic ecosystem which result to stress on the aquatic biota are caused by either physical, chemical, or biological processes. Numerous nanomaterials have the capability of escaping into the purified water during the purification process. Thus, the rate of production of some nanomaterials such as TiO_2 and Ag nanoparticles keeps increasing annually [27]. These nanomaterials are regarded as a threat to the aquatic biota as reported by Kim et al. [55] in their study on the impact of TiO_2 and Ag nanoparticles on the aquatic plant duckweed growth. The concentration of nanomaterials in some aquatic environments such as surface water is assessed to be micro or nanogram per liter. This concentration rises with the increase in nanomaterials production. Nanomaterials such as ZnO, TiO_2, and Ag are commonly used for the purification of water due to their antibacterial properties, hence replacing the chemical's antiseptics [56]. Similarly, impregnated filter ceramics with Ag nanoparticles are used for water purification. The filtrate water pass via the coated nano filter papers, however, the study showed some traceable quantity of the Ag nanoparticles which are substantially below the WHO permissible limit guidelines [57]. Report on the investigation studies of these leaching nanomaterials in the aquatic ecosystem and their harm to the aquatic biota is scanty. Thus, some investigations reported damage to some vital organs and the destruction of DNA strands in rats that consumed water containing TiO_2 nanoparticles. Also, nanomaterials that are engineered are reported to induce some adverse health effects, which include, circulatory defect, carcinogenicity, genotoxicity, and inflammation of the pulmonary arteries [58].

6.   Nanomaterials pathways to aquatic ecosystem

The vicinity of the water bodies is known to receive higher volumes of nanomaterials loads from the main point sources as compared to those farther away from the point sources of the nanomaterials [59]. The amount of the nanomaterials load can be elucidated by studying the complete life cycle of the release nanomaterials. An outline of the release nanomaterials patterns at various stages comprising of (i) the production or state of the nanomaterials, (ii) compounding of the nanomaterials into the product of consumers, (iii) utilization of the product by consumers, and (iv) disposal of the end used products, as depicted in Fig. 1. Some of the nanomaterials can inadvertently escape to the environment during the stage (1) and (ii) before the release of the products to the consumers, especially the powdered engineered nanomaterials and other products that can escape into the water via cleaning the equipment which contaminated the water [60]. The discharge of the nanomaterials on the water surface reacts with other contaminants in the water bodies such as organic and inorganic materials, sediments, etc. from which they can be stuck to the sediment, travel across the water, or undergo irreparable transformation [61].

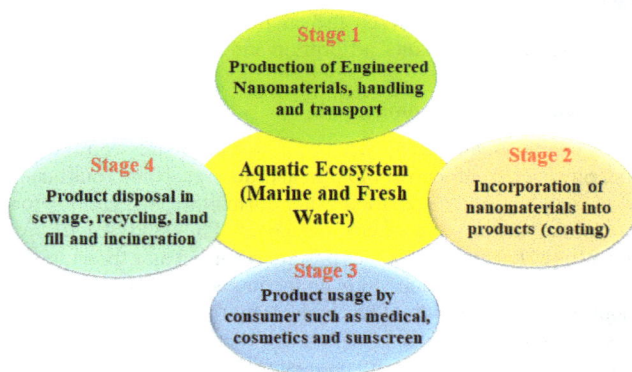

Figure 1: Nanomaterial life cycle stages and their potential release into the aquatic environment

7.   Nanomaterials toxicity in aquatic ecosystem

Natural nanomaterials have been in existence within the ecosystem from time immemorial and have been in interaction with various organisms. However, synthesized nanomaterials have been increasing and accumulating in our everyday lives unknowingly. Manufactured nanomaterials tend to possess peculiar surface structures and properties which gave them unique physical, chemical, and toxicological characteristics that differ from natural nanomaterials [62]. Similarly, the proliferation of nanomaterials in both the terrestrial and aquatic environment necessitated the in-depth comprehension of their behavioural and characteristics. The quantity of the nanomaterial in the aquatic environment determines its possible carriage and subsequent toxicity. However, the conventional distribution of hydrophobic contaminants of organic is largely dependent on the phenomenon of portioning equilibrium, moreover, other processes may be required for the nanomaterials [63]. The knowledge of colloids chemistry is important in knowing the fate of nanomaterial because the nano-scale materials are similar to the system of colloidal mixtures, in which by characterization the nanomaterial fall within the ranges of colloids substances that are described as particles that range between 1 and 1000 nanometer [64]; [65].   The aggregation of colloids and their sedimentation are the two significant chemical processes that determine the dispersion and behavioural features of nanomaterials. The aggregate formation is largely base on the attached efficiency and frequency of the collision. The attached efficiency designates the chances of a particle to be attach by forming an aggregate after they collide, this can occur as a result of many influences such as; hydration, and magnetic forces, steric hindrance, Van der Waals and electrostatic forces [28]. The frequency of collision among the particles is basically dependent on the differential settling, fluid and Brownian motions [66]. Aggregation of nanomaterials can occur with other engineered nanoparticles to form homogenous aggregation, it can also aggregate or associate with organic substances or other different particles found in the aquatic ecosystem such as dissolved solids, suspended solids, and natural colloids to form heterogeneous aggregation [66]. The formation of the sedimentation depends on the frequency of the collision and efficiency of the attachments which result in the formation of large aggregates.

Moreover, the transformation of the nanomaterials via chemical processes can be possible, including biological degradation, hydrolysis, dissolution, oxidation, and reduction processes [67]. The composition of the nanomaterials is the key determinant factor that determines its transformation [Table 3]. Also, the kind of water body that receives the nanomaterial influence the status of the nanomaterials, this is due to the variation in the composition of the substances present in the water, such as dissolved molecules, salinity, turbulences, pH, natural colloids, organic matters, and other pollutants. Model organisms are of great significant in determining the basis for predicting and assessing the hazardous nature of nanomaterial, however among the drawback is the inadequate experimental tools and the current challenges associated with the analysis of the nanomaterials in their complex forms. To ascertain the behaviour of nanomaterials, it is substantial to examine some precise processes such as dissolution and sedimentation. Enumeration of these precise processes varies from the categories of conventional contaminants like organic micropollutants and heavy metals [68,69]. Aggregation and sedimentation data from the experimental process are necessarily required for the improvement of the parameters input of the nanomaterial fate model [70,71]. Thus, the rate of heterogeneous aggregation and sedimentation data of nanomaterials are mostly lacking for natural water over varying geochemical effects. This is because the present analytical tools generally cannot separate precisely among the unused nanomaterials such as the natural nanoparticles, generated combustion or bulk nanoparticles, engineered nanoparticles, and measurement of the concentration; from which the evaluated concentration are likely to be the higher concentration compare to the predicted one because the fraction of the total nano scale size is the engineered part mostly considered for the prediction model [72]. Hence, enhancement of the analytical techniques for complex matrices in the environment will immensely pave way for better comprehension of the transporting system and fate model of the nanomaterials and also enhance the evaluation process from the exposed concentration [70].

Table 3: Influence of nanomaterials (NMs) toxicity on the aquatic biota

Variable	Nanomaterial toxicity	Examples	References
Charges on the NMs Surface	Nanomaterial rate of agglomeration is influenced by surface charges which regulate the toxicity	Silver nanoparticles (Ag-NPs) toxicity was reported to be dependent on the charges on the surface of the particles	[73]
Nanomaterials sizes	The size of the nanomaterials is a key determinant in the toxicity of the particles	Aluminium oxide (Al_2O_3) nanosized was reported to exhibit less bacterial toxicity in comparison with the same nanomaterial that has a size of less than 50 nm.	[74,75]
Structural crystals	Cytotoxicity and genotoxicity have been described to be associated with the structure of the nanomaterial crystal	Higher toxicity (oxidative stress) was reported in Anastase ($nTiO_2$) as compared to rutile ($nTiO_2$)	[42,76]
Nanomaterials morphology	Morphology of NMs was reported to influence their toxicity	Higher toxicity of silver (Ag) nanoplates was reported in the embryo of zebrafish and epithelial cell	[77,78]

		of fish gill as compared to the spheres and wires	
Concentration and time of exposure	The concentration and time of exposure influence the toxicity of NMs within the aquatic environment	The nZnO NPs exposed to Snails (*Lymnaea luteola)* were reported to be time and concentration-dependent	[79,80]
coating Surface of NMs	The NMs interaction with the experimental organism tends to increase or decrease depending on the chemistry nature of the NMs coating	The polyethylene glycol-(PEG) coated nAg was reported to be lower in toxicity in algae species as compared to nAg coated-Polyvinylpyrrolidone (PVP) and citrates	[81] [82]
Other contaminants in the aquatic environment	Information on the interaction between nanomaterials and other chemical contaminants in the aquatic ecosystem is scarce	Higher chromosomal aberration was reported in the *Mytilus edulis* (blue mussels) when expose to $nTiO_2$ combine with benzo (a) pyrene, while less chromosomal abnormality was observed when expose to nTiO2 or benzo (a) pyrene	[83] [83]
Effect of photoactive	Light dissolution of NMs and reactive oxygen species production were reported after exposing NMs to ultraviolet (UV) and visible light	Damages in chromosomes were reported to increase after exposing nTiO2 to irradiation of ultraviolet A on the cell line of fish as compared to exposing to only nTiO2 nanoparticles	[84]

7.1   Factors affecting nanomaterials toxicity

The toxicity of nanomaterials in the aquatic environment is mainly affected by three different factors, this includes (i) physicochemical characteristics of the nanomaterials, (ii) functional groups behaviour of the nanomaterials, and (iii) nanomaterial and co-pollutant interactions within the aqueous solution [81]. The dissolution of nanomaterials into some metallic ions within the aquatic ecosystem and the generation of reactive oxygen species (ROS) are among the examples of functional group behaviour which can influence the nanomaterial's toxicity [85]. Among the nanomaterials that dissolve completely in aquatic

Emerging Nanomaterials and Their Impact on Society in the 21st Century   Materials Research Forum LLC
Materials Research Foundations 135 (2023) 304-340   https://doi.org/10.21741/9781644902172-13

media include silver (Ag) surfaces and copper [Cu] related nanoparticles, which result in the production of peroxide radicals and Ag^{2+} and Cu^{2+} respectively. Similarly, many studies reveal that ionic forms of metal are more noxious than their nanoparticle counterparts [86,62]. For the determination of entire aquatic toxicity, nanomaterials concentration is considered a significant parameter. Little concentration of the nanomaterials such as 5 to 50 μg/L can result in changes in the organism's physiology, oxidative stress, and chromosomal aberration. Nevertheless, higher dose concentration of the nanomaterials such as 1 mg/L was revealed to directly cause mortality [87]. Furthermore, the toxicity of nanomaterials in aquatic biota is precise on the particles and majorly depends on the mode of ingestion into the cell of an organism [88]. The mode of ingestion into the organism's cell starts by adhering to the pores of the membrane in the cells, which is followed by the complete entering into the cell through the system of ion transport or endocytosis (Fig. 2). The meddling with the process of electron transport or the generation of reactive oxygen species [ROS] caused as a result of nanomaterials ingress has varied adverse effects, such as damaging of the membrane of the cell, functional organelles and nucleic acid [89,90].

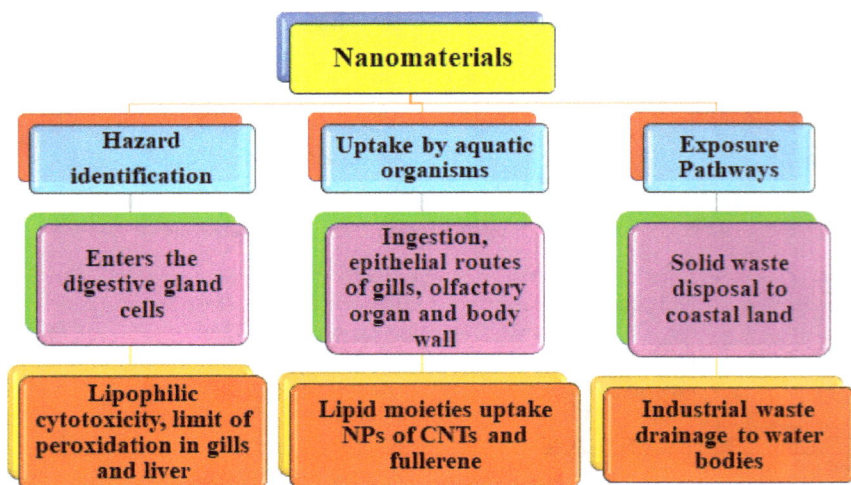

Figure 2: Toxicity and Exposure of Nanomaterials to Aquatic Biota

7.2   Nanomaterials toxicity on algae and aquatic plants

The exposure or entry of nanomaterials to the aquatic biota is at increases, due to the mass production and their application in various sectors, for example, *a* study on the *Spirodela polyrrhiza* plant tissue revealed various adverse results depending on the nature and type of the nanomaterials. Studies have shown that different concentrations of nanomaterials cause a decrease in the parameters of plant growths as well as a reduction in the plant pigments used for photosynthesis, thereby increasing the activity of some oxidative stress enzymes and also generating more reactive oxygen species [91,86]. The phytotoxicity of some nanomaterials is identified to inhibit different developmental stages, especially at a higher concentration [92]. Similarly, the architectural anatomy of the plant roots also influences the toxicity of the nanomaterials on the plant. The nanomaterial was reported to affect microalgae *Raphidocelis subcapitata* which is a single-celled alga, thereby causing aggregations that result in depletion of nutrients and shading [66]. The toxicity of nanomaterials in the plant is associated with the surface properties and surface area at both the cells and seedlings level [93]. Nevertheless, other conducted investigations revealed the significance of the surface characteristics in the nanomaterial toxicity [94]. Also, different pH range was used to evaluate the influence of the pH on the algae, and the study was shown to be dependent on pH, from which pH 4 was observed to be the range with high solubility. Exposure of dissolved nanomaterials to alga species results in the generation of reactive oxygen species (ROS) which leads to a drastic reduction in the content of chlorophyll, viability, and growth of the algal cells [95,96]. Furthermore, damage to membranes of an alga and depletion in their nutrient were reported as a result of the negative influence of the nanomaterial [97,97]. Similarly, the nanomaterial accumulation of ionic form in the cells of an alga was reported to be the cause of the toxicity mechanisms [98,99]. this hypothesis does not correspond with that found in the seedling of plants [100]. Thus, it is imperative to mention that the growth of the plant is not only impeded by the phytotoxicity chemicals of nanomaterials, but also because of the physical nanomaterial interaction and the pathways in the transportation to the plant cells. Among the adverse examples of physical nanomaterials, the interaction includes the impediment of the spaces within the intercellular component of the plant cells and a decrease in the conduction of hydraulic and plasmodesmata nanosized impediment [101].

7.3   Plant and nanomaterials interactions

Interaction between the plants and nanomaterials occurs during the plant uptake, translocation, and accumulation process (UTA). The UTA was observed to be precise for the plant type and nanomaterials, for instance, iron oxide uptake observed in the root of *Cucurbita maxima* (pumpkin) which translocated via the plant was observed to reach the plant leaves in a small quantity [102]. The uptake process and translocation pattern of the nanomaterials start with infiltration into the cell walls thereby reaching the membrane's plasma and the root vascular conducting tissue (xylem). The pore size of a cell wall ranges from 3-8 nm, which is considered to be smaller compares to most nanomaterials [103]. Similarly, in another hypothesis, the typical diameter of a cell wall ranges between 5 to 20 nm, hence, the accumulation of nanomaterials may pass only via a plant provided their size is not bigger compared to the largest cell wall as indicated above. Investigations have revealed the ability of the nanomaterial to induce some changes within the pore sizes of the cell wall thereby making a large entrance for easy passage of the nanomaterials [104]. Furthermore, the passage of nanomaterials via the plant cell walls has been observed to be analogous to that of the pathways of endocytic cells in mammals [101]. The intercellular plant organelles (plasmodesmata) analysis result from green fluorescent protein (GFP) and transmission electron microscopy (TEM) revealed dilation of sizes after being exposed to nanomaterials, thereby increasing the passage rate of the nanomaterials. This dilation is known as a limit of exclusion from which the largest molecule can pass via the plasmodesmata [105,106,105,107].

7.4   Bacteria and aquatic microbe nanomaterials toxicity

The bacteria are mostly considered less impacted by nanomaterial toxicity as compared to other aquatic organisms, this may be due to their capability to withstand stressful situations thereby developing defense mechanisms [108]. Nevertheless, bacteria are regarded as model organisms mainly used for the examination of nanomaterial toxicological effects because of their short lifespan, and easy and rapid growth [109]. The interaction of nanomaterials with bacteria occurs via four different channels: (i) damage of protein or DNA, (ii) agglomeration of bacteria, (iii) toxicity impurity, and (iv) toughness of membrane lining. The toxicity of nanomaterials generally differs between and or amongst the strains of bacteria depending on the composition contains in the lipid membranes. It was observed that gram-positive bacteria are more sensitive than that gram-negative bacteria [108]. Different parameters like changes in the bacteria morphology, generation of reactive oxygen species (ROS), growth and viability of the bacteria are mainly used to evaluate the influence of the nanomaterial exposure to the bacteria (Fig. 3). The response of bacteria is affected by several factors starting with the conditions of the surrounding environment such as pH, UV irradiation, interaction with the nanomaterials; agglomeration state, concentration, surface modification, etc. [110]. The long- and short-term toxicity effect of nanomaterials is well influenced by their transformation from the environment which can be instigated by the interaction with sulfidation, chloride ions, and inorganic and organic materials [111]. Furthermore, nanomaterials can influence the integrity of the bacteria cell membranes thereby causing uptake of the nanomaterials into the cell via osmotic pressure which takes place in the absence of a light source (i.e., without the illumination of UV) hence, distressing the activities of some proteins and genes [112]. Different investigations were carried out to evaluate the toxicity influence of nanomaterials on the *Escherichia coli* organisms, from which the result revealed that the toxicity of nanomaterials relies more on the oxidative stress radicals or reactive oxygen species generation (ROS) as compared to the size of the particle or surface area under the irradiation light sources [113]. Also, the nanomaterial's toxicity in the *Escherichia coli* was reported to be dependent on pH which is impacted during the oxidative stress by some generation of ions, such as silver ion (Ag^+) that tend to bind to the cytoplasm of a cell thereby resulting into lysis of the cells [114].

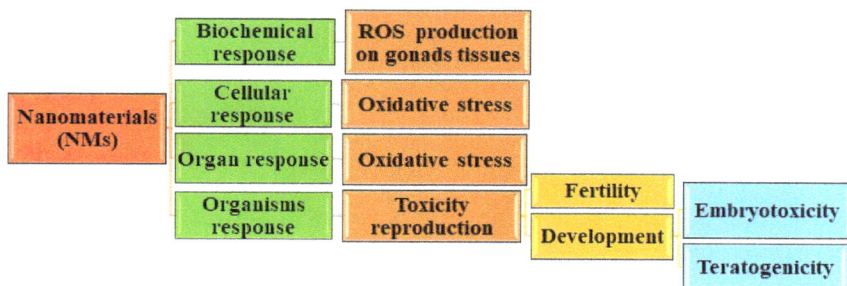

Figure 3: Nanomaterials route uptake at different stages of aquatic organisms

7.5   Semiaquatic nanomaterial toxicity

Semiaquatic animals are referred to as different types of organisms that partially spend some part of their life in water for example the amphibians. The amphibians are categorized into caecilians, urodelans, and anurans. The amphibian larval stage depends on an aquatic environment with well-adapted physiology and anatomy; they possess gills for breathing and the presence of loops and elongated guts for an herbivory diet. Their mode of excretion in the aquatic environment is well adapted, they possess a strong mechanism for water removal and they have caudal fin and tail used for swimming to different locations [16]. Once they leave the aquatic environment they metamorphose to adapt to the terrestrial environment, by modifying their respiratory organs which involve the developing of the lungs and fading of the gills. Similarly, the circulatory system is also modified, in which gills fade, and arches of the aorta became dedicated to both the large and small circulations. The amphibians are found to be excellent animals used as models for the determination of the toxic influence of nanomaterials. The nanomaterial's effects on the thyroid and pituitary gland have been examined using an in vitro tadpole (*Lithobates catesbeianus)* cultured caudal fins [115,116]. From the tadpole examination, effects of different nanomaterials such as zinc oxides (ZnO) nanoparticles, silver (Ag) nanoparticles, and aggregate of Ag nanoparticles on the various genes containing 3- 3-'5'- triiodothyronine (T3) were observed via the quantitative polymerase chain reaction (qPCR). The results obtained were subjected to the comparison between those nanomaterials like $AgNO_3,$ and cadmium telluride particles both measured in several micrometers (μm). The silver nanoparticles (Ag-NPs) and the small aggregate influence the transcript expression associated with the T3, causing multiple stressed molecules. The zinc oxide nanoparticles (ZnO-NPs) show no effects, while T3 signal disruption was reported from a small dosage of silver nanoparticles without causing any stress [117]. An experimental study using the strains of epithelial *Xenopus laevis* revealed copper toxicity under three various nanomaterial forms, that is aggregate of 100 nm CuO nanoparticles, 6nm CuO nanoparticles, and Cu^{++}. The impact of these cytotoxics differs based on the type of compounds and the phase of the cell's cycle. The mitotic cell division was observed to stop after subjecting the three nanomaterials to the epithelia strains of the *xenopus laevis*, thereby resulting t The knowledge of colloids chemistry is important in knowing the fate of nanomaterials because the nano-scale materials are similar to the system of colloidal mixtures, in which by characterization the nanomaterial fall within the ranges of colloids substances that are described as particles that range between 1 and 1000 nanometer [64]o a substantial decrease in proliferation of the cells and also increase in the number of apoptotic cells at different time intervals such as 48 h, 144 h and 168 h using nanomaterial's treatment of 100nm CuO-NPs, 6nm CuO-NPs and Cu^{++} respectively. Other treatments remained similar on separated cells after exposure to a short period [118].

A frail lethal impact on an embryo's development using different concentrations of ZnO, CuO, and TiO_2 nanoparticles was reported in an amphibian. These nanoparticles show teratogenic effects especially on the intestine at a higher concentration of 50 mg/L. The ZnO nanoparticles prompted the most severe influence on the barrier of the intestine, thereby allowing the nanomaterials to get into the connective tissue. However, TiO_2 nanoparticles were reported to be dimly teratogenic with possibly unnoticed physiological impacts. The dissolved CuO nanoparticles could be more adverse as compared to the ZnO [119]. A mortality rate lower than 11% was reported in tadpole (*Pelophylax perezi*) after exposure to a concentration less than 50 nm of $TiSiO_4$ nanoparticles. Similarly, adverse effects were reported on the melanin and lactate with subsequent oxidative stress increase, hence, $TiSiO_4$ nanomaterials, therefore, result in long-term consequences on the tadpole [120].

7.6   Nanomaterial toxicity on aquatic invertebrates

Studies on nanomaterial toxicity use various kinds of organisms as a model for evaluating their different impact. For instance, model organisms were used from macroinvertebrates for the toxicity determination of heavy metals on aquatic biota [121]. Generally, aquatic invertebrates comprise very wide animal groups constituting up to 30 phyla and more than 1000 different species. The largest phylum in the invertebrate group is the Arthropoda, which comprises two major dominant groups, the Insecta and the crustaceans [122]. The crustaceans are considered the most significant group among the invertebrates, they are typically used as ecotoxicological models. The planktonic or sedimental crustaceans are significant because they play a major role within the aquatic ecosystem food chain. Also, the sedimental crustaceans help in the biodegradation of numerous contaminants that are adsorbed at the bottom layer of the aquatic environment which includes the accumulated nanoparticles [12]. The peroxidation of lipid and reduction in the sodium ion content from the entire organism's body *Gammarux purex* was depicted after exposing the macroinvertebrates to a concentration of 11 Cu µg/L and 3.8 Agµg/L of copper and silver nanoparticles respectively [123,124]. The Nanotoxicity of metals was also reported in an investigation study carried out on zebrafish [125]. The eggs of the zebrafish were exposed to copper [Cu] nanoparticles for 48 hr from which the result showed delayed and hatching malformation while the exposure to gold [Au] nanoparticles did not reveal any teratogenic impacts for the same zebrafish eggs on the same 48 hr [126].

Among the invertebrate, *Daphnia magna* is regarded as the most common model used to evaluate the potential risk associated with nanomaterials [15]. The Daphnia genus members are also used as a determinant of exposure to toxicants, this is due to their possession of a significant number of genes that are common to that human as well as the availability of the entire sequence genome [127]. The *Daphnia magna* which is also called the filter feeder is known to consume different particles such as inorganic, organic, algae, and bacteria which gave them the capability of filtering much quantity of water [82,128]. The *Daphnia magna* for example shows an increase in the rate of mortality after increasing the exposure concentration of TiO_2 nanoparticles [129]. However, the *Daphnia magna* exposure to copper (Cu) related nanoparticles revealed that the chemistry of the water and the sizes are determinant factors in the natural freshwater toxicity [130]. In addition, the coated tyrosine silver nanoparticles (tyrosine-AgNPs) and silver ions were studied on three different macroinvertebrates. The tyrosine -silver nanoparticles toxicities were observed in the following increasing magnitude; daphnia greater than shrimps followed by hydra while the and silver ion toxicities were observed to be, the shrimp greater than that of daphnia followed by hydra [131]. Furthermore, the *Daphnia Magna* can take up an aggregate of nanomaterials through ingestion or filtration. Studies have shown that ingested nanomaterials in the digestive tract of the *Daphnia* sp can be excreted afterward 48 h or can be translocated to another different part of the body thereby causing bioaccumulation of the nanomaterials [132,129]. The nanomaterial's influence on the *Daphnia magna* is not only limited to reproduction abnormality and mortality, but it can also cause various physiological and behavioral changes starting with the rate of the heart, the velocity of swimming, vertical migration, and rate of alga feeding [133,134,135].

Another group of aquatic biotas comprising bivalves [molluscs] and echinoderms are also used as a good model for toxicological studies of nanomaterials in aquatic ecosystems. The influence of metallic oxides such as Fe_3O_4, CeO_2, and SnO_2 nanoparticles was evaluated on sea urchin (*Paracentrotus lividus)* immune cells. The nanomaterials were seen in the coelomocytes which are made up of the sea urchin immune cells. These cells depicted a well-morphological variable alteration of endoplasmic reticulum lysosomes, cholinesterase reduced activity, and under-regulating stress proteins [136,137]. In the molluscs (bivalves), their immune system is mainly influenced by nanomaterials, for example, *Elliptio complanate* was reported in 2009 to show immunomodulation which is caused by some nanomaterials [138]. Different experiments were conducted using *Mytilus galloprovincialis* to evaluate the nanomaterial's influence on their immune system. Various nanomaterials were absorbed through the heamocytes thereby influencing different parameters such as apoptotic activity (Fig. 4), free radical production, the capacity of phagocytosis, and activity of the lysosomal [139]. Similarly, various nanomaterials exerted an adverse effect on the immune system for example, in the *Crassostrea gigas* [an Oyster], the exposure of different types of nanomaterials causes an adverse effect on the activity of the lysosomes, oxidative stress, and apoptosis activity, particularly in the development of the embryo, digestive gland and the gills [140]. Nanomaterials accumulation was also reported in many tissues of an invertebrate, depending on the type of nanomaterials. For instance, in the *Mytilus edulis*, several effects on immunomodulators were reported after exposure to TiO_2 nanoparticles. These nanoparticles [TiO_2] were reported to accumulate in the water followed by pasting onto the gills of the *Mytilus edulis* alongside passing directly into the digestive gland thereby penetrating the cells and resulting in the perturbation of the lysosomes, modification of the genes expression, causing of antioxidant stress and influencing the immune system, more especially antimicrobial and lysozymes peptides [141]. However, the nanomaterials may pass out of the digestive tract through the haemolymph and or heamocytes from which they tend to cause modification in their functions, particularly in phagocytosis, functioning of the lysosomes, reactive oxygen species (ROS), and nitric oxide (NO) production for inducing oxidative stress, initiation of apoptosis, in addition to mitochondria and membranes [142]. Nevertheless, substantial variations were reported in the expression of the genes of immunity and antioxidant, yet, the nanoparticles (TiO_2) initiated a hypo and hyper-regulation in the immune genes within the digestive gland and hemocytes respectively [143]. The RNA quantification analysis (meta-transcriptomic) extracted from the plankton's samples, which comprise small aquatic communities of organisms revealed some toxic influences of silver ions (Ag^+) and silver nanoparticles (Ag-NPs) of 6 nm on the plankton's composition and their metabolism [144].

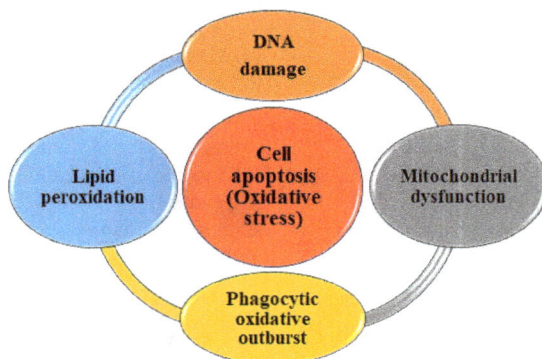

Figure 4: Nanomaterials influence the molecular level of an aquatic biota

7.7   Nanomaterial Toxicity on the Aquatic Vertebrates

The nanomaterials can be formed singly or in a combination of single metal oxide or multiple metals. Therefore, characteristics of the chemicals and the metallic behaviour are associated with both the metal ions equilibrium, aggregation dynamics, and also the level of metal nanoparticles toxicity. The biological influence of nanomaterials can be associated with the availability of ion carriers and the endocytosis level. Other experiments reported that nanomaterials can lead to the death of the fishes depending on the concentration of the particles which ranges from $\mu g/L$ to mg/L. In some fish species, the metallic nanomaterials can be more lethal as compared to the dissolved state. For example, in zebrafish [*Danio rerio*] the LC-50 was attained at 0.71mg/L for copper nanoparticles (CU-NPs) while 1.78mg/L was reported for the dissolved copper (Cu) ions. These nanoparticles induced sublethal impacts such as oxidative stress, sodium–potassium pump inhibition (Na^+/K^+-ATPase), tissue elemental disruption, and toxic breath [145]; [146]. Thus, various organs such as the brain, intestines, liver, and gills can be affected by different concentrations of metals and nanoparticles. Silver nanoparticles were also reported to pass through the egg envelope (chorion) or the evolving embryo. The ZnO and Cu nanoparticles are reported to be more noxious for juveniles and embryos as compared to their salt [147].

Different experimental investigations on the toxicological impact of nanomaterials on aquatic vertebrates were conducted majorly on fish species. For example, the largemouth bass (*Micropterus salmoides)* juvenile was used to evaluate the toxicity influence of some nanomaterials, from which the result shows peroxidation of lipid in the fish brain [148]; [149]. Furthermore, other kinds of fish species used for the nanomaterial's toxicity include Japanese rice fish (*Oryzias latipes*) and fathead minnow (*Pimephales promelas*) which were exposed to nanomaterials for 72 hr, from which a decrease in the peroxisomal membrane protein (PMP 70) was recorded in the fathead minnow fish [149]. Similarly, exposure of fathead minnow to tetrahydrofuran (THF) suspension with fullerene (nC60) revealed a 100% mortality rate within 6 -18 hr. On the other hand, zero mortality was recorded when fullerene (nC60) was exposed without the THF suspension in the water. However, peroxidation of lipids was observed when compared with the control fish [150].

In another experiment conducted on the rainbow trout (*Oncorhynchus mykiss*), it was observed that the dosage of the nanomaterials determines their toxicity effects, in which it influences the liver and gills after increasing the glutathione total level to 18% and 28% respectively. The impact observed affected the brain, liver, and gills which lead to the mortality of some of the fish [151]. The zebrafish (*Danio rerio)* embryo is also used for toxicological experiments because of its peculiar characteristics such as rapid development and small size [126,152]. For instance, in an experiment conducted on the *Cyprinus carpio* fish organs, the result showed that nanomaterials exposure over a long period causes damage to the cells leading to necrosis [153,125]. Furthermore, oxidative stress leading to liver and gills damage in the *Danio rerio* fish was also reported during the nanomaterial's long-term exposure [154,15]. The young salmon fish was also exposed to suspension of commercial and synthesized silver nanoparticles [Ag-NPs] with silver nitrate ($AgNO_3$) after reduction with Sodium borohydride ($NaBH_4$). The sizes of both the synthesized and commercial colloidal silver nanoparticles range between 3 and 220 nm. The gills of the fish were reported to accumulate the silver nanoparticles in the entire experiments, with the exception of least concentration of the nanomaterials that is 1 $\mu g/L$. The silver nanoparticles induced gill lamellae necrosis at the maximum dose (100 $\mu g/L$) leading to the mortality of 73% individual fishes [155].

Studies on the mechanical influence of nanomaterials on fishes focused more on the excretion, metabolism, distribution, and absorption of the materials in the fishes as compared to other chemical substances. For instance, examination of fullerene C60 and TiO_2 nanoparticles show an adverse effect on different organs such as the adrenal gland, liver, digestive tract, and gills. These nanomaterials infiltrated the tissues intensively via endocytosis as compared with the ionic carriers, like that of the corresponding metal ions, or through the diffusion process. Nanomaterials in the fish hardly have been excreted via the kidneys, however, they can be removed via the bile [156]. The influence of silver nanoparticles was also examined on zebrafish, and the results revealed that the Ag-NPs caused neurotoxic effects as compared to the silver ions metal. The effects caused by the silver nanoparticles of various sizes ranging from 12 and 28 nm, silver nanoparticles coated with Polyvinylpyrrolidone (PVP) ranging between 324, 65, 63, and 45 nm, and silver ions, were reported to affect the development of the embryo. The silver ion (Ag^+) was reported to slow down the swim bladder development and also different malformation. The silver nanoparticles (Ag-NPs) show adverse malformation also, and the behaviour of the fish was modified in the presence of light stimuli, the least PVP-coated nanoparticles induced hyperactivity while the highest PVP-coated induced hypoactivity [157]. Similarly, nickel nanoparticles (Ni-NPs) were reported to induce malformation and mortality in the embryo of zebrafish. Their intestine was reported to become thin and the muscle of the skeleton was affected by the soluble nickel, 100, 60, and 30 nm of the nickel nanoparticles [158]. Consequently, the nickel conformation (nanoparticles, ions, or aggregate) was considered more significant than their sizes [65]. Gold nanoparticles (Au-NPs) were also reported to induce toxicity in the embryo of zebrafish. The Au-NPs accumulated in various sizes to form aggregates which determine the concentration rate. However, the Au-NPs influence on the development of the embryo was disproportionate to its concentrations. The deformities observed in the embryo during its developmental stages could be a result of random dispersal of the Au-NPs within its cells, meanwhile, the toxicity depends on the chemical characteristics. The gold nanoparticles (Au-NPs) were reported to be less toxic compared to the silver nanoparticles (Ag-NPs); hence, the embryo of the zebrafish can be used as a suitable model for toxicological studies, particularly relating to the biocompatibility of the nanomaterials [159]. The trout fishes [*Oncorhynchus* sp] are also used as models for toxicological studies, for example, the influence of carbon nanoparticles and TiO_2 nanoparticles were evaluated on the hepatocytes of trout fish. The carbon nanomaterials used comprise Single-walled carbon nanotubes (SWCNTs), fullerene (C60) and multi-walled carbon nanotubes (MWNTs), and TiO_2 nanoparticles ranging between 5 to 200 nm. Experimental studies show that these nanomaterials pose some toxicological effects on trout fish [160]. The influence of Copper (II) sulfate ($CuSO_4$) salt and that of copper nanoparticles [Cu-NPs 87 nm] were investigated on the *Oncorhynchus mykiss* from which an aggregate Cu-NPs and $CuSO_4$ salt were detected in the juvenile *O. mykiss* gills but with varying proportions. However, both the materials (Cu-NPs and $CuSO_4$) were not detected in the muscle, brain, or spleen. Yet, a reduction of sodium-potassium enzymes (Na^+/K^+ ATPase) was seen in the intestine and brain [146]. Furthermore, exposure to TiO_2 (20 nm) to *Oncorhynchus mykiss* induced adverse effects on its gills, pathologically characterized by edema gill lamellae thickness [161]. Generally, fish are considered at low risk as compared to other aquatic organisms because of their specialized defensive immune system [155].

8.   Challenges and regulations of nanomaterials

The expanding growth and production of nanomaterials have led to more complications because of their noxious effect on the aquatic biota, human health, and the environment at large. A lot of efforts have been ongoing to investigate the nanotoxicity effect of these nanomaterials alongside their interactions with aquatic biota. However, it is difficult to evaluate and confirm the nanotoxicity effects of nanomaterials in the biological system [162]. Yet, owing to the various hazardous nature of some nanomaterials to aquatic biota, regulations of these materials from their fabrications and commercial uses are of significant importance. One of the major challenges associated with the synthesis of nanomaterial is the use of less noxious chemicals and reductions in the required energy for acquiring pure, stable, smaller, and less harmful nanoparticles with well-improved properties [4]. However, presently, no wholistic standard techniques are available for assessment of the toxic nature of the fabricated nanomaterials, this complicated the classification and commercialization of the nanomaterials for industrial application. Different countries have their regulatory bodies for nanomaterials which are responsible for evaluating the toxic nature of the nanomaterials before approving commercial applications. Nevertheless, the worldwide standard organization for the monitoring of the nanomaterial's toxicity is lacking, which is crucial in the deployment of some essential products of nanomaterials in the international market [1].

Conclusion and future outlooks

The recent development in nanotechnology has resulted in the emergence of new nanomaterials that are used in various aspects of human endeavour which comprise medical, agriculture, and environmental remediation among others. However, despite their wide application in these areas, they still pose some threat to the human, aquatic biota, and the environment at large. Thus, less toxic nanomaterials required much attention due to their harmless application. Understanding their biological influence and environmental fate is very substantial for synthesizing a biocompatible material. Nanomaterial toxicity has attracted much interest in the field of scientific research, because of its intrinsic and extrinsic properties, which involve their peculiar behaviour in water bodies and their possible damage to wildlife, human health, and the ecosystem. In recent years, studies have been conducted to assess the toxicity effects of nanomaterials on the aquatic biota alongside their interaction with other pollutants along with the adverse effect on the different biota. Nanomaterials are potential noxious substances; they can cause adverse effects directly or indirectly due to their chemical constituents such as other industrial substances like pesticides. The influence of nanomaterials is gaining more attention recently, especially in using different organisms as bioindicators. However, several investigations regarding nanomaterials recently focus on knowing their influence on the biota living both in the semi-aquatic; such as amphibians and aquatic comprising of marine and freshwater organisms, examples include, algae, planktons, invertebrates [bivalves, insects, etc.] and vertebrates such as fish species. A novel research area is currently developing, focusing on the measurement of the precise possible adverse influence of the nanomaterials thereby resulting in effective environmental protection. collaborative investigation of the nanomaterial's toxicity is more required nowadays due to the high increase in the fabrication of these nanomaterials. Furthermore, multidisciplinary research will lead to more findings on the new physiological mechanisms of interaction between the biological organism and the nanomaterials, which can result in providing new knowledge in the field of bio-nanotoxicology. Nanomaterials characterizations provide vital information about the substances or constituents used in the formulation of the nanomaterials. Similarly, it also established the interaction between the nanostructure and biological functions of a biota [163,164]. Thus, some of the basic models for predicting the properties of nanomaterials and their biological impacts comprise (i) nanotoxicological change mechanism from the aspect of correlative to aspect of causative, (ii) conquering the interference of nanomaterials for a clear invitro study, (iii) provision of a cellular (single-celled level) response for the interaction of nanomaterials and (iv) knowing the interference of kinetic models' parameter of biological and nanomaterials coherence [165,166,167].

References

[1] Rastogi S, Sharma G, Kandasubramanian B. Nanomaterials and the environment. In: The ELSI Handbook of Nanotechnology: Risk, Safety, ELSI and Commercialization. 2020. https://doi.org/10.1002/9781119592990.ch1

[2] Albalawi F, Hussein MZ, Fakurazi S, Masarudin MJ. Engineered nanomaterials: The challenges and opportunities for nanomedicines. International Journal of Nanomedicine. 2021. https://doi.org/10.2147/IJN.S288236

[3] Challagulla NV, Rohatgi V, Sharma D, Kumar R. Recent developments of nanomaterial applications in additive manufacturing: a brief review. Current Opinion in Chemical Engineering. 2020. https://doi.org/10.1016/j.coche.2020.03.003

[4] Allam BK, Musa N, Debnath A, Usman UL, Banerjee S. Recent developments and application of bimetallic based materials in water purification. Environ Challenges [Internet]. 2021;5[July]:100405. Available from: https://doi.org/10.1016/j.envc.2021.100405 https://doi.org/10.1016/j.envc.2021.100405

[5] Chitranshi M, Pujari A, Ng V, Chen D, Chauhan D, Hudepohl R, et al. Carbon nanotube sheet-synthesis and applications. Nanomaterials. 2020; https://doi.org/10.3390/nano10102023

[6] Potbhare AK, Chouke PB, Mondal A, Thakare RU, Mondal S, Chaudhary RG, Rai AR. Rhizoctonia solani assisted biosynthesis of silver nanoparticles for antibacterial

assay. Materials Today: Proceedings. 2020.
https://doi.org/10.1016/j.matpr.2020.05.419

[7] Yashwant Singh Sahu. Nano zinc oxide market by Application [Paints & Coatings, Cosmetics, and Others] - Global Opportunity Analysis and Industry Forecast, 2014-2022. Allied Market Research. 2016.

[8] Saleem H, Zaidi SJ, Alnuaimi NA. Recent advancements in the nanomaterial application in concrete and its ecological impact. Materials. 2021. https://doi.org/10.3390/ma14216387

[9] Laux P, Tentschert J, Riebeling C, Braeuning A, Creutzenberg O, Epp A, et al. Nanomaterials: certain aspects of application, risk assessment and risk communication. Archives of Toxicology. 2018. https://doi.org/10.1007/s00204-017-2144-1

[10] Opinion F. Scientific Committee on Consumer Safety SCCS OPINION on Titanium dioxide [TiO 2] used in cosmetic products that lead to exposure by inhalation. 2020. 1-61 p.

[11] Liu J, Williams PC, Goodson BM, Geisler-Lee J, Fakharifar M, Gemeinhardt ME. TiO2 nanoparticles in irrigation water mitigate impacts of aged Ag nanoparticles on soil microorganisms, Arabidopsis thaliana plants, and Eisenia fetida earthworms. Vol. 172, Environmental Research. 2019. 202-215 p. https://doi.org/10.1016/j.envres.2019.02.010

[12] Bundschuh M, Filser J, Lüderwald S, McKee MS, Metreveli G, Schaumann GE, et al. Nanoparticles in the environment: where do we come from, where do we go to? Environmental Sciences Europe. 2018. https://doi.org/10.1186/s12302-018-0132-6

[13] Musial J, Krakowiak R, Mlynarczyk DT, Goslinski T, Stanisz BJ. Titanium dioxide nanoparticles in food and personal care products-what do we know about their safety? Nanomaterials. 2020. https://doi.org/10.3390/nano10061110

[14] Joner E, Hartnik T, Amundsen C. Environmental fate and ecotoxicity of engineered nanoparticles. Norwegian Pollution Control Authority Report no. TA 2304/2007. 2007.

[15] Khoshnamvand M, Hao Z, Fadare OO, Hanachi P, Chen Y, Liu J. Toxicity of biosynthesized silver nanoparticles to aquatic organisms of different trophic levels. Chemosphere. 2020; https://doi.org/10.1016/j.chemosphere.2020.127346

[16] Exbrayat JM, Moudilou EN, Lapied E. Harmful Effects of Nanoparticles on Animals. Journal of Nanotechnology. 2015. https://doi.org/10.1155/2015/861092

[17] Chouke PB, Potbhare AK, Meshram NP, Rai MM, Dadure KM, Chaudhary K, Rai AR, Desimone MF, Chaudhary RG, Masram DT. Bioinspired NiO nanospheres: Exploring in vitro toxicity using Bm-17 and L. rohita liver cells, DNA degradation, docking, and proposed vacuolization mechanism. ACS omega. 2022. https://doi.org/10.1021/acsomega.1c06544

[18] Vimercati L, Cavone D, Caputi A, De Maria L, Tria M, Prato E, et al. Nanoparticles: An Experimental Study of Zinc Nanoparticles Toxicity on Marine Crustaceans. General Overview on the Health Implications in Humans. Front Public Heal. 2020; https://doi.org/10.3389/fpubh.2020.00192

[19] Lee Y-L, Shih Y-S, Chen Z-Y, Cheng F-Y, Lu J-Y, Wu Y-H, et al. Toxic Effects and Mechanisms of Silver and Zinc Oxide Nanoparticles on Zebrafish Embryos in Aquatic Ecosystems. Nanomaterials. 2022;12[4]:717. https://doi.org/10.3390/nano12040717

[20] Valdiglesias V. Cytotoxicity and Genotoxicity of Nanomaterials. Nanomaterials. 2022;12[4]:12-4. https://doi.org/10.3390/nano12040634

[21] Kong IC, Ko KS, Koh DC, Chon CM. Comparative effects of particle sizes of cobalt nanoparticles to nine biological activities. Int J Mol Sci. 2020;21[18]:1-19. https://doi.org/10.3390/ijms21186767

[22] Chouke PB, Potbhare AK, Dadure KM, Mungole AJ, Meshram NP, Chaudhary RR, Rai AR, Chaudhary RG. An antibacterial activity of Bauhinia racemosa assisted ZnO nanoparticles during lunar eclipse and docking assay. Materials Today: Proceedings. 2020. https://doi.org/10.1016/j.matpr.2020.04.758

[23] Sahu SC, Hayes AW. Toxicity of nanomaterials found in human environment. Toxicol Res Appl. 2017; https://doi.org/10.1177/2397847317726352

[24] Accomasso L, Cristallini C, Giachino C. Risk assessment and risk minimization in nanomedicine: A need for predictive, alternative, and 3Rs strategies. Frontiers in Pharmacology. 2018. https://doi.org/10.3389/fphar.2018.00228

[25] Kahru A, Dubourguier HC. From ecotoxicology to nanoecotoxicology. Toxicology. 2010. https://doi.org/10.1016/j.tox.2009.08.016

[26] Kong B, Seog JH, Graham LM, Lee SB. Experimental considerations on the cytotoxicity of nanoparticles. Nanomedicine. 2011. https://doi.org/10.2217/nnm.11.77

[27] Barhoum A, García-Betancourt ML, Jeevanandam J, Hussien EA, Mekkawy SA, Mostafa M, et al. Review on Natural, Incidental, Bioinspired, and Engineered Nanomaterials: History, Definitions, Classifications, Synthesis, Properties, Market, Toxicities, Risks, and Regulations. Nanomaterials. 2022;12[2]. https://doi.org/10.3390/nano12020177

[28] Jeevanandam J, Barhoum A, Chan YS, Dufresne A, Danquah MK. Review on nanoparticles and nanostructured materials: History, sources, toxicity and regulfile:///C:/Users/Usuario.Usuario-PC/Documents/INVESTIGACION/1-NANOPARTÍCULAS/4-ATM/Bibliografía/Aplicaciones/Dang2019_Effect of Gold Nanoparticle Size on Their Prope. Beilstein J Nanotechnol. 2018;

[29] Hochella MF, Spencer MG, Jones KL. Nanotechnology: Nature's gift or scientists' brainchild? Environmental Science: Nano. 2015. https://doi.org/10.1039/C4EN00145A

[30] Konishi H, Xu H, Guo H. Nanostructures of natural iron oxide nanoparticles. In: Nature's Nanostructures. 2012.

[31] Holmes-Maybank KT, Schreiner AD, Houchens N, Brzezinski WA. Dust in the wind. J Hosp Med. 2017; https://doi.org/10.12788/jhm.2845

[32] Sapkota A, Symons JM, Kleissl J, Wang L, Parlange MB, Ondov J, et al. Impact of the 2002 Canadian forest fires on particulate matter air quality in Baltimore City. Environ Sci Technol. 2005; https://doi.org/10.1021/es035311z

[33] Buseck PR, Pósfai M. Airborne minerals and related aerosol particles: Effects on climate and the environment. Proc Natl Acad Sci U S A. 1999; https://doi.org/10.1073/pnas.96.7.3372

[34] Sioutas C, Delfino RJ, Singh M. Exposure assessment for atmospheric Ultrafine Particles [UFPs] and implications in epidemiologic research. Environmental Health Perspectives. 2005. https://doi.org/10.1289/ehp.7939

[35] Westerdahl D, Fruin S, Sax T, Fine PM, Sioutas C. Mobile platform measurements of ultrafine particles and associated pollutant concentrations on freeways and residential streets in Los Angeles. Atmos Environ. 2005; https://doi.org/10.1016/j.atmosenv.2005.02.034

[36] Ning Z, Cheung CS, Fu J, Liu MA, Schnell MA. Experimental study of environmental tobacco smoke particles under actual indoor environment. Sci Total Environ. 2006; https://doi.org/10.1016/j.scitotenv.2006.02.017

[37] Monteiro-Riviere NA, Wiench K, Landsiedel R, Schulte S, Inman AO, Riviere JE. Safety evaluation of sunscreen formulations containing titanium dioxide and zinc oxide nanoparticles in UVB sunburned skin: An In vitro and in vivo study. Toxicol Sci. 2011; https://doi.org/10.1093/toxsci/kfr148

[38] Stefani D, Wardman D, Lambert T. The implosion of the calgary general hospital: Ambient air quality issues. J Air Waste Manag Assoc. 2005; https://doi.org/10.1080/10473289.2005.10464605

[39] De Matteis V. Exposure to inorganic nanoparticles: Routes of entry, immune response, biodistribution and in vitro/In vivo toxicity evaluation. Toxics. 2017. https://doi.org/10.3390/toxics5040029

[40] Chouke PB, Potbhare AK, Bhusari G, Somkuwar S, Shaik D, Mishra RK, Chaudhary RG. Green fabrication of zinc oxide nanospheres by Aspidopterys cordata for effective antioxidant and antibacterial activity. Advanced Materials Letters. 10(2019)355-360.

[41] Sirelkhatim A, Mahmud S, Seeni A, Kaus NHM, Ann LC, Bakhori SKM, et al. Review on zinc oxide nanoparticles: Antibacterial activity and toxicity mechanism. Nano-Micro Letters. 2015. https://doi.org/10.1007/s40820-015-0040-x

[42] Lv X, Tao J, Chen B, Zhu X. Roles of temperature and flow velocity on the mobility of nano-sized titanium dioxide in natural waters. Sci Total Environ. 2015; https://doi.org/10.1016/j.scitotenv.2016.03.001

[43] Bandaru PR. Electrical properties and applications of carbon nanotube structures. Journal of Nanoscience and Nanotechnology. 2007. https://doi.org/10.1166/jnn.2007.307

[44] Bosi S, Da Ros T, Spalluto G, Prato M. Fullerene derivatives: An attractive tool for biological applications. European Journal of Medicinal Chemistry. 2003. https://doi.org/10.1002/chin.200410254

[45] Ali H, Khan E, Ilahi I. Environmental chemistry and ecotoxicology of hazardous heavy metals: Environmental persistence, toxicity, and bioaccumulation. J Chem. 2019;2019[Cd]. https://doi.org/10.1155/2019/6730305

[46] Khan I, Saeed K, Khan I. Nanoparticles: Properties, applications and toxicities. Arabian Journal of Chemistry. 2019. https://doi.org/10.1016/j.arabjc.2017.05.011

[47] Abbasi E, Aval SF, Akbarzadeh A, Milani M, Nasrabadi HT, Joo SW, et al. Dendrimers: Synthesis, applications, and properties. Nanoscale Research Letters. 2014. https://doi.org/10.1186/1556-276X-9-247

[48] Goswami L, Kim KH, Deep A, Das P, Bhattacharya SS, Kumar S, et al. Engineered nano particles: Nature, behavior, and effect on the environment. Journal of Environmental Management. 2017. https://doi.org/10.1016/j.jenvman.2017.01.011

[49] Dimapilis EAS, Hsu CS, Mendoza RMO, Lu MC. Zinc oxide nanoparticles for water disinfection. Sustainable Environment Research. 2018. https://doi.org/10.1016/j.serj.2017.10.001

[50] Sukhanova A, Bozrova S, Sokolov P, Berestovoy M, Karaulov A, Nabiev I. Dependence of Nanoparticle Toxicity on Their Physical and Chemical Properties. Nanoscale Research Letters. 2018. https://doi.org/10.1186/s11671-018-2457-x

[51] Zuber A, Purdey M, Schartner E, Forbes C, Van Der Hoek B, Giles D, et al. Detection of gold nanoparticles with different sizes using absorption and fluorescence based method. Sensors Actuators, B Chem. 2016; https://doi.org/10.1016/j.snb.2015.12.044

[52] Biola-Clier M, Gaillard JC, Rabilloud T, Armengaud J, Carriere M. Titanium dioxide nanoparticles alter the cellular phosphoproteome in A549 cells. Nanomaterials. 2020; https://doi.org/10.3390/nano10020185

Materials Research Foundations 135 (2023) 304-340   https://doi.org/10.21741/9781644902172-13

[53] Wu Z, Hu T, Hu W, Shao L, Sun Y, Xue F, et al. Evolution in physicochemical properties of fine particles emitted from residential coal combustion based on chamber experiment. Gondwana Res. 2021; https://doi.org/10.1016/j.gr.2021.10.017

[54] Gatoo MA, Naseem S, Arfat MY, Mahmood Dar A, Qasim K, Zubair S. Physicochemical properties of nanomaterials: Implication in associated toxic manifestations. BioMed Research International. 2014. https://doi.org/10.1155/2014/498420

[55] Kim E, Kim SH, Kim HC, Lee SG, Lee SJ, Jeong SW. Growth inhibition of aquatic plant caused by silver and titanium oxide nanoparticles. Toxicol Environ Health Sci. 2011; https://doi.org/10.1007/s13530-011-0071-8

[56] Wang L, Hu C, Shao L. The antimicrobial activity of nanoparticles: Present situation and prospects for the future. International Journal of Nanomedicine. 2017. https://doi.org/10.2147/IJN.S121956

[57] Zhu Y, Liu X, Hu Y, Wang R, Chen M, Wu J, et al. Behavior, remediation effect and toxicity of nanomaterials in water environments. Environmental Research. 2019. https://doi.org/10.1016/j.envres.2019.04.014

[58] Kan H, Pan D, Castranova V. Engineered nanoparticle exposure and cardiovascular effects: the role of a neuronal-regulated pathway. Inhalation Toxicology. 2018. https://doi.org/10.1080/08958378.2018.1535634

[59] Gottschalk F, Ort C, Scholz RW, Nowack B. Engineered nanomaterials in rivers - Exposure scenarios for Switzerland at high spatial and temporal resolution. Environ Pollut. 2011; https://doi.org/10.1016/j.envpol.2011.08.023

[60] Caballero-Guzman A, Nowack B. A critical review of engineered nanomaterial release data: Are current data useful for material flow modeling? Environmental Pollution. 2016. https://doi.org/10.1016/j.envpol.2016.02.028

[61] Bathi JR, Moazeni F, Upadhyayula VKK, Chowdhury I, Palchoudhury S, Potts GE, et al. Behavior of engineered nanoparticles in aquatic environmental samples: Current status and challenges. Science of the Total Environment. 2021. https://doi.org/10.1016/j.scitotenv.2021.148560

[62] Klaine SJ, Alvarez PJJ, Batley GE, Fernandes TF, Handy RD, Lyon DY, et al. Nanomaterials in the environment: Behavior, fate, bioavailability, and effects. Environmental Toxicology and Chemistry. 2008. https://doi.org/10.1897/08-090.1

[63] Sanchez-Martin MJ, Rodriguez-Cruz MS, Andrades MS, Sanchez-Camazano M. Efficiency of different clay minerals modified with a cationic surfactant in the adsorption of pesticides: Influence of clay type and pesticide hydrophobicity. Appl Clay Sci. 2006; https://doi.org/10.1016/j.clay.2005.07.008

[64] Baig N, Kammakakam I, Falath W, Kammakakam I. Nanomaterials: A review of synthesis methods, properties, recent progress, and challenges. Materials Advances. 2021. https://doi.org/10.1039/D0MA00807A

[65] Mourdikoudis S, Pallares RM, Thanh NTK. Characterization techniques for nanoparticles: Comparison and complementarity upon studying nanoparticle properties. Nanoscale. 2018. https://doi.org/10.1039/C8NR02278J

[66] Boies AM, Hoecker C, Bhalerao A, Kateris N, de La Verpilliere J, Graves B, et al. Agglomeration Dynamics of 1D Materials: Gas-Phase Collision Rates of Nanotubes and Nanorods. Small. 2019; https://doi.org/10.1002/smll.201900520

[67] Saleh TA. Trends in the sample preparation and analysis of nanomaterials as environmental contaminants. Trends in Environmental Analytical Chemistry. 2020. https://doi.org/10.1016/j.teac.2020.e00101

[68] Guerra FD, Attia MF, Whitehead DC, Alexis F. Nanotechnology for environmental remediation: Materials and applications. Molecules. 2018. https://doi.org/10.3390/molecules23071760

[69] Kumar L, Ragunathan V, Chugh M, Bharadvaja N. Nanomaterials for remediation of contaminants: a review. Environmental Chemistry Letters. 2021. https://doi.org/10.1007/s10311-021-01212-z

[70] Shiu RF, Vazquez CI, Tsai YY, Torres G V., Chen CS, Santschi PH, et al. Nano-plastics induce aquatic particulate organic matter [microgels] formation. Sci Total Environ. 2020; https://doi.org/10.1016/j.scitotenv.2019.135681

[71] Markus AA, Parsons JR, Roex EWM, de Voogt P, Laane RWPM. Modelling the transport of engineered metallic nanoparticles in the river Rhine. Water Res. 2016; https://doi.org/10.1016/j.watres.2016.01.003

[72] Ellis LJA, Valsami-Jones E, Lead JR, Baalousha M. Impact of surface coating and environmental conditions on the fate and transport of silver nanoparticles in the aquatic environment. Sci Total Environ. 2016; https://doi.org/10.1016/j.scitotenv.2016.05.199

[73] El Badawy AM, Silva RG, Morris B, Scheckel KG, Suidan MT, Tolaymat TM. Surface charge-dependent toxicity of silver nanoparticles. Environ Sci Technol. 2011; https://doi.org/10.1021/es1034188

[74] Chan VSW. Nanomedicine: An unresolved regulatory issue. Regul Toxicol Pharmacol. 2006; https://doi.org/10.1016/j.yrtph.2006.04.009

[75] Bhuvaneshwari M, Bairoliya S, Parashar A, Chandrasekaran N, Mukherjee A. Differential toxicity of Al2O3 particles on Gram-positive and Gram-negative sediment bacterial isolates from freshwater. Environ Sci Pollut Res. 2016; https://doi.org/10.1007/s11356-016-6407-9

[76] Jin C, Tang Y, Yang FG, Li XL, Xu S, Fan XY, et al. Cellular toxicity of TiO2 nanoparticles in anatase and rutile crystal phase. Biol Trace Elem Res. 2011; https://doi.org/10.1007/s12011-010-8707-0

[77] Simon-Deckers A, Loo S, Mayne-L'Hermite M, Herlin-Boime N, Menguy N, Reynaud C, et al. Size-, composition- and shape-dependent toxicological impact of metal oxide nanoparticles and carbon nanotubes toward bacteria. Environ Sci Technol. 2009; https://doi.org/10.1021/es9016975

[78] George S, Lin S, Ji Z, Thomas CR, Li L, Mecklenburg M, et al. Surface defects on plate-shaped silver nanoparticles contribute to its hazard potential in a fish gill cell line and zebrafish embryos. ACS Nano. 2012; https://doi.org/10.1021/nn204671v

[79] Wise JP, Goodale BC, Wise SS, Craig GA, Pongan AF, Walter RB, et al. Silver nanospheres are cytotoxic and genotoxic to fish cells. Aquat Toxicol. 2010;

[80] Ali D, Alarifi S, Kumar S, Ahamed M, Siddiqui MA. Oxidative stress and genotoxic effect of zinc oxide nanoparticles in freshwater snail Lymnaea luteola L. Aquat Toxicol. 2012; https://doi.org/10.1016/j.aquatox.2012.07.012

[81] Mahaye N, Thwala M, Cowan DA, Musee N. Genotoxicity of metal based engineered nanoparticles in aquatic organisms: A review. Mutation Research - Reviews in Mutation Research. 2017. https://doi.org/10.1016/j.mrrev.2017.05.004

[82] Kalman J, Paul KB, Khan FR, Stone V, Fernandes TF. Characterisation of bioaccumulation dynamics of three differently coated silver nanoparticles and aqueous silver in a simple freshwater food chain. In: Environmental Chemistry. 2015. https://doi.org/10.1071/EN15035

[83] Farkas J, Bergum S, Nilsen EW, Olsen AJ, Salaberria I, Ciesielski TM, et al. The impact of TiO2 nanoparticles on uptake and toxicity of benzo[a]pyrene in the blue mussel [Mytilus edulis]. Sci Total Environ. 2015; https://doi.org/10.1016/j.scitotenv.2014.12.084

[84] Reeves JF, Davies SJ, Dodd NJF, Jha AN. Hydroxyl radicals [{radical dot}OH] are associated with titanium dioxide [TiO2] nanoparticle-induced cytotoxicity and oxidative DNA damage in fish cells. Mutat Res - Fundam Mol Mech Mutagen. 2008; https://doi.org/10.1016/j.mrfmmm.2007.12.010

[85] Kadiyala NK, Mandal BK, Ranjan S, Dasgupta N. Bioinspired gold nanoparticles decorated reduced graphene oxide nanocomposite using Syzygium cumini seed extract: Evaluation of its biological applications. Mater Sci Eng C. 2018; https://doi.org/10.1016/j.msec.2018.07.075

[86] Rastogi A, Zivcak M, Sytar O, Kalaji HM, He X, Mbarki S, et al. Impact of metal and metal oxide nanoparticles on plant: A critical review. Frontiers in Chemistry. 2017. https://doi.org/10.3389/fchem.2017.00078

[87] Turan NB, Erkan HS, Engin GO, Bilgili MS. Nanoparticles in the aquatic environment: Usage, properties, transformation and toxicity-A review. Process Safety and Environmental Protection. 2019. https://doi.org/10.1016/j.psep.2019.08.014

[88] Gupta R, Xie H. Nanoparticles in daily life: Applications, toxicity and regulations. J Environ Pathol Toxicol Oncol. 2018; https://doi.org/10.1615/JEnvironPatholToxicolOncol.2018026009

[89] Rani M, Shanker U, Jassal V. Recent strategies for removal and degradation of persistent & toxic organochlorine pesticides using nanoparticles: A review. Journal of Environmental Management. 2017. https://doi.org/10.1016/j.jenvman.2016.12.068

[90] Buzea C, Pacheco II, Robbie K. Nanomaterials and nanoparticles: Sources and toxicity. Biointerphases. 2007; https://doi.org/10.1116/1.2815690

[91] Etesami H, Fatemi H, Rizwan M. Interactions of nanoparticles and salinity stress at physiological, biochemical and molecular levels in plants: A review. Ecotoxicol Environ Saf. 2021; https://doi.org/10.1016/j.ecoenv.2021.112769

[92] Ma X, Geiser-Lee J, Deng Y, Kolmakov A. Interactions between engineered nanoparticles [ENPs] and plants: Phytotoxicity, uptake and accumulation. Science of the Total Environment. 2010. https://doi.org/10.1016/j.scitotenv.2010.03.031

[93] Usman UL, Sajad S, Musa N, Banerjee S. Plant Mediated Nanomaterials for Water. 2020;10[December]:9-23.

[94] Warheit DB. Hazard and risk assessment strategies for nanoparticle exposures: How far have we come in the past 10 years? F1000Research. 2018. https://doi.org/10.12688/f1000research.12691.1

[95] Sharma P, Jha AB, Dubey RS, Pessarakli M, Afsar T, Razak S, et al. Reactive Oxygen Species, Oxidative Damage, and Antioxidative Defense Mechanism in Plants under Stressful Conditions. BMC Complementary and Alternative Medicine. 2016.

[96] Khan MN, Mobin M, Abbas ZK, AlMutairi KA, Siddiqui ZH. Role of nanomaterials in plants under challenging environments. Plant Physiology and Biochemistry. 2017. https://doi.org/10.1016/j.plaphy.2016.05.038

[97] Akter M, Sikder MT, Rahman MM, Ullah AKMA, Hossain KFB, Banik S, et al. A systematic review on silver nanoparticles-induced cytotoxicity: Physicochemical properties and perspectives. Journal of Advanced Research. 2018. https://doi.org/10.1016/j.jare.2017.10.008

[98] Nolte TM, Hartmann NB, Kleijn JM, Garnæs J, van de Meent D, Jan Hendriks A, et al. The toxicity of plastic nanoparticles to green algae as influenced by surface modification, medium hardness and cellular adsorption. Aquat Toxicol. 2017; https://doi.org/10.1016/j.aquatox.2016.12.005

[99] Wang F, Guan W, Xu L, Ding Z, Ma H, Ma A, et al. Effects of nanoparticles on algae: Adsorption, distribution, ecotoxicity and fate. Applied Sciences [Switzerland]. 2019. https://doi.org/10.3390/app9081534

[100] Yamamoto Y. Aluminum toxicity in plant cells: Mechanisms of cell death and inhibition of cell elongation. Soil Science and Plant Nutrition. 2019. https://doi.org/10.1080/00380768.2018.1553484

[101] Banerjee K, Pramanik P, Maity A, Joshi DC, Wani SH, Krishnan P. Methods of Using Nanomaterials to Plant Systems and Their Delivery to Plants [Mode of Entry, Uptake, Translocation, Accumulation, Biotransformation and Barriers]. Advances in Phytonanotechnology: From Synthesis to Application. 2019. 123-152 p. https://doi.org/10.1016/B978-0-12-815322-2.00005-5

[102] Zhu H, Han J, Xiao JQ, Jin Y. Uptake, translocation, and accumulation of manufactured iron oxide nanoparticles by pumpkin plants. J Environ Monit. 2008; https://doi.org/10.1039/b805998e

[103] Mittal D, Kaur G, Singh P, Yadav K, Ali SA. Nanoparticle-Based Sustainable Agriculture and Food Science: Recent Advances and Future Outlook. Front Nanotechnol. 2020; https://doi.org/10.3389/fnano.2020.579954

[104] Pérez-de-Luque A. Interaction of nanomaterials with plants: What do we need for real applications in agriculture? Front Environ Sci. 2017; https://doi.org/10.3389/fenvs.2017.00012

[105] Zhai G, Walters KS, Peate DW, Alvarez PJJ, Schnoor JL. Transport of Gold Nanoparticles through Plasmodesmata and Precipitation of Gold Ions in Woody Poplar. Environ Sci Technol Lett. 2014; https://doi.org/10.1021/ez400202b

[106] Kurczynska E, Sala K, Godel-Jedrychowska K, Milewska-Hendel A. applied sciences Nanoparticles - Plant Interaction : What We Know , Where We Are ? Appl Sci. 2021;11[5473]:1-12. https://doi.org/10.3390/app11125473

[107] Sevilem I, Yadav SR, Helariutta Y. Plasmodesmata: Channels for intercellular signaling during plant growth and development. Methods Mol Biol. 2015; https://doi.org/10.3389/fpls.2014.00044

[108] Freixa A, Acuña V, Sanchís J, Farré M, Barceló D, Sabater S. Ecotoxicological effects of carbon based nanomaterials in aquatic organisms. Science of the Total Environment. 2018. https://doi.org/10.1016/j.scitotenv.2017.11.095

[109] Jackson P, Jacobsen NR, Baun A, Birkedal R, Kühnel D, Jensen KA, et al. Bioaccumulation and ecotoxicity of carbon nanotubes. Chem Cent J. 2013; https://doi.org/10.1186/1752-153X-7-154

[110] Maurer-Jones MA, Gunsolus IL, Murphy CJ, Haynes CL. Toxicity of engineered nanoparticles in the environment. Anal Chem. 2013; https://doi.org/10.1021/ac303636s

[111] Levard C, Hotze EM, Lowry G V., Brown GE. Environmental transformations of silver nanoparticles: Impact on stability and toxicity. Environmental Science and Technology. 2012. https://doi.org/10.1021/es2037405

[112] Sohm B, Immel F, Bauda P, Pagnout C. Insight into the primary mode of action of TiO2 nanoparticles on Escherichia coli in the dark. Proteomics. 2015; https://doi.org/10.1002/pmic.201400101

[113] Potbhare AK, Chaudhary RG, Chouke PB, Yerpude S, Mondal A, Sonkusare VN, Rai AR, Juneja HD. Phytosynthesis of nearly monodisperse CuO nanospheres using Phyllanthus reticulatus/Conyza bonariensis and its antioxidant/antibacterial assays. Materials Science and Engineering: C. 2019. https://doi.org/10.1016/j.msec.2019.02.010

[114] Slavin YN, Asnis J, Häfeli UO, Bach H. Metal nanoparticles: Understanding the mechanisms behind antibacterial activity. Journal of Nanobiotechnology. 2017. https://doi.org/10.1186/s12951-017-0308-z

[115] Thambirajah AA, Koide EM, Imbery JJ, Helbing CC. Contaminant and environmental influences on thyroid hormone action in amphibian metamorphosis. Frontiers in Endocrinology. 2019. https://doi.org/10.3389/fendo.2019.00276

[116] Thambirajah AA, Koide EM, Imbery JJ, Helbing CC. Corrigendum: Contaminant and Environmental Influences on Thyroid Hormone Action in Amphibian Metamorphosis. Front Endocrinol [Lausanne]. 2019; https://doi.org/10.3389/fendo.2019.00405

[117] Faizan M, Bhat JA, Noureldeen A, Ahmad P, Yu F. Zinc oxide nanoparticles and 24-epibrassinolide alleviates Cu toxicity in tomato by regulating ROS scavenging, stomatal movement and photosynthesis. Ecotoxicol Environ Saf. 2021; https://doi.org/10.1016/j.ecoenv.2021.112293

[118] Wongrakpanich A, Mudunkotuwa IA, Geary SM, Morris AS, Mapuskar KA, Spitz DR, et al. Size-dependent cytotoxicity of copper oxide nanoparticles in lung epithelial cells. Environ Sci Nano. 2016; https://doi.org/10.1039/C5EN00271K

[119] Al Sharie AH, El-Elimat T, Darweesh RS, Swedan S, Shubair Z, Al-Qiam R, et al. Green synthesis of zinc oxide nanoflowers using Hypericum triquetrifolium extract: characterization, antibacterial activity and cytotoxicity against lung cancer A549 cells. Appl Organomet Chem. 2020; https://doi.org/10.1002/aoc.5667

[120] Salvaterra T, Alves MG, Domingues I, Pereira R, Rasteiro MG, Carvalho RA, et al. Biochemical and metabolic effects of a short-term exposure to nanoparticles of titanium silicate in tadpoles of Pelophylax perezi [Seoane]. Aquat Toxicol. 2013; https://doi.org/10.1016/j.aquatox.2012.12.014

[121] Gheorghe S, Stoica C, Vasile GG, Nita-Lazar M, Stanescu E, Lucaciu IE. Metals Toxic Effects in Aquatic Ecosystems: Modulators of Water Quality. In: Water Quality. 2017. https://doi.org/10.5772/65744

Emerging Nanomaterials and Their Impact on Society in the 21st Century   Materials Research Forum LLC

Materials Research Foundations 135 (2023) 304-340   https://doi.org/10.21741/9781644902172-13

[122] Giribet G, Edgecombe GD. The Phylogeny and Evolutionary History of Arthropods. Current Biology. 2019. https://doi.org/10.1016/j.cub.2019.04.057

[123] Felten V, Charmantier G, Mons R, Geffard A, Rousselle P, Coquery M, et al. Physiological and behavioural responses of Gammarus pulex [Crustacea: Amphipoda] exposed to cadmium. Aquat Toxicol. 2008; https://doi.org/10.1016/j.aquatox.2007.12.002

[124] Zilberg D, Munday BL, Gross A, Findlay VL, Girling P, Zhao Z, et al. The consequences of escaped farmed salmon in the Pacific North West. Dis Aquat Organ. 2001;

[125] Haque E, Ward AC. Zebrafish as a model to evaluate nanoparticle toxicity. Nanomaterials. 2018. https://doi.org/10.3390/nano8070561

[126] Verma SK, Nandi A, Sinha A, Patel P, Jha E, Mohanty S, et al. Zebrafish [Danio rerio] as an ecotoxicological model for Nanomaterial induced toxicity profiling. Precis Nanomedicine. 2021; https://doi.org/10.33218/001c.21978

[127] Hu Y, Wang Y, Lin D. A review of the toxicity of nanoparticles to Daphnia magna. Kexue Tongbao/Chinese Science Bulletin. 2017.

[128] Seabra AB, Durán N. Nanotoxicology of metal oxide nanoparticles. Metals. 2015. https://doi.org/10.3390/met5020934

[129] Martinez DST, Ellis LJA, Da Silva GH, Petry R, Medeiros AMZ, Davoudi HH, et al. Daphnia magna and mixture toxicity with nanomaterials - Current status and perspectives in data-driven risk prediction. Nano Today [Internet]. 2022;43:101430. Available from: https://doi.org/10.1016/j.nantod.2022.101430 https://doi.org/10.1016/j.nantod.2022.101430

[130] Lead JR, Batley GE, Alvarez PJJ, Croteau MN, Handy RD, McLaughlin MJ, et al. Nanomaterials in the environment: Behavior, fate, bioavailability, and effects-An updated review. Environmental Toxicology and Chemistry. 2018. https://doi.org/10.1002/etc.4147

[131] Lekamge S, Miranda AF, Abraham A, Li V, Shukla R, Bansal V, et al. The toxicity of silver nanoparticles [AgNPs] to three freshwater invertebrates with different life strategies: Hydra vulgaris, Daphnia carinata, and Paratya australiensis. Front Environ Sci. 2018; https://doi.org/10.3389/fenvs.2018.00152

[132] Skjolding LM, Kern K, Hjorth R, Hartmann N, Overgaard S, Ma G, et al. Uptake and depuration of gold nanoparticles in Daphnia magna. Ecotoxicology. 2014; https://doi.org/10.1007/s10646-014-1259-x

[133] Dong F, Koodali RT, Wang H, Ho WK. Nanomaterials for environmental applications. J Nanomater. 2014; https://doi.org/10.1155/2014/276467

134] Corsi I, Winther-Nielsen M, Sethi R, Punta C, Della Torre C, Libralato G, et al. Ecofriendly nanotechnologies and nanomaterials for environmental applications: Key issue and consensus recommendations for sustainable and ecosafe nanoremediation. Ecotoxicol Environ Saf. 2018; https://doi.org/10.1016/j.ecoenv.2018.02.037

[135] Yang J, Hou B, Wang J, Tian B, Bi J, Wang N, et al. Nanomaterials for the removal of heavy metals from wastewater. Nanomaterials. 2019. https://doi.org/10.3390/nano9030424

[136] Eid R, Arab NTT, Greenwood MT. Iron mediated toxicity and programmed cell death: A review and a re-examination of existing paradigms. Biochimica et Biophysica Acta - Molecular Cell Research. 2017. https://doi.org/10.1016/j.bbamcr.2016.12.002

[137] Butt SA. Toxicity of Engineered Nanoparticles to Marine Bivalve Molluscs. 2018;[July]. Available from: http://theses.ncl.ac.uk/jspui/handle/10443/4330

[138] Bouchard B, Gagné F, Fortier M, Fournier M. An in-situ study of the impacts of urban wastewater on the immune and reproductive systems of the freshwater mussel Elliptio complanata. Comp Biochem Physiol - C Toxicol Pharmacol. 2009; https://doi.org/10.1016/j.cbpc.2009.04.002

[139] Behzadi S, Serpooshan V, Tao W, Hamaly MA, Alkawareek MY, Dreaden EC, et al. Cellular uptake of nanoparticles: Journey inside the cell. Chemical Society Reviews. 2017. https://doi.org/10.1039/C6CS00636A

[140] González-Fernández C, Tallec K, Le Goïc N, Lambert C, Soudant P, Huvet A, et al. Cellular responses of Pacific oyster [Crassostrea gigas] gametes exposed in vitro to polystyrene nanoparticles. Chemosphere. 2018; https://doi.org/10.1016/j.chemosphere.2018.06.039

[141] Malhotra S, Hayes D, Wozniak DJ. Cystic fibrosis and pseudomonas aeruginosa: The host-microbe interface. Clinical Microbiology Reviews. 2019. https://doi.org/10.1128/CMR.00138-18

[142] Zorov DB, Juhaszova M, Sollott SJ. Mitochondrial reactive oxygen species [ROS] and ROS-induced ROS release. Physiological Reviews. 2014. https://doi.org/10.1152/physrev.00026.2013

[143] Wang Y, Hu M, Li Q, Li J, Lin D, Lu W. Immune toxicity of TiO2 under hypoxia in the green-lipped mussel Perna viridis based on flow cytometric analysis of hemocyte parameters. Sci Total Environ. 2014; https://doi.org/10.1016/j.scitotenv.2013.09.060

[144] Baptista MS, Miller RJ, Halewood ER, Hanna SK, Almeida CMR, Vasconcelos VM, et al. Impacts of Silver Nanoparticles on a Natural Estuarine Plankton Community. Environ Sci Technol. 2015; https://doi.org/10.1021/acs.est.5b03285

[145] Katuli KK, Massarsky A, Hadadi A, Pourmehran Z. Silver nanoparticles inhibit the gill Na+/K+-ATPase and erythrocyte AChE activities and induce the stress response in adult zebrafish [Danio rerio]. Ecotoxicol Environ Saf. 2014; https://doi.org/10.1016/j.ecoenv.2014.04.001

[146] Pirahanchi Y, Aeddula NR. Physiology, Sodium Potassium Pump [Na+ K+ Pump]. StatPearls. 2019.

[147] Vicario-Parés U, Castañaga L, Lacave JM, Oron M, Reip P, Berhanu D, et al. Comparative toxicity of metal oxide nanoparticles [CuO, ZnO and TiO 2] to developing zebrafish embryos. J Nanoparticle Res. 2014; https://doi.org/10.1007/s11051-014-2550-8

[148] Oberdörster E. Manufactured nanomaterials [fullerenes, C60] induce oxidative stress in the brain of juvenile largemouth bass. Environ Health Perspect. 2004; https://doi.org/10.1289/ehp.7021

[149] Zhu S, Oberdörster E, Haasch ML. Toxicity of an engineered nanoparticle [fullerene, C60] in two aquatic species, Daphnia and fathead minnow. Mar Environ Res. 2006; https://doi.org/10.1016/j.marenvres.2006.04.059

[150] Zhang X dong, Wu T xing, Cai L sheng, Zhu Y fei. Influence of fasting on muscle composition and antioxidant defenses of market-size Sparus macrocephalus. J Zhejiang Univ Sci B. 2007; https://doi.org/10.1631/jzus.2007.B0906

[151] Ostaszewska T, Chojnacki M, Kamaszewski M, Sawosz-Chwalibóg E. Histopathological effects of silver and copper nanoparticles on the epidermis, gills, and liver of Siberian sturgeon. Environ Sci Pollut Res. 2016; https://doi.org/10.1007/s11356-015-5391-9

[152] Rahman Khan F, Sulaiman Alhewairini S. Zebrafish [Danio rerio] as a Model Organism . In: Current Trends in Cancer Management. 2019. https://doi.org/10.5772/intechopen.81517

[153] Lee B, Duong CN, Cho J, Lee J, Kim K, Seo Y, et al. Toxicity of citrate-capped silver nanoparticles in common carp [Cyprinus carpio]. J Biomed Biotechnol. 2012; https://doi.org/10.1155/2012/262670

[154] Souza JP, Baretta JF, Santos F, Paino IMM, Zucolotto V. Toxicological effects of graphene oxide on adult zebrafish [Danio rerio]. Aquat Toxicol. 2017; https://doi.org/10.1016/j.aquatox.2017.02.017

[155] Mabrouk MM, Mansour AT, Abdelhamid AF, Abualnaja KM, Mamoon A, Gado WS, et al. Impact of aqueous exposure to silver nanoparticles on growth performance, redox status, non-specific immunity, and histopathological changes of Nile Tilapia, Oreochromis niloticus, challenged with Aeromonas hydrophila. Aquac Reports. 2021; https://doi.org/10.1016/j.aqrep.2021.100816

[156] Prabhu S, Poulose EK. Silver nanoparticles: mechanism of antimicrobial action, synthesis, medical applications, and toxicity effects. Int Nano Lett. 2012; https://doi.org/10.1186/2228-5326-2-32

[157] Ahlberg S, Antonopulos A, Diendorf J, Dringen R, Epple M, Flöck R, et al. PVP-coated, negatively charged silver nanoparticles: A multi-center study of their

physicochemical characteristics, cell culture and in vivo experiments. Beilstein Journal of Nanotechnology. 2014. https://doi.org/10.3762/bjnano.5.205

[158] Imran Din M, Rani A. Recent advances in the synthesis and stabilization of nickel and nickel oxide nanoparticles: A green adeptness. International Journal of Analytical Chemistry. 2016. https://doi.org/10.1155/2016/3512145

[159] Lee KY, Jang GH, Byun CH, Jeun M, Searson PC, Lee KH. Zebrafish models for functional and toxicological screening of nanoscale drug delivery systems: Promoting preclinical applications. Bioscience Reports. 2017. https://doi.org/10.1042/BSR20170199

[160] Yen HJ, Horng JL, Yu CH, Fang CY, Yeh YH, Lin LY. Toxic effects of silver and copper nanoparticles on lateral-line hair cells of zebrafish embryos. Aquat Toxicol. 2019; https://doi.org/10.1016/j.aquatox.2019.105273

[161] Varol M, Kaya GK, Alp A. Heavy metal and arsenic concentrations in rainbow trout [Oncorhynchus mykiss] farmed in a dam reservoir on the Firat [Euphrates] River: Risk-based consumption advisories. Sci Total Environ. 2017; https://doi.org/10.1016/j.scitotenv.2017.05.052

[162] Chrishtop V V., Prilepskii AY, Nikonorova VG, Mironov VA. Nanosafety vs. nanotoxicology: adequate animal models for testing in vivo toxicity of nanoparticles. Toxicology. 2021. https://doi.org/10.1016/j.tox.2021.152952

[163] Stroscio M, Dutta M. Biological Nanostructures and Applications of Nanostructures in Biology: Electrical, Mechanical, and Optical Properties. Mater Today. 2004; https://doi.org/10.1007/b112184

[164] Lin Z, Lu Y, Huang J. A hierarchical Ag2O-nanoparticle/TiO2-nanotube composite derived from natural cellulose substance with enhanced photocatalytic performance. Cellulose. 2019; https://doi.org/10.1007/s10570-019-02573-z

[165] Tirumala MG, Anchi P, Raja S, Rachamalla M, Godugu C. Novel Methods and Approaches for Safety Evaluation of Nanoparticle Formulations: A Focus Towards In Vitro Models and Adverse Outcome Pathways. Frontiers in Pharmacology. 2021. https://doi.org/10.3389/fphar.2021.612659

[166] Sayes CM, Warheit DB. Characterization of nanomaterials for toxicity assessment. Wiley Interdisciplinary Reviews: Nanomedicine and Nanobiotechnology. 2009. https://doi.org/10.1002/wnan.58

[167] Jagdale SC, Hude RU, Chabukswar AR. Zebrafish: A laboratory model to evaluate nanoparticle toxicity. In: Model Organisms to Study Biological Activities and Toxicity of Nanoparticles. 2020. https://doi.org/10.1007/978-981-15-1702-0_18

Emerging Nanomaterials and Their Impact on Society in the 21st Century   Materials Research Forum LLC
Materials Research Foundations 135 (2023) 341-365   https://doi.org/10.21741/9781644902172-14

Chapter 14

Nanomaterials and Safety Concerns to End Users

A. Ale[1*], S. Municoy[2], J. Cazenave[1,3], M.F. Desimone[2]

[1]Instituto Nacional de Limnología, CONICET, UNL, Santa Fe, Argentina. Paraje El Pozo, Ciudad Universitaria UNL, Santa Fe, Argentina

[2]Universidad de Buenos Aires. Instituto de la Química y Metabolismo del Fármaco (IQUIMEFA), CONICET, Facultad de Farmacia y Bioquímica. Buenos Aires, Argentina

[3]Facultad de Humanidades y Ciencias, UNL, Paraje El Pozo, Ciudad Universitaria UNL, Santa Fe, Argentina

* aale@inali.unl.edu.ar

Abstract

Nanomaterials (NM) are part of the daily life since decades, and the advantages that nanotechnology has brought to humanity in relation to the novel nanoproducts with unique properties are undeniable. However, concern about engineered NM has raised because of the lack of both regulation to the increasing production and safety assessments in users (with emphasis on workers). As robust evidence suggests toxic effects of NM on human health in case of exposure, this chapter aimed to describe not only the main applications and benefits of NM but also their fate, exposure pathway, and ultimate toxicological effects.

Keywords

Applications, Engineered nanomaterials, Environmental fate, Exposure pathway, Nanoparticles, Nanotoxicology

Contents

Nanomaterials and Safety Concerns to End users.......................................341

1.   **Introduction**...342

2.   **Applications**..343

3.   **Global market and production levels** 345

4.   **Environmental fate** 347

5.   Exposure pathway 348

    5.1   Inhalation 348

    5.2   Penetration through skin...348

    5.3   Ingestion...349

6.   Toxicological effects..350

Conclusions 354

References 354

1.   Introduction

Humans have been using nanomaterials (NM) since prehistoric times and first civilizations without knowing their existence [1]. For example, prehistoric humans used combustion products of fire, like carbon NM, for cave paintings. Ancient Egyptians synthesized palladium disulfide nanoparticles (NP) of 5 nm for the formulation of hair colorants. However, many years passed until in 1914, Richard Adolf Zsigmondy first introduced the term of nanometer. Since then, the "nano-universe" has been growing unlimited. In 1959, Richard Feynman installed the new concept of nanotechnology [2], and in 1974, Norio Taniguchi was the first Japanese researcher to use the term "nanotechnology" [3]. Over the years, the use of nanotechnology has expanded dramatically. Nowadays, it could be found in almost every field of science, such as cosmetics, wastewater treatment, environmental remediation, sensors, electronics, medicine, catalyst and energy- storage devices [4].

The fundamental constituents of nanotechnology are the NM. These materials of 10^{-9} m in size have at least one dimension between 1-100 nm. They can be of different shapes; like NP, nanofibers, nanoplatelets, nanorods and nanosheets, and different composition such as inorganic, organic, carbon and hybrid composites NM [5]. At the nanoscale, these materials exhibit different physicochemical properties when compared to their bulk, and can contribute to increase or improve the resistance, lightness, cleanliness and intelligence of the surfaces and systems were they are included [6].

Having been widely proved the advantageous properties of NM in comparison with the micrometric counterparts, the increased amount of production of these novel materials has raised concern regarding both their fate when released and potential exposure to humans. However, the knowledge about their safety remains under development despite the fact that there is updated evidence suggesting toxic effects in case of exposure (with emphasis in case of occupational one). In this sense, this chapter aimed to describe the main NM applications, the values towards the global market and production levels, and also to provide information about their potential fate and toxicological effects on humans.

Emerging Nanomaterials and Their Impact on Society in the 21st Century   Materials Research Forum LLC
Materials Research Foundations 135 (2023) 341-365   https://doi.org/10.21741/9781644902172-14

2. Applications

Nanomaterials are now being applied in the manufacture of a wide variety of daily use products, ranging from cosmetics, paints and coating, food, automotive, toys, textile, electronic, plastic and sport industry to biomedical and pharmaceutical sector (Figure 1). In fact, according to a BCC report, there are more than three thousand commercial applications based on NM [7].

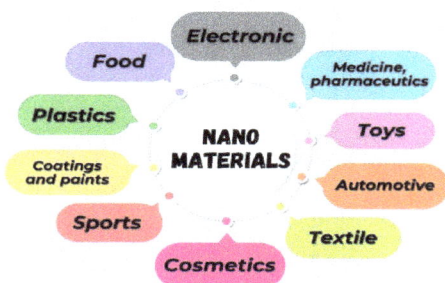

Figure 1. Nanomaterials are applied in a wide range of consumer nanoproducts from different sectors.

For example, in the cosmetic industry, nanosized materials can be found in moisturizers, hair care products, make up and sunscreens [8]. The incorporation of NM in personal care products seeks to prolong their useful life by increasing the stability of the cosmetic components and improving the delivery of the ingredients through the skin. Cosmetic industry also aims to develop new products whose nanometric elements can target the right amount of the active ingredients to the required part of the body and promote a controlled release of them. At the same time, NM are used to give new color, transparency, and UV protection. This is the case of titanium dioxide (TiO_2NP) and zinc oxide nanoparticles ($ZnONP$) that are incorporated in modern sunscreens as an UV filter and to make creams and lotions transparent [9]. Nanometric lipid vesicles, liposomes, have been another protagonist of cosmetic products since the 1980s to improve the topical delivery of cosmeceuticals [10]. The interest of cosmetics brands in incorporating liposomes into their products arose from the many advantages they offer: biodegradability, improved moisturization, recovery action, biocompatibility and controlled dermal release. Moreover, due to their similarity to cell membranes, they can penetrate epidermal barrier in contrast to other delivery systems, increasing cosmeceuticals accumulation in the skin [11]. Taking advantage of all these properties of liposomes, the L'Oréal group for Lancôme and Dior LVMH were the two pioneer brands to launch "Niosomes®" and "Capture®" onto the market in 1986, respectively.

Paints and coatings are other sectors that mainly include, for example, TiO_2NP, nanosized antimony tin oxide (ATONP), silver NP (AgNP) and nano silicon dioxide NP (SiO_2NP) to enhance durability and incorporate new functions to the products [12,13]. While TiO_2NP [14] and AgNP [15] provide antimicrobial properties against Gram positive and Gram negative bacteria, silica NP (SiNP) are included in paints to improve their physic-mechanical properties, and increase durability and resistance to abrasions, scratches and weather [16,17]. ATONP provides coatings thermal insulation and transparent protection against UV radiation, which was not seen with the bulk material [18].

Nanotechnology is widely used in various fields of food industry as well [19,20]. Production, processing, safety, and packaging of food have included NM to guarantee contamination-free products and provide end consumers with food of higher quality and enhanced properties. In this sense, nanostructured components aim to change textures and colors of food, give new tastes, and deliver and control the release of food ingredients, flavors, and additives. Nanofillers can also improve mechanical properties, increase durability, control gas permeability and provide antimicrobial action to food packaging, ensuring a well preserved product [21]. It is also possible to develop food nanosensors or nanobiosensors by using thin films, nanorods, NP, and nanofibers to detect pathogens in processing stages or in the food material, advising consumers and distributors about the condition of the product [22].

The electronic industry takes advantage of the magnetic, electrical, optical and chemical properties of NM to increase the capabilities of electronics devices, such as recording media, data storage devices, and computer components [23]. For example, nanoelectronics have allowed to reduce power consumption, weigh and thickness of the computer and television screens, but also helped to increase the computers response speed and the capacity of hard disk drive storage. Carbon nanotubes (CNT) [24] are incorporated in transistors to magnify the processing power of silicon chips, reducing the use of energy, the heat generated, and the size and weight of the components [25]. Additionally, nanotechnology is applied to conserve power and prolong battery life of small portable electronic gadgets. Magnetic ferrite-based NP with high saturation magnetization and electrical resistivity can be applied in magnetic recording media [26,27].

Sports technology also includes different nanoscale materials, specially CNT, to produce lighter but stiffer equipment and enhance strength and durability of conventional sporting equipment [28]. In this sense, CNT are used to improve the performance of tennis, badminton, golf, kayaking and archery equipment. TiO_2NP, SiNP and fullerenes [29] are included in tennis rackets; nanonickel [30] is used in golf clubs to increase their moment of inertia and stability; carbon nanofibers help to reduce the weight and increase stiffness of bicycles; and nanoclay enhances speed of water-boats [31]. In the case of textile industry, nanotechnology is used to improve specific functions in clothing, like water repellence, antimicrobial action, conductivity, and antistatic and antiwrinkle characteristics [32]. AgNP provide bactericidal protection to baby clothing, TiO_2NP acts as UV protector in swimwears, synthetic amorphous SiNP increases hydrophobicity of waterproof mountain jackets and tablecloths, and nano aluminium oxide and CNT enhance resistance to scratches. It is noteworthy that these advantages can be achieved without altering breathability or texture of textiles, and this is only possible with the application of the nanoengineered materials [33]

Nanocomposites made of polymers entrapping several varieties of NM, are widely used to improve physical properties and the efficiency of plastics manufacturing [34]. As well as this, toys containing antimicrobial nanostructures, like AgNP, are already available in the market [35]. Automotive sector appeals to nanotechnology to increase strength and improve adhesion of tires to the road [36]. In fact, carbon black and silica fillers, nanoclay, CNT, and graphene are incorporated in rubber of automobile tires to enhance security, fuel efficiency, control, and handling performance. But also, NM are present in the surface coating, fluids, interior and exterior parts, electronics, and batteries of the cars. Pharmaceutics, biology and medicine are other fields where the use of nanotechnology has grown remarkably [37–39]. In fact, NM are applied in different areas like drug delivery, imaging, gene therapy, tissue engineering, diagnosis, nanoscale biochips, and alternative therapeutics. Nanoparticles can be used to miniaturize biosensors and develop more sophisticated biochemical and diagnostic analysis approaches. Currently, the use of NP for diagnosis has extended to treatment of different diseases like cancer and diabetes. For example, iron oxide NP are widely employed in Magnetic Resonance Imaging [40], gold NP are used in cancer treatment, drug delivery and as antibacterial material [41], and AgNP are used as antibacterial agents for dressing wounds [42–44]. Inorganic, polymeric and lipid NP can help with targeted drug delivery, increasing bioavailiablity of the drug [45,46], and biomaterials and nanofiber based scaffolds are favorable to stimulate growth, proliferation, and cell attachment for tissue engineering [47,48].

Although these are just some examples, there are many other NM and applications that are currently used to improve the quality and properties of countless products. In fact, it is considered that NM have been introduced in consumerization category, as they migrated from the laboratory research to enterprise and then into the market. As a result, there is a constant exposure to different nanostructures. However, it has been reported that 70% of the products do not include information about the NM employed [49]. This is particularly the case for the automotive, cosmetics and electronics sectors, where 92%, 60% and 88% of the products, respectively, do not declare and identify a specific NM. Although this information should be available for final users, the determination and quantification of NM content in commercial products lead to many technical problems for companies, as they require special techniques and regulations to characterize them [50]. All this is especially problematic since the adverse physicochemical effects that NM can have on living beings remain ambiguous [51]. Not only the final users but also exposed workers and the general public could experience side effects against NM.

3.   Global market and production levels

Nowadays, the scientific community has successfully achieved the ability to integrate the novel properties of nanoscale materials into new products and systems. In addition, nanotechnology has generated considerable support and funding from Governments and industrial companies which drive the growth of the nanotechnology market. For example, in the United States, the National Nanotechnology Initiative enhances interagency coordination of nanotechnology research and development from early-stage fundamental science through applications-driven activities. In this way, new developments are transferred and manufactured into products to benefit society [52]. The United States President's Budget provided $1.2 billion for the National Nanotechnology Initiative to support innovation competitiveness, economic growth, as well as national security [53]. In the year 2014, the incorporation of nanotechnology into the different industries of the United States increased the U.S. gross domestic product and at the same time created nearly 1.8 million new jobs in the process [54]. Indeed, the global nanotechnology market is supposed to overcome a net value of $124 billion by the year 2024 [55].

In parallel, the United State Food and Drug Administration (FDA) received an augmented demand for regulatory review of products containing NM. Furthermore, a Nanotechnology Task Force was created in 2006 to help FDA's regulatory authorities. For example, since 1970, the Center for Drug Evaluation and Research of the FDA has received more than 600 applications for products containing NM. However, half of them were received during the last 10 years [56]. The nanomedicine's market had an estimated value of $53 billion in 2009, while by 2025 the business of nanomedicines is expected to reach a total market value of approximately $334 billion [57]. Undoubtedly, nanoscience and nanotechnology are important players of economic growth and industrial revolution.

Products that result from nanotechnology take advantages of the better and new properties to the nanoscale, in relation to larger size materials. However, the same interesting properties that make NP useful from a technological point of view would be involved in toxic effects for human health and the environment. In this context, specific regulations that ensure the safety use of NM are of paramount importance in order to present technical and accurate information about NM. Until now, the number of publications that address the nanotechnology applications together with regulatory issues and environmental management of NP is scare, but it is a rapidly growing field [51,58,59].

Due to the different applications of nanotechnology, the impacts on society are many and diverse. The top three nanotechnological applications are electronics, energy and biomedical, altogether they reach *ca.* 70% of the global nanotechnology market [60]. It is worth to mention that NP represent more than 85% of the global nanotechnological product market [60]. The United State National Institute for Occupational Safety and Health (NIOSH) reported that more than 2 million persons were exposed to large amounts of NP, and this number would rise fast and reach 4 million soon, thus is vital to evaluate the exposure and develop efficient monitoring methods and control techniques [61].

The production of NM is expected to keep increasing through the years. In 2014 (with predictions that would last up to 2025), it was estimated that the global market for NM was from 300,000 tons up to 1.6 million tons. The Asian region accounts for largest market share (34%), followed by North America (31%) and Europe (30%). Among the most produced NM, the SiO_2NP were between 185,000 and 1,400,000; TiO_2NP between 60,000 and 150,000; and CNT between 1,550 and 1,950 annuals tons [62].

Even though, recent surveys revealed that the majority of the population is aware about the nanotechnology and most of them recognizes its beneficial effects in medical and technological applications. Indeed, only a small fraction of the population declared aversion towards nanotechnology. The public perception about nanotechnology has a direct implication on both the implementation and regulation of NM or products that contains them [63].

4.   Environmental fate

Massive production of nano-enabled products inevitably leads to the release of engineered NM into the environment. Along with the product life cycle, there could be potential emissions and wastes during production processes, usage, and final disposal of consumer nanoproducts [64–66]. For example, Kaegi et al. [67] confirmed TiO_2NP emission from exterior paints into runoff waters. Similarly, several studies have investigated the release of Ag^+ from silver nanoproducts such as textiles (shirts, towels, socks), medical devices (mask, commercial dressing, medical cloth), personal and health care products (detergent, shampoo, toothpaste, toothbrushes) and paints [44,68–73]. The presence of AgNP was confirmed in most products and Ag release was quantified in the wash water samples [73–75].

In the environment, emitted NM can be deposited on land and remain in the soil or can be transported by wind or rainwater runoff to aquatic systems [76]. In turn, homo- and heteroaggregation lead to an accelerated air or sediment bed deposition [77]. Resuspension of NM is possible via wind, water currents, or biological processes (e.g., disturbance of sedimentary deposits by living organisms). In addition, particles could slowly migrate into groundwaters [78].

Wastewater treatment plants (WWTP), landfills and waste incineration plants (atmospheric emissions) are the main sources of NP emissions. Thus, NP are expected to be found in various environmental compartments (air, soil, water). Several authors have made attempts to calculate final concentrations of NM in environmental matrices (surface waters, wastewater treatment plant effluents, biosolids, sediments, soils, and air), even taking into consideration dynamic material flow analysis models (i.e., the flow of NM from their synthesis to final disposal) [64,79–81]. Despite the predicted environmental concentrations (PEC) that have been modelled for several NM, their field validation has not been confirmed yet [65,82]. The order of magnitude of the modeled PEC in surface waters was estimated in the range of mg L^{-1} (for TiO_2NP) or ng L^{-1} (for ZnONP, AgNP, fullerenes, CNT, and CeO_2NP) [64], while the concentrations in waste incineration residues are at the mg kg^{-1} level [82]. Analytical methods are currently under development, and the quantification of NM in environmental samples in the field is still not possible. Likely, this is due to the low concentrations and the technical difficulties in distinguishing the naturally occurring NM from engineered NM.

5.   Exposure pathway

Exposure to NM could take place through various routes. Due to their small size, nanosized materials can easily enter into the human body by inhalation, ingestion, through the skin, and by injection, crossing numerous biological barriers and distributing throughout the organism. Thus, the immune system reacts and initiates a clearing process of the body. If these biological responses are not controlled, NM are considered toxic [83]. Ultimately, those NP will be accumulated in tissues meanwhile there is some evidence suggesting their biomagnification through the trophic web [84,85]. In addition, other research includes NP routes through both reproductive and circulatory systems [86]. The cycle of life of different NM in the human body, their residence times, and fate to each organ could be dissimilar and highly dependent on the exposure pathway apart from the chemical and physical properties of the particles [78]. Overall, the main documented routes of exposure are those described below and represented in Figure 2.

5.1   Inhalation

Toxicity through airborne exposure to NP has been a major concern as it was considered one of the main routes of exposures for humans. Intrinsic properties of the particles such as their size and spreading pattern will ultimately determine the area of the respiratory tract in which the NP will settle. However, once settled, the NP are able to enter the blood circulation and reach different parts of the body including the brain [87]. Particularly, a complete review carried out by Garcés et al. [86] explained that AgNP caused lung toxicity by altering O_2 metabolisms and generating oxidative damage in mice, which was related to alveolar epithelial injury after an acute exposure. In addition, the presence of NM in the air has been associated with low quality indexes and related to a cause of a variety of cancers depending on the particle density. Urban air can contain up to 10,000-50,000 NP cm^{-3}. For example, in the case of AgNP having a density > 6 g cm^3, the proposed exposure threshold is 20,000 NP cm^{-3} [78]. Urban air could have up to 50,000 NP despite the real concern regarding work safety is increased in the case of the proximity of industries which process NM (like leather tanneries) as the level increases in multi-fold ways [88].

5.2   Penetration through skin

Despite that inhalation has been considered the primary route of exposure to NP for humans, there is an ongoing debate about the potentiality of NM passage through the skin and its significant health hazard in humans [89]. The increasing number of products containing NM or nanoproducts has reached an amount of 3,000 and they belong to various fields of applications with emphasis on personal care category (including baby products). Therefore, the current concern lies in the fact that those nanoproducts are being applied directly to the skin, being the most used the TiO_2NP, ZnONP, and AgNP [90,91]. In addition, the aforementioned particles are also applied in textile industries (for deodorizing and antibacterial purposes), antibacterial products, and even some NP has a direct use like nanosilver as biocide [92]. For example, a toxicological study about a nanocrystalline silver-impregnated coated dressing showed high amounts of silver ions release when dispersed in saline solution (simulating the humans body's fluids) up to 150 µg ml^{-1} [44]. NP are able to cross over the skin and reach the basal layers, and the smaller they are, the easier they cross given the size of skin pores (< 4 nm) [93]. Therefore, people who suffer from dermatitis or chronic eczema can have severe toxic effects as it has been proved regarding SiO$_2$NP [94]. A review carried out by De Matteis [90] explained that the particle shape also will determine the ultimate skin penetration as rods have more ability to penetrate in comparison with sphere and triangles. Lastly, the NP ultimate distribution will depend on the application as it was proved in mice that exposure to gold NP (AuNP) induced a flux in the blood vessel only after an intravenous administration, and it did not in the case of the topical one [95].

5.3   Ingestion

Food and beverages widely contain NM, therefore, gastrointestinal tract is a vital route of exposure [78]. As examples, TiO$_2$NP can act as a food coloring agent in sweets, which concentrations generally are between 1-5 µg mg^{-1}, and they are also present in chewing gums, confectionery, sauces and dressings, creamers, and dietary supplements. Average consumption was estimated as 0.2-0.7 mg kg weight per day in the United States and 1 mg kg body weight per day in the United Kingdom and Germany [96]. Particularly, concern about TiO$_2$NP ingestion raised in relation to the highest exposure scenario in the case of the children (1-2 mg kg^{-1} body weight per day, < 10 years old) [97]. Unluckily, there is still a lack of studies addressing the safety of TiO$_2$NP consumption [78]. On another hand, AgNP use is also a matter of concern. Wijnhoven et al. [98] highlighted the addition of AgNP into the food production chain as their application occurs at all stages: during the processing of food (the particles are applied in food preparation equipment), for conservation purposes (the refrigerators storage containers are applied with AgNP), and at the ultimate food consumption (nanosilver is applied as a supplement). In all cases, the authors stated that the AgNP application has antibacterial purposes and avoids pathogens growth although the expected consumer exposure remains low and hard to estimate. Lastly, another study showed that commercially available plastic food containers were proved to release up to 3.1 ng of AgNP cm^{-2} after 10 days, and that the particles were capable of migrating to food [72]. Regarding SiNP, Dekkers et al. [99] explained in those particles widely present in food products could led to a daily oral intake of the particles at 1.8 mg kg^{-1} bw per day for an adult of 70 kg. Notably, the authors listed a diverse variety of food-based products where they detected SiNP such as mix for lasagna sauce, pancake mix, coffee creamer, instant soups, and steak and vegetable rubs.

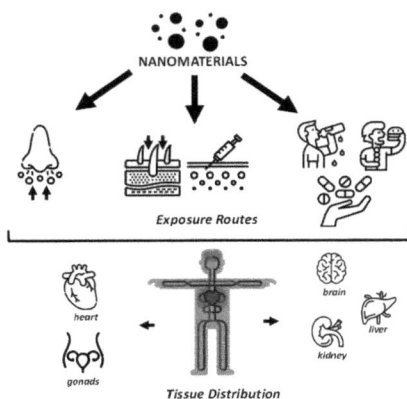

Figure 2. Main routes of exposure to nanomaterials and distribution.

6.   Toxicological effects

The benefits that NM have brought to humanity are countless, so their widespread use is expected to keep growing through the years. However, consumer knowledge regarding their functionality, benefits, and potential dangers of nanotechnology to final users is still modest [63]. According to Rodríguez-Ibarra et al. [100], the PubMed database showed 25,000 papers published related to engineered NM applications while less than 3% included occupational safety. In this sense, an updated search in Scopus database showed that 4,700 articles included the keyword "engineered nanomaterials" (ENM), but only the 43% contemplated their applications, and just a 20-30% included their toxicity or risks. Unluckily, less than 8% took into account occupational-exposure implications (Figure 3).

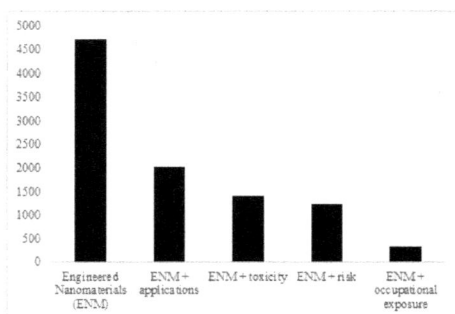

Figure 3. Articles available in Scopus database. Last update: May 2022. Modified from Rodríguez-Ibarra et al. [100]. Published by Elsevier. License number: 5315960975965.

Technically, a product including nanosized materials or implicating the use of nanotechnology cannot be considered as strictly dangerous. The harmfulness and toxicity of a certain new NM should be determined by its applications, the production, and the final disposal. By considering the wide variety of possible application sectors, the list of NM that are spread in the environment and represent a risk for human health is very extensive. However, they have different impact on the market. Inorganic non-metallic NM, metal NP, and carbon-based NM are the groups of nanostructures with the largest market volume that require an immediate regulatory attention.

Some effort was made experimentally in order to assess and, in the best-case scenario, prevent deleterious effect to users and consumers of nanoproducts. In this sense, the research field of "nano-ecotoxicology" emerged and gained importance in the last decades [65]. Nevertheless, despite several discussions and reviews about such field of study, official descriptions of health implications and environmental risk to users exposed to NM are severely lacking [101].

For example, a global report called "EMERGNANO" attempted to identify and assess worldwide progress in relation to nanotechnology risks and highlighted the need for a better characterization of nanotechnology constructs and for the production of "reagentgrade" NM. This matter of "nanosafety" has become under the spotlight since some unfortunate events such as a fatal one regarding women who worked for a paint factory in China and NP found in their lungs were associated with disease [102]. Therefore, it results of vital importance the availability of detailed information about the NM (chemical composition, morphology, aspect ratio, particle size, zeta potential, solubility, known hazards) as well as the complete description of how the NM should be used and how they were produced [101].

The most common toxic effects produced by nanostructured materials include oxidative stress caused by reactive oxygen species (ROS) release, protein denaturation, and alteration of mitochondrial and phagocytic functions. Other mechanisms are also related to physiological parameters that can go from physiological implications to reproductive failure by modifying hormones or hatching enzymes [65]. Additionally, when NM are deposited and accumulated in specific organs during an extended period of time, they can generate organelle damage and organ dysfunction, and lead to chronic health diseases, like asthma and cancer. However, these toxicity effects depends on many factors, like the chemical composition, morphology, core material, size and aging of each nanostructure [91]. For example, it has been demonstrated that nanotoxicity increases as size decreases [103]. Smaller materials can penetrate cells more easily and, due to their high specific surface area, they are more reactive and can have more interactions with cellular components. This increases the risk of suffering a prolonged inflammation, which can lead to the disruption of cellular machinery and finally to cellular death.

Despite short-term exposures to human were already considered to have a low risk, long periods of time at low dose of NM are a current matter of concern regarding safe implications. In this sense, a review carried out by Cronin et al. [104] explained that the immune system is particularly affected (with emphasis on the innate immune system). The authors highlighted that the exposure to nanosized particles could be related to deleterious effects in immune functions like penetrating into neutrophils, disrupting epithelial barriers, influencing the form of active uptake used by phagocytic cells, and even deleterious effects were analyzed on fetus in case of exposure.

Interestingly, Akçan et al. [87] detailed the main chemical composition of NP and the main reported toxic effects. For example, carbon-based NM were related to cytotoxicity on alveolar macrophages, meanwhile quantum dots core metalloid complexes were proved to cross the blood-brain barriers and placenta and also distribute to all body tissues (being the liver and kidney the target organs of toxicity). Other metallic NP (made of silver, molybdenum, iron, aluminum, etc.) reduced cell proliferation and even cause death. Lastly, other kinds of NP like the ones doped with cerium, showed low toxicity in relation to cell proliferation. The charge of the NP, apart from their main component, will also have influence on the ultimate effects to users. In this sense, positively charged poly amidoamine dendrimers showed higher deposition in comparison with the neutral ones.

Morphology and crystal structure also affect nanotoxicity levels. As an example, amorphous TiO_2 generates more ROS than the two crystalline polymorphs, anatase and rutile [105]. Furthermore, rutile TiO_2NP can cause DNA damage in the dark, while anatase TiO_2NP do not disturb genetic material in the same conditions. As well as this, rod-shaped iron oxide NM have shown much higher cytotoxic effect in a murine macrophage cell line compared to spherical NM [106]. In addition to this, concentration, aggregation, and time exposure are other factors that condition the degree of nanotoxicity [107–109]. It is considered that the amount of NM that penetrates into the cells keeps a directly proportional relationship with its concentration and period exposure. Nevertheless, at the same time, the higher the concentration, the greater the possibility that the nanomaterial aggregates and cannot cross the biological barriers.

Surface functionalization is also crucial to modulate the nanotoxicity since it determines uptake, the body's immune response and biocompatibility of the nanosized material [110]. Biochemical modification of the outside part of nanostructures with specific targeting, such as small proteins, peptides, polymers, and oligosaccharides, also improve their stability and increase specificity and efficacy. For example, polyethyleneglycol (PEG), dextran, chitosan and lipids are commonly used to reduce cytotoxicity and enhance biocompatibility of NM, while hydrophilic polymers can reduce the response of the immune system [111].

The application of NM in consumer products and the use of nanotechnology are constantly growing, but the available information about their possible effects in living beings is not enough. Therefore, more studies of toxicity, health impact, biological interaction of NM, is urgently necessary, as well as regulations to control the potential risks of nanosized structures. Overall, given the variety of NM regarding their composition, and together with the different kinds of exposure (via airborne, skin penetration or digestion), it results a challenge to characterize the toxicity response in human/user in case exposure. The Table 1 summarizes some interesting findings based on previous evidence. Notably, there is robust evidence suggesting the toxicity of carbon-based NM meanwhile there is a lack of research into other kinds such as AgNP despite their wide use and multiple applications.

Table 1. Cases of studies based on NM exposure to workers. Modified from Schulte et al. [112]. Published by Scandinavian Journal of Work, Environment & Health. License under Creative Commons Attribution 4.0.

Nanomaterial (NM)	Exposure	Main findings	References
Carbon-based NM	Production, manufacturing, and packing.	Reduction of lung functions (but not pathologically). Development of respiratory allergies. Increased inflammatory cytokines from sputum samples. Altered inflammatory biomarkers in serum. Increased number of eosinophiles in peripheral blood. Depression of neutrophils and elevation of monocytes, mean platelet volume, immature platelet fraction, and immature reticulocytes fraction. Altered antioxidant enzymatic activities in blood. Oxidative damage in lipids, genotoxicity in DNA, and pro-inflammatory responses from exhaled breath condensates. Altered gene expression related to cell cycle regulation, apoptosis, and proliferation. Identified pathways and signaling networks related to potential trigger of pulmonary and cardiovascular effects and carcinogenic outcomes. Changes in methylation of gens.	[113–119]
Silica-based NM	Spray painting, manufacturing.	Shortness of breath, pleural and pericardial effusion, and pulmonary inflammation. Anomalies in cellular cytoplasmatic components. Decreased global DNA methylation. Lower blood and white blood cells.	[102,120,121]

Titanium dioxide-based NM	Process and handle of NM and manufacturing.	Suspected effects on heart rate variability. Oxidative damage of nucleic acids and proteins. Augmented lipid peroxidation levels and altered antioxidant enzymatic activities. Altered biomarkers related to lung damage and cardiovascular disease. Decreased in blood and white blood cells number.	[120,122–124]
Silver-based NM	Manufacturing.	Detectable concentrations of silver in blood but not in urine. Altered activities of antioxidant enzymes (with great alterations after 6 months of exposure). Increased cardiovascular makers.	[125–127]

Conclusions

Nanotechnology has become a vital tool for human being daily life and a key field of study for many subjects, from pharmaceutics to automotive industry. In this sense, the development of novel materials or NM inevitably has increased in an exponentially manner, leading to concern about the unregulated production and consumption of nano-based products, together with their release into environment as exposure to humans also may occur. From that concern, other fields of research such as "nanotoxicology" have emerged, despite the fact that the latter is still under development and available and robust information remains scarce. Since "nanosafety" is a matter of key importance, more studies aiming to guarantee both the correct manipulation by users and discharge of NM are mandatory. Only after an appropriate balance between the design and application of further NM and their side effects and risks (with emphasis on occupational exposure), nanotechnology will probably be one of the most promising fields of study worldwide in the long run.

References

[1] A. Barhoum, M.L. García-Betancourt, J. Jeevanandam, E.A. Hussien, S.A. Mekkawy, M. Mostafa, M.M. Omran, M.S. Abdalla, M. Bechelany, Review on Natural, Incidental, Bioinspired, and Engineered Nanomaterials: History, Definitions, Classifications, Synthesis, Properties, Market, Toxicities, Risks, and Regulations, Nanomaterials. 12 (2022). https://doi.org/10.3390/nano12020177

[2] Editorial, "Plenty of room" revisited, Nat. Nanotechnol. 4 (2009) 781. https://doi.org/10.1038/nnano.2009.356

[3] S. Bayda, M. Adeel, T. Tuccinardi, M. Cordani, F. Rizzolio, The history of nanoscience and nanotechnology: From chemical-physical applications to nanomedicine, Molecules. 25 (2020) 1-15. https://doi.org/10.3390/molecules25010112

[4] N. Baig, I. Kammakakam, W. Falath, I. Kammakakam, Nanomaterials: A review of synthesis methods, properties, recent progress, and challenges, Mater. Adv. 2 (2021) 1821-1871. https://doi.org/10.1039/D0MA00807A

[5] L.A. Kolahalam, I. V. Kasi Viswanath, B.S. Diwakar, B. Govindh, V. Reddy, Y.L.N. Murthy, Review on nanomaterials: Synthesis and applications, Mater. Today Proc. 18 (2019) 2182-2190. https://doi.org/10.1016/j.matpr.2019.07.371

[6] J. Jeevanandam, A. Barhoum, Y.S. Chan, A. Dufresne, M.K. Danquah, Review on nanoparticles and nanostructured materials: History, sources, toxicity and regulations, Beilstein J. Nanotechnol. 9 (2018) 1050-1074. https://doi.org/10.3762/bjnano.9.98

[7] J. Highsmith, Global Markets: A BCC Research Report: Nanoparticles in Biotechnology, Drug Development and Drug Delivery Systems, 2021.

[8] G. Fytianos, A. Rahdar, G.Z. Kyzas, Nanomaterials in cosmetics: Recent updates, Nanomaterials. 10 (2020) 1-16. https://doi.org/10.3390/nano10050979

[9] B. Dréno, A. Alexis, B. Chuberre, M. Marinovich, Safety of titanium dioxide nanoparticles in cosmetics, J. Eur. Acad. Dermatology Venereol. 33 (2019) 34-46. https://doi.org/10.1111/jdv.15943

[10] Y. Rahimpour, H. Hamishehkar, Liposomes in cosmeceutics, Expert Opin. Drug Deliv. 9 (2012) 443-455. https://doi.org/10.1517/17425247.2012.666968

[11] M.F. Peralta, M.L. Guzmán, A.P. Pérez, G.A. Apezteguia, M.L. Fórmica, E.L. Romero, M.E. Olivera, D.C. Carrer, Liposomes can both enhance or reduce drugs penetration through the skin, Sci. Rep. 8 (2018) 13253. https://doi.org/10.1038/s41598-018-31693-y

[12] J. Vidales-Herrera, I. López, Chapter 3 - Nanomaterials in coatings: an industrial point of view, in: C.M.B.T.-H. of N. for M.A. Hussain (Ed.), Micro Nano Technol., Elsevier, 2020: pp. 51-77. https://doi.org/10.1016/B978-0-12-821381-0.00003-X

[13] E.R. Sadiku, O. Agboola, O. Agboola, I.D. Ibrahim, P.A. Olubambi, B. Avabaram, M. Bandla, W.K. Kupolati, J. Tippabattini, J. Tippabattini, K. Varaprasad, K. Varaprasad, S.C. Agwuncha, S.C. Agwuncha, J. Mochane, O.O. Daramola, B. Oboirien, T.A. Adegbola, C. Nkuna, S.J. Owonubi, S.J. Owonubi, V.O. Fasiku, B. Aderibigbe, V. Ojijo, R. Dunne, K. Selatile, G. Makgatho, C. Khoathane, W. Mhike, O.F. Biotidara, M.K. Dludlu, A. Adeboje, O.A. Adeyeye, A. Ndamase, S. Sanni, G.F. Molelekwa, P. Selvam, R. Nambiar, A.B. Perumal, J. Jayaramudu, J. Jayaramudu, N. Iheaturu, I. Diwe, B. Chima, Nanotechnology in Paints and Coatings, in: L. Li, Q. Yang (Eds.), Adv. Coat. Mater., Scrivener Publishing LLC, 2018: pp. 175-233. https://doi.org/10.1002/9781119407652.ch7

[14] S. Arango-Santander, A. Pelaez-Vargas, S.C. Freitas, C. García, A novel approach to create an antibacterial surface using titanium dioxide and a combination of dip-pen

nanolithography and soft lithography, Sci. Rep. 8 (2018) 15818.
https://doi.org/10.1038/s41598-018-34198-w

[15] T. Bruna, F. Maldonado-Bravo, P. Jara, N. Caro, Silver Nanoparticles and Their
Antibacterial Applications, Int. J. Mol. Sci. 22 (2021) 7202.
https://doi.org/10.3390/ijms22137202

[16] D. Zhi, H. Wang, D. Jiang, I.P. Parkin, X. Zhang, Reactive silica nanoparticles turn
epoxy coating from hydrophilic to super-robust superhydrophobic, RSC Adv. 9 (2019)
https://doi.org/10.1039/C8RA10046B

[17] S. Mahović Poljaček, T. Tomašegović, M. Leskovšek, U. Stanković Elesini, Effect
of sio2 and tio2 nanoparticles on the performance of uv visible fluorescent coatings,
Coatings. 11 (2021) 1-18. https://doi.org/10.3390/coatings11080928

[18] H. Sun, B. Liu, X. Liu, Z. Yin, Dispersion of antimony doped tin oxide nanopowders
for preparing transparent thermal insulation water-based coatings, J. Mater. Res. 32
(2017) 2414-2422. https://doi.org/10.1557/jmr.2017.211

[19] T. Singh, S. Shukla, P. Kumar, V. Wahla, V.K. Bajpai, I.A. Rather, Application of
Nanotechnology in Food Science: Perception and Overview, Front. Microbiol. 8
(2017) 1-7. https://doi.org/10.3389/fmicb.2017.01501

[20] X. He, H. Deng, H. Hwang, The current application of nanotechnology in food and
agriculture, J. Food Drug Anal. 27 (2019) 1-21.
https://doi.org/10.1016/j.jfda.2018.12.002

[21] M.A. Emamhadi, M. Sarafraz, M. Akbari, V.N. Thai, Y. Fakhri, N.T.T. Linh, A.
Mousavi Khaneghah, Nanomaterials for food packaging applications: A systematic
review, Food Chem. Toxicol. 146 (2020) 111825.
https://doi.org/10.1016/j.fct.2020.111825

[22] P. Chandra, P.S. Panesar, eds., Nanosensing and Bioanalytical Technologies in Food
Quality Control, Springer Singapore, 2022. https://doi.org/10.1007/978-981-16-7029-9

[23] M. Chethipuzha, A.R. Abraham, N. Kalarikkal, S. Thomas, S. Sreeja, Chapter 16 -
Embracing nanotechnology concepts in the electronics industry, in: S. Thomas, N.
Kalarikkal, A.R.B.T.-F. and P. of M.N. Abraham (Eds.), Micro Nano Technol.,
Elsevier, 2021: pp. 405-421. https://doi.org/10.1016/B978-0-12-822352-9.00004-3

[24] T. Maruyama, Carbon nanotubes, in: S. Thomas, C. Sarathchandran, S.A. Ilangovan,
J.C.B.T.-H. of C.-B.N. Moreno-Piraján (Eds.), Micro Nano Technol., Elsevier, 2021:
pp. 299-319. https://doi.org/10.1016/B978-0-12-821996-6.00009-9

[25] N. Gupta, A. Dixit, Carbon Nanotube Field-Effect Transistors (CNFETs): Structure,
Fabrication, Modeling, and Performance, in: A. Hazra, R. Goswami (Eds.), Carbon
Nanomater. Electron. Devices Appl., Springer Singapore, Singapore, 2021: pp. 199-
214. https://doi.org/10.1007/978-981-16-1052-3_9

[26] D.E. El-Nashar, M.A. Aly, M.A. Ashmawy, W.R. Agami, Enhancing the
magnetization, electric resistivity and mechanical properties of silicone rubber loaded

by Co-Zn ferrite nanoparticles as filler, J. Magn. Magn. Mater. 553 (2022) 169252. https://doi.org/10.1016/j.jmmm.2022.169252

[27] K.K. Kefeni, T.A.M. Msagati, B.B. Mamba, Ferrite nanoparticles: Synthesis, characterisation and applications in electronic device, Mater. Sci. Eng. B Solid-State Mater. Adv. Technol. 215 (2017) 37-55. https://doi.org/10.1016/j.mseb.2016.11.002

[28] M. Ćibo, A. Šator, A. Kazlagić, E. Omanović-Mikličanin, Application and Impact of Nanotechnology in Sport, in: M. Brka, E. Omanović-Mikličanin, L. Karić, V. Falan, A. Toroman (Eds.), 30th Sci. Conf. Agric. Food Ind., Springer International Publishing, Cham, 2020: pp. 349-362. https://doi.org/10.1007/978-3-030-40049-1_44

[29] S.F.A. Acquah, A. V Penkova, D.A. Markelov, A.S. Semisalova, B.E. Leonhardt, J.M. Magi, The Beautiful Molecule: 30 Years of C60 and Its Derivatives, ECS J. Solid State Sci. Technol. 6 (2017) M3155-M3162. https://doi.org/10.1149/2.0271706jss

[30] A. Sharma, J. Hickman, N. Gazit, E. Rabkin, Y. Mishin, Nickel nanoparticles set a new record of strength, Nat. Commun. 9 (2018) 4102. https://doi.org/10.1038/s41467-018-06575-6

[31] M. Shettar, M. Doshi, A.K. Rawat, Study on mechanical properties and water uptake of polyester-nanoclay nanocomposite and analysis of wear property using RSM, J. Mater. Res. Technol. 14 (2021) 1618-1629. https://doi.org/10.1016/j.jmrt.2021.07.034

[32] A.K. Yetisen, H. Qu, A. Manbachi, H. Butt, M.R. Dokmeci, J.P. Hinestroza, M. Skorobogatiy, A. Khademhosseini, S.H. Yun, Nanotechnology in Textiles, ACS Nano. 10 (2016) 3042-3068. https://doi.org/10.1021/acsnano.5b08176

[33] M.A. Shah, B.M. Pirzada, G. Price, A.L. Shibiru, A. Qurashi, Applications of nanotechnology in smart textile industry: A critical review, J. Adv. Res. 38 (2022) 55-75. https://doi.org/10.1016/j.jare.2022.01.008

[34] T. Lindström, F. Österberg, Evolution of biobased and nanotechnology packaging - a review, Nord. Pulp Pap. Res. J. 35 (2020) 491-515. https://doi.org/10.1515/npprj-2020-0042

[35] M.E. Quadros, R. Pierson, N.S. Tulve, R. Willis, K. Rogers, T.A. Thomas, L.C. Marr, Release of Silver from Nanotechnology-Based Consumer Products for Children, Environ. Sci. Technol. 47 (2013) 8894-8901. https://doi.org/10.1021/es4015844

[36] R. Asmatulu, P. Nguyen, E. Asmatulu, Nanotechnology Safety in the Automotive Industry, in: Nanotechnol. Saf., 1st ed., © 2013 Elsevier B.V. All rights reserved., 2013: pp. 57-72. https://doi.org/10.1016/B978-0-444-59438-9.00005-9

[37] J. Damodharan, Nanomaterials in medicine - An overview, Mater. Today Proc. 37 (2021) 383-385. https://doi.org/10.1016/j.matpr.2020.05.380

[38] R. Mitarotonda, E. Giorgi, T. Eufrasio-da-Silva, A. Dolatshahi-Pirouz, Y.K. Mishra, A. Khademhosseini, M.F. Desimone, M. De Marzi, G. Orive, Immunotherapeutic nanoparticles: From autoimmune disease control to the development of vaccines, Biomater. Adv. (2022) 212726. https://doi.org/10.1016/j.bioadv.2022.212726

[39] S. Sargazi, Z. Ahmadi, M. Barani, A. Rahdar, S. Amani, M.F. Desimone, S. Pandey, G.Z. Kyzas, Can nanomaterials support the diagnosis and treatment of human infertility? A preliminary review, Life Sci. 299 (2022) 120539. https://doi.org/10.1016/j.lfs.2022.120539

[40] M.O. Besenhard, L. Panariello, C. Kiefer, A.P. LaGrow, L. Storozhuk, F. Perton, S. Begin, D. Mertz, N.T.K. Thanh, A. Gavriilidis, Small iron oxide nanoparticles as MRI T1 contrast agent: scalable inexpensive water-based synthesis using a flow reactor, Nanoscale. 13 (2021) 8795-8805. https://doi.org/10.1039/D1NR00877C

[41] Y. Zheng, H. Chen, X.-P. Liu, J.-H. Jiang, Y. Luo, G.-L. Shen, R.-Q. Yu, An ultrasensitive chemiluminescence immunosensor for PSA based on the enzyme encapsulated liposome, Talanta. 77 (2008) 809-814. https://doi.org/10.1016/j.talanta.2008.07.038

[42] S. Municoy, P.E. Antezana, C.J. Pérez, M.G. Bellino, M.F. Desimone, Tuning the antimicrobial activity of collagen biomaterials through a liposomal approach, J. Appl. Polym. Sci. n/a (2020) 50330. https://doi.org/10.1002/app.50330

[43] P.E. Antezana, S. Municoy, C.J. Pérez, M.F. Desimone, Collagen Hydrogels Loaded with Silver Nanoparticles and Cannabis Sativa Oil, Antibiotics. 10 (2021) 1420. https://doi.org/10.3390/antibiotics10111420

[44] A. Ayech, M.E. Josende, J. Ventura-Lima, C. Ruas, M.A. Gelesky, A. Ale, J. Cazenave, J.M. Galdopórpora, M.F. Desimone, M. Duarte, P. Halicki, D. Ramos, L.M. Carvalho, G.C. Leal, J.M. Monserrat, Toxicity evaluation of nanocrystalline silver-impregnated coated dressing on the life cycle of worm Caenorhabditis elegans, Ecotoxicol. Environ. Saf. 197 (2020) 110570. https://doi.org/10.1016/j.ecoenv.2020.110570

[45] J.K. Patra, G. Das, L.F. Fraceto, E.V.R. Campos, M. del P. Rodriguez-Torres, L.S. Acosta-Torres, L.A. Diaz-Torres, R. Grillo, M.K. Swamy, S. Sharma, S. Habtemariam, H.-S. Shin, Nano based drug delivery systems: recent developments and future prospects, J. Nanobiotechnology. 16 (2018) 71. https://doi.org/10.1186/s12951-018-0392-8

[46] G.S. Alvarez, C. Helary, A.M. Mebert, X. Wang, T. Coradin, M.F. Desimone, Antibiotic-loaded silica nanoparticle-collagen composite hydrogels with prolonged antimicrobial activity for wound infection prevention, J. Mater. Chem. B. 2 (2014) 4660-4670. https://doi.org/10.1039/c4tb00327f

[47] S. Municoy, M.I. Álvarez Echazú, P.E. Antezana, J.M. Galdopórpora, C. Olivetti, A.M. Mebert, M.L. Foglia, M. V. Tuttolomondo, G.S. Alvarez, J.G. Hardy, M.F. Desimone, Stimuli-responsive materials for tissue engineering and drug delivery, Int. J. Mol. Sci. 21 (2020) 1-39. https://doi.org/10.3390/ijms21134724

[48] M.F. Desimone, C. Hélary, I.B. Rietveld, I. Bataille, G. Mosser, M.M. Giraud-Guille, J. Livage, T. Coradin, Silica-collagen bionanocomposites as three-dimensional

358

scaffolds for fibroblast immobilization, Acta Biomater. 6 (2010) 3998-4004. https://doi.org/10.1016/j.actbio.2010.05.014

[49] S.F. Hansen, O.F.H. Hansen, M.B. Nielsen, Advances and challenges towards consumerization of nanomaterials, Nat. Nanotechnol. 15 (2020) 964-965. https://doi.org/10.1038/s41565-020-00819-7

[50] C. Contado, Nanomaterials in consumer products: a challenging analytical problem, Front. Chem. 3 (2015) 1-20. https://doi.org/10.3389/fchem.2015.00048

[51] A. Ale, M.F. Gutierrez, A.S. Rossi, C. Bacchetta, M.F. Desimone, J. Cazenave, Ecotoxicity of silica nanoparticles in aquatic organisms: An updated review, Environ. Toxicol. Pharmacol. 87 (2021) 103689. https://doi.org/10.1016/j.etap.2021.103689

[52] The National Nanotechnology Initiative (NNI), (n.d.).

[53] NNI Supplement to the President's 2018 Budget, (2017).

[54] Benedette Cuffari, Nanotechnology in the USA: Market Report, (2018).

[55] Global Nanotechnology Market 2018-2024: Market is Expected to Exceed US$ 125 Billion, (2018).

[56] The U.S. Food and drug administration, Nanotechnology-Over a Decade of Progress and Innovation, 2020.

[57] R. Bosetti, S.L. Jones, Cost-effectiveness of nanomedicine: estimating the real size of nano-costs, Nanomedicine. 14 (2019) 1367-1370. https://doi.org/10.2217/nnm-2019-0130

[58] L. Almeida, I. Felzenszwalb, M. Marques, C. Cruz, Nanotechnology activities: environmental protection regulatory issues data, Heliyon. 6 (2020) e05303-e05303. https://doi.org/10.1016/j.heliyon.2020.e05303

[59] I. Corsi, M.F. Desimone, J. Cazenave, Building the Bridge From Aquatic Nanotoxicology to Safety by Design Silver Nanoparticles, Front. Bioeng. Biotechnol. 10 (2022) 28. https://doi.org/10.3389/fbioe.2022.836742

[60] Global Nanotechnology Market, (2018).

[61] E.M. Osman, Environmental and Health Safety Considerations of Nanotechnology, Nano Safety, Biomed. J. Sci. Tech. Res. 19 (2019) 14501-14515. https://doi.org/10.26717/BJSTR.2019.19.003346

[62] J. Pulit-Prociak, M. Banach, Silver nanoparticles - A material of the future...?, Open Chem. 14 (2016) 76-91. https://doi.org/10.1515/chem-2016-0005

[63] I.A. Joubert, M. Geppert, S. Ess, R. Nestelbacher, G. Gadermaier, A. Duschl, A.C. Bathke, M. Himly, Public perception and knowledge on nanotechnology: A study based on a citizen science approach, NanoImpact. 17 (2020) 100201. https://doi.org/10.1016/j.impact.2019.100201

[64] F. Gottschalk, T. Sun, B. Nowack, Environmental concentrations of engineered nanomaterials: Review of modeling and analytical studies, Environ. Pollut. 181 (2013) 287-300. https://doi.org/10.1016/j.envpol.2013.06.003

[65] M. Bundschuh, J. Filser, S. Lüderwald, M.S. McKee, G. Metreveli, G.E. Schaumann, R. Schulz, S. Wagner, Nanoparticles in the environment: where do we come from, where do we go to?, Environ. Sci. Eur. 30 (2018). https://doi.org/10.1186/s12302-018-0132-6

[66] Q. Abbas, B. Yousaf, Amina, M.U. Ali, M.A.M. Munir, A. El-Naggar, J. Rinklebe, M. Naushad, Transformation pathways and fate of engineered nanoparticles (ENPs) in distinct interactive environmental compartments: A review, Environ. Int. 138 (2020) 105646. https://doi.org/10.1016/j.envint.2020.105646

[67] R. Kaegi, A. Ulrich, B. Sinnet, R. Vonbank, A. Wichser, S. Zuleeg, H. Simmler, S. Brunner, H. Vonmont, M. Burkhardt, M. Boller, Synthetic TiO 2 nanoparticle emission from exterior facades into the aquatic environment, Environ. Pollut. 156 (2008) 233-239. https://doi.org/10.1016/j.envpol.2008.08.004

[68] T. Benn, B. Cavanagh, K. Hristovski, J.D. Posner, P. Westerhoff, The Release of Nanosilver from Consumer Products Used in the Home, J. Environ. Qual. 39 (2010) 1875-1882. https://doi.org/10.2134/jeq2009.0363

[69] R. Kaegi, A. Voegelin, B. Sinnet, S. Zuleeg, H. Hagendorfer, M. Burkhardt, H. Siegrist, Behavior of metallic silver nanoparticles in a pilot wastewater treatment plant, Environ. Sci. Technol. 45 (2011) 3902-3908. https://doi.org/10.1021/es1041892

[70] C. Lorenz, L. Windler, N. von Goetz, R.P. Lehmann, M. Schuppler, K. Hungerbühler, M. Heuberger, B. Nowack, Characterization of silver release from commercially available functional (nano)textiles, Chemosphere. 89 (2012) 817-824. https://doi.org/10.1016/j.chemosphere.2012.04.063

[71] E.M. Sussman, P. Jayanti, B.J. Dair, B.J. Casey, Assessment of total silver and silver nanoparticle extraction from medical devices, Food Chem. Toxicol. 85 (2015) 10-19. https://doi.org/10.1016/j.fct.2015.08.013

[72] A. Mackevica, M.E. Olsson, S.F. Hansen, Silver nanoparticle release from commercially available plastic food containers into food simulants, J. Nanoparticle Res. 18 (2016) 1-11. https://doi.org/10.1007/s11051-015-3313-x

[73] T.M. Benn, P. Westerhoff, Nanoparticle silver released into water from commercially available sock fabrics, Environ. Sci. Technol. 42 (2008) 7025-7026. https://doi.org/10.1021/es801501j

[74] L. Geranio, M. Heuberger, B. Nowack, The behavior of silver nanotextiles during washing, Environ. Sci. Technol. 43 (2009) 8113-8118. https://doi.org/10.1021/es9018332

[75] J. Farkas, H. Peter, P. Christian, J.A. Gallego Urrea, M. Hassellöv, J. Tuoriniemi, S. Gustafsson, E. Olsson, K. Hylland, K.V. Thomas, Characterization of the effluent from a nanosilver producing washing machine, Environ. Int. 37 (2011) 1057-1062. https://doi.org/10.1016/j.envint.2011.03.006

[76] P.C. Ray, H. Yu, P.P. Fu, Toxicity and environmental risks of nanomaterials: Challenges and future needs, in: J. Environ. Sci. Heal. - Part C Environ. Carcinog. Ecotoxicol. Rev., 2009: pp. 1-35. https://doi.org/10.1080/10590500802708267

[77] A.A. Keller, N. Parker, Chapter 7. Innovation in procedures for human and ecological health risk assessment of engineered nanomaterials, Elsevier Inc., 2019. https://doi.org/10.1016/B978-0-12-814835-8.00007-8

[78] A. Malakar, S.R. Kanel, C. Ray, D.D. Snow, M.N. Nadagouda, Nanomaterials in the environment, human exposure pathway, and health effects: A review, Sci. Total Environ. 759 (2021) 143470. https://doi.org/10.1016/j.scitotenv.2020.143470

[79] A.A. Keller, W. Vosti, H. Wang, A. Lazareva, Release of engineered nanomaterials from personal care products throughout their life cycle, J Nanopart Res. 16 (2014). https://doi.org/10.1007/s11051-014-2489-9

[80] T.Y. Sun, F. Gottschalk, K. Hungerbühler, B. Nowack, Comprehensive probabilistic modelling of environmental emissions of engineered nanomaterials, Environ. Pollut. 185 (2014) 69-76. https://doi.org/10.1016/j.envpol.2013.10.004

[81] R. Song, Y. Qin, S. Suh, A.A. Keller, Dynamic Model for the Stocks and Release Flows of Engineered Nanomaterials, 2017. https://doi.org/10.1021/acs.est.7b01907

[82] T.Y. Sun, N.A. Bornhöft, K. Hungerbühler, B. Nowack, Dynamic Probabilistic Modeling of Environmental Emissions of Engineered Nanomaterials, Environ. Sci. Technol. 50 (2016) 4701-4711. https://doi.org/10.1021/acs.est.5b05828

[83] S.C. Sahu, A.W. Hayes, Toxicity of nanomaterials found in human environment, Toxicol. Res. Appl. 1 (2017) 1-13. https://doi.org/10.1177/2397847317726352

[84] M. Yoo-iam, R. Chaichana, T. Satapanajaru, Toxicity, bioaccumulation and biomagnification of silver nanoparticles in green algae (Chlorella sp.), water flea (Moina macrocopa), blood worm (Chironomus spp.) and silver barb (Barbonymus gonionotus), Chem. Speciat. Bioavailab. 26 (2014) 257-265. https://doi.org/10.3184/095422914X14144332205573

[85] X. Chen, Y. Zhu, K. Yang, L. Zhu, D. Lin, Nanoparticle TiO2 size and rutile content impact bioconcentration and biomagnification from algae to daphnia, Environ. Pollut. 247 (2019) 421-430. https://doi.org/10.1016/j.envpol.2019.01.022

[86] M. Garcés, N.D. Magnani, A. Pecorelli, V. Calabró, T. Marchini, L. Cáceres, E. Pambianchi, J. Galdoporpora, T. Vico, J. Salgueiro, M. Zubillaga, M.A. Moretton, M.F. Desimone, S. Alvarez, G. Valacchi, P. Evelson, Alterations in oxygen metabolism are associated to lung toxicity triggered by silver nanoparticles exposure, Free Radic. Biol. Med. 166 (2021) 324-336. https://doi.org/10.1016/j.freeradbiomed.2021.02.008

[87] R. Akçan, H.C. Aydogan, M.Ş. Yildirim, B. Taştekin, N. Sağlam, Nanotoxicity: A challenge for future medicine, Turkish J. Med. Sci. 50 (2020) 1180-1196. https://doi.org/10.3906/sag-1912-209. https://doi.org/10.3906/sag-1912-209

[88] P. Oberbek, P. Kozikowski, K. Czarnecka, P. Sobiech, S. Jakubiak, T. Jankowski, Inhalation exposure to various nanoparticles in work environment-contextual information and results of measurements, J. Nanoparticle Res. 21 (2019). https://doi.org/10.1007/s11051-019-4651-x

[89] F. Sarwar, R.N. Malik, C.W. Chow, K. Alam, Occupational exposure and consequent health impairments due to potential incidental nanoparticles in leather tanneries: An evidential appraisal of south Asian developing countries, Environ. Int. 117 (2018) 164-174. https://doi.org/10.1016/j.envint.2018.04.051

[90] V. De Matteis, Exposure to inorganic nanoparticles: Routes of entry, immune response, biodistribution and in vitro/In vivo toxicity evaluation, Toxics. 5 (2017). https://doi.org/10.3390/toxics5040029

[91] Y.W. Huang, M. Cambre, H.J. Lee, The Toxicity of Nanoparticles Depends on Multiple Molecular and Physicochemical Mechanisms, Int. J. Mol. Sci. 18 (2017). https://doi.org/10.3390/ijms18122702

[92] P. Mantecca, K. Kasemets, A. Deokar, I. Perelshtein, A. Gedanken, Y.K. Bahk, B. Kianfar, J. Wang, Airborne Nanoparticle Release and Toxicological Risk from Metal-Oxide-Coated Textiles: Toward a Multiscale Safe-by-Design Approach, Environ. Sci. Technol. 51 (2017) 9305-9317. https://doi.org/10.1021/acs.est.7b02390

[93] F. Larese Filon, M. Mauro, G. Adami, M. Bovenzi, M. Crosera, Nanoparticles skin absorption: New aspects for a safety profile evaluation, Regul. Toxicol. Pharmacol. 72 (2015) 310-322. https://doi.org/10.1016/j.yrtph.2015.05.005

[94] F. Rancan, Q. Gao, C. Graf, S. Troppens, S. Hadam, S. Hackbarth, C. Kembuan, U. Blume-Peytavi, E. Rühl, J. Lademann, A. Vogt, Skin penetration and cellular uptake of amorphous silica nanoparticles with variable size, surface functionalization, and colloidal stability, ACS Nano. 6 (2012) 6829-6842. https://doi.org/10.1021/nn301622h

[95] E.A. Sykes, Q. Dai, K.M. Tsoi, D.M. Hwang, W.C.W. Chan, Nanoparticle exposure in animals can be visualized in the skin and analysed via skin biopsy, Nat. Commun. 5 (2014). https://doi.org/10.1038/ncomms4796

[96] M. Ropers, H. Terrisse, M. Mercier-bonin, B. Humbert, Titanium Dioxide as Food Additive, INTECH Open Sci. Open Minds. (n.d.) 3-22.

[97] H.C. Winkler, T. Notter, U. Meyer, H. Naegeli, Critical review of the safety assessment of titanium dioxide additives in food, J. Nanobiotechnology. (2018) 1-19. https://doi.org/10.1186/s12951-018-0376-8

[98] S.W.P. Wijnhoven, W.J.G.M. Peijnenburg, C.A. Herberts, W.I. Hagens, A.G. Oomen, E.H.W. Heugens, B. Roszek, J. Bisschops, I. Gosens, D. Van De Meent, S. Dekkers, W.H. De Jong, M. Van Zijverden, A.J.A.M. Sips, R.E. Geertsma, Nano-silver - A review of available data and knowledge gaps in human and environmental risk assessment, Nanotoxicology. 3 (2009) 109-138. https://doi.org/10.1080/17435390902725914

[99] S. Dekkers, P. Krystek, R.J.B. Peters, D.P.K. Lankveld, B.A.S.G.H. Bokkers, P.H.V.A.N. Hoeven-arentzen, H. Bouwmeester, A.G. Oomen, Presence and risks of nanosilica in food products, 5 (2011) 393-405. https://doi.org/10.3109/17435390.2010.519836

[100] C. Rodríguez-Ibarra, A. Déciga-Alcaraz, O. Ispanixtlahuatl-Meráz, E.I. Medina-Reyes, N.L. Delgado-Buenrostro, Y.I. Chirino, International landscape of limits and recommendations for occupational exposure to engineered nanomaterials, Toxicol. Lett. 322 (2020) 111-119. https://doi.org/10.1016/j.toxlet.2020.01.016

[101] A. Groso, A. Petri-Fink, A. Magrez, M. Riediker, T. Meyer, Management of nanomaterials safety in research environment, Part. Fibre Toxicol. 7 (2010) 1-8. https://doi.org/10.1186/1743-8977-7-40

[102] Y. Song, X. Li, X. Du, Exposure to nanoparticles is related to pleural effusion, pulmonary fibrosis and granuloma, Eur Respir J. 34 (2009) 559-567. https://doi.org/10.1183/09031936.00178308

[103] L. Truong, T. Zaikova, B.L. Baldock, M. Balik-Meisner, K. To, D.M. Reif, Z.C. Kennedy, J.E. Hutchison, R.L. Tanguay, Systematic determination of the relationship between nanoparticle core diameter and toxicity for a series of structurally analogous gold nanoparticles in zebrafish, Nanotoxicology. 13 (2019) 879-893. https://doi.org/10.1080/17435390.2019.1592259

[104] J.G. Cronin, N. Jones, C.A. Thornton, G.J.S. Jenkins, S.H. Doak, M.J.D. Clift, Nanomaterials and innate immunity: A perspective of the current status in nanosafety, Chem. Res. Toxicol. 33 (2020) 1061-1073. https://doi.org/10.1021/acs.chemrestox.0c00051

[105] M.S.Umekar, R.G. Chaudhary, G.S. Bhusari, A.Mondal, A.K. Potbhare, M. Sami, Phytoreduced graphene oxide-titanium dioxide nanocomposites using Moringa oleifera stick extract, Mater. Today: Procs, 29 (2020) 709-714. https://doi.org/10.1016/j.matpr.2020.04.169

[106] J.H. Lee, J.E. Ju, B. Il Kim, P.J. Pak, E.-K. Choi, H.-S. Lee, N. Chung, Rod-shaped iron oxide nanoparticles are more toxic than sphere-shaped nanoparticles to murine macrophage cells, Environ. Toxicol. Chem. 33 (2014) 2759-2766. https://doi.org/10.1002/etc.2735

[107] P.B. Chouke, A.K. Potbhare, N.P. Meshram, M.M. Rai, K.M. Dadure, K Chaudhary, A.R. Rai, M. Desimone, R. Chaudhary, D.T. Masram, Bioinspired NiO nanospheres: Exploring in-vitro toxicity using Bm-17 and L. rohita liver cells, DNA degradation, docking and proposed vacuolization mechanism, ACS Omega. 7 (2022) 6869−6884. https://doi.org/10.1021/acsomega.1c06544

[108] R. Mitarotonda, M. Saraceno, M. Todone, E. Giorgi, E.L. Malchiodi, M.F. Desimone, M.C. De Marzi, Surface chemistry modification of silica nanoparticles alters the activation of monocytes, Ther. Deliv. 12 (2021) 443-459. https://doi.org/10.4155/tde-2021-0006

[109] M.C. De Marzi, M. Saraceno, R. Mitarotonda, M. Todone, M. Fernandez, E.L. Malchiodi, M.F. Desimone, Evidence of size-dependent effect of silica micro- and nano-particles on basal and specialized monocyte functions, Ther. Deliv. 8 (2017). https://doi.org/10.4155/tde-2017-0053

[110] I. Paatero, E. Casals, R. Niemi, E. Özliseli, J.M. Rosenholm, C. Sahlgren, Analyses in zebrafish embryos reveal that nanotoxicity profiles are dependent on surface-functionalization controlled penetrance of biological membranes, Sci. Rep. 7 (2017) 8423. https://doi.org/10.1038/s41598-017-09312-z

[111] G. Sanità, B. Carrese, A. Lamberti, Nanoparticle Surface Functionalization: How to Improve Biocompatibility and Cellular Internalization, Front. Mol. Biosci. 7 (2020). https://doi.org/10.3389/fmolb.2020.587012

[112] P.A. Schulte, V. Leso, M. Niang, I. Iavicoli, Current state of knowledge on the health effects of engineered nanomaterials in workers: A systematic review of human studies and epidemiological investigations, Scand. J. Work. Environ. Heal. 45 (2019) 217-238. https://doi.org/10.5271/sjweh.3800

[113] R. Zhang, Y. Dai, X. Zhang, Y. Niu, T. Meng, Y. Li, H. Duan, P. Bin, Reduced pulmonary function and increased pro-inflammatory cytokines in nanoscale carbon black-exposed workers, Part. Fibre Toxicol. (2014) 1-14. https://doi.org/10.1186/s12989-014-0073-1

[114] Y. Dai, Y. Niu, H. Duan, B.A. Bassig, M. Ye, X. Zhang, T. Meng, P. Bin, X. Jia, M. Shen, R. Zhang, W. Hu, X. Yang, R. Vermeulen, D. Silverman, N. Rothman, Q. Lan, S. Yu, Effects on Occupational Exposure to Carbon Black on Peripheral White Blood Cell Counts and Lymphocyte Subsets, Environ. Mol. Mutagen. (2016) 515-622. https://doi.org/10.1002/em.22036

[115] L.M. Fatkhutdinova, T.O. Khaliullin, O.L. Vasil, R.R. Zalyalov, I.G. Musta, E.R. Kisin, M.E. Birch, N. Yanamala, A.A. Shvedova, Fibrosis biomarkers in workers exposed to MWCNTs, Toxicol. Appl. Pharmacol. 299 (2016) 125-131. https://doi.org/10.1016/j.taap.2016.02.016

[116] J. Vlaanderen, A. Pronk, N. Rothman, A. Hildesheim, D. Silverman, H.D. Hosgood, S. Spaan, E. Kuijpers, P. Hoet, Q. Lan, R. Vermeulen, J. Vlaanderen, A. Pronk, N. Rothman, A. Hildesheim, H.D. Hosgood, S. Spaan, E. Kuijpers, L. Godderis, P. Hoet, A cross-sectional study of changes in markers of immunological effects and lung health due to exposure to multi-walled carbon nanotubes, Nanotoxicology. 5390 (2017). https://doi.org/10.1080/17435390.2017.1308031

[117] J.D. Beard, A. Erdely, M.M. Dahm, M.A. De Perio, M.E. Birch, D.E. Evans, J.E. Fernback, T. Eye, V. Kodali, R.R. Mercer, S.J. Bertke, M.K. Schubauer-berigan, Carbon nanotube and nano fi ber exposure and sputum and blood biomarkers of early e ff ect among U . S . workers, Environ. Int. 116 (2018) 214-228. https://doi.org/10.1016/j.envint.2018.04.004

[118] M.K. Schubauer-berigan, M.M. Dahm, A. Erdely, J.D. Beard, M.E. Birch, D.E. Evans, J.E. Fernback, R.R. Mercer, S.J. Bertke, T. Eye, M.A. De Perio, Association of pulmonary , cardiovascular , and hematologic metrics with carbon nanotube and nanofiber exposure among U.S. workers : a cross-sectional study, Part. Fibre Toxicol. (2018) 1-14. https://doi.org/10.1186/s12989-018-0258-0

[119] E. Di Ianni, P. Møller, U.B. Vogel, N.R. Jacobsen, Pro-inflammatory response and genotoxicity caused by clay and graphene nanomaterials in A549 and THP-1 cells, Mutat. Res. - Genet. Toxicol. Environ. Mutagen. 872 (2021). https://doi.org/10.1016/j.mrgentox.2021.503405

[120] S. Liou, W. Wu, H. Liao, C. Chen, C. Tsai, W. Jung, H. Lee, Global DNA methylation and oxidative stress biomarkers in workers exposed to metal oxide nanoparticles, J. Hazard. Mater. 331 (2017) 329-335. https://doi.org/10.1016/j.jhazmat.2017.02.042

[121] M. Song, Yuguo; Xue, Li; Liying, Wang; Yon, Rojanasakul; Vincent, Castranova; Huiling, Li; Jing, Nanomaterials in Humans: Identification, Characteristics, and Potential Damage, Toxicol. Pathol. 39 (2011) 841-849. https://doi.org/10.1177/0192623311413787

[122] S. Ichihara, W. Li, S. Omura, Y. Fujitani, Y. Liu, Q. Wang, Y. Hiraku, N. Hisanaga, K. Wakai, X. Ding, T. Kobayashi, G. Ichihara, Exposure assessment and heart rate variability monitoring in workers handling titanium dioxide particles : a pilot study, J. Nanoparticle Res. 18 (2016) 1-14. https://doi.org/10.1007/s11051-016-3340-2

[123] D. Pelclova, V. Zdimal, P. Kacer, N. Zikova, Z. Fenclova, S. Vlckova, J. Schwarz, K. Syslova, T. Navratil, F. Turci, I. Corazzari, Markers of oxidative damage of nucleic acids and proteins among workers exposed to TiO2 (nano) particles, Nanotoxicology. 0 (2016) 000. https://doi.org/10.1080/17435390.2016.1262921

[124] L. Zhao, L. Zhao, Y. Zhu, Z. Chen, H. Xu, J. Zhou, Cardiopulmonary effects induced by occupational exposure to titanium dioxide nanoparticles Cardiopulmonary effects induced by occupational exposure to titanium dioxide nanoparticles, Nanotoxicology. 0 (2018) 1-16. https://doi.org/10.1080/17435390.2018.1425502

[125] J.H. Lee, M. Kwon, J.H. Ji, C.S. Kang, K.H. Ahn, J.H. Han, I.J. Yu, Exposure assessment of workplaces manufacturing nanosized TiO 2 and silver, Inhal. Toxicol. 23 (2011) 226-236. https://doi.org/10.3109/08958378.2011.562567

[126] J.H. Lee, J. Mun, J.D. Park, I.J. Yu, A health surveillance case study on workers who manufacture silver nanomaterials, Nanotoxicology. 6 (2012) 667-669. https://doi.org/10.3109/17435390.2011.600840

[127] H. Liao, Y. Chung, C. Lai, S. Wang, H. Chiang, L. Li, T. Tsou, W. Li, H. Lee, W. Wu, M. Lin, J. Hsu, J. Ho, C. Chen, T. Shih, C. Lin, S. Liou, Six-month follow-up study of health markers of nanomaterials among workers handling engineered nanomaterials, Nanotoxicology. 5390 (2014) 100-110. https://doi.org/10.3109/17435390.2013.858793

Keyword Index

Additives   23
Allotropes   40
Anti-Microbial Resistance   284
Applications   40, 341
Aquatic Ecosystem   304

Biomedical Application   152, 265
Bone Fillers   265

Carbon Dots   72
Carbon Nanotubes   72, 178
Carbon   40
Characteristics   304
Classification   152
Concrete   226
Cosmetic Industry   200
Cytotoxicity   265

Diagnostic   125
Disease Control   125
Drug Delivery   100, 200, 284

Electrodes   178
Engineered Nanomaterials   341
Environmental Fate   341
Exposure Pathway   341
Exposure   304

Friction   23
Functionalized CNMs   72

Geopolymer   226
Graphene Oxide   72

Healthcare   284
Historical Development   152
Hole-transporting Materials   178
Hybrid Nanomaterials   152
Hydrogel   125

LC3 Cement   226
Lipoprotein   125
Liposomes   125
Lubricant   23
Lycurgus Cup   1
Mechanism   23
Medical Science   125
Metal Oxide   284
Micelles   125

Nanoemulsion   200
Nanoparticles   23, 40, 125, 341
Nanoscience   1
Nanosize   1
Nanotechnology   1
Nanotoxicology   341

Perovskite Solar Cells   178
Phase Change Materials   226
Photovoltaic Applications   178
Portland Cement   226
Potential Applications of CNMs   72

Sensors   100
Smart Nanomaterials   100
Soy Protein Isolate   265
Stability   200
Strategies for Synthesis   152

Targeted Drug Delivery   265
Tissue Engineering   100
Toxicity   304
Tumor   125

Ultrasonication   200

Wear   23
Wound Dressing   265

About the Editors

Prof. N.B. Singh, former Head Chemistry Department and Former Dean Faculty of Science, DDU Gorakhpur University, Gorakhpur is presently Emeritus professor at Department of Chemistry and Biochemistry and Research Development Cell, Sharda University, Greater Noida, UP. Dr. Singh received the most prestigious Alexander von Humboldt fellowship (Germany) in 1977 and has worked at different universities in Germany. His main areas of research are Solid state chemistry, materials science, cement chemistry, thermodynamics, and water purification. He has published more than 300 research articles, 8 books, and more than 50 book chapters. Prof. Singh is president of Indian Association for Solid State Chemists and Allied Scientists Jammu and received many awards. His name appeared twice in the list of 2% scientists of world declared by Stanford University, USA.

Dr. Md. Abu Bin Hasan Susan, a Professor of Chemistry at Dhaka University, Bangladesh, received his Ph.D. degree from Yokohama National University, Japan in 2000 as a MEXT scholar and was awarded VBL, JSPS, CREST, and Bridge postdoctoral fellowships. He has 176 articles including 18 book chapters. He collaborates with 14 renowned laboratories and visited eleven countries. He was nominated for awards from FACS, in 2005 and ISESCO, in 2010. He received the Dean's Award of the Faculty of Science, Dhaka University (2011), the UGC Awards of Bangladesh in 2011 and 2013, the United Group Paper Awards in 2016 and 2017, and the Bangladesh Academy of Sciences

Gold Medal in 2012. He received the Silver Medal of the Society of Promotion of Education and Science, India in 2019. He is an *Associate Editor* of Spectrum of Emerging Sciences, an *Editorial Board Member* of the Journal of Bangladesh Academy of Sciences and Universal Journal of Electrochemistry, and the *Journal of Scientific and Technical Research*. He served as the National Representative of Division I of IUPAC. He has been a Fellow (2018) of the Bangladesh Academy of Sciences and IUPAC (2016). His citations are 10,901, *h*-index- 30 and *i*-10 index- 79.

Dr. Ratiram Gomaji Chaudhary is presently working as Associate Professor and Head, Department of Chemistry, Seth Kesarimal Porwal College of Arts, Science and Commerce, Kamptee. His research areas are *Biogenic Synthesis*, *Phytosynthesis*, *Metal oxide/Graphene-based nanohybrids*, *Toxic Dyes Degradations* etc. He has been awarded two times with *'Rajiv Gandhi National Fellowship Award'* as JRF by UGC, New Delhi for pursuing MPhil and PhD degrees. He has published 3 books, 19 book chapters, and 102 regular articles, and 12 review articles in peer-reviewed SCI/Scopus indexing journals having Total Citations of 1351 with h_index 20. He is a recipient of several awards like 'Best Researcher Award', 'Young Scientist Award', 'Outstanding Reviewer Award by IOP-VAST', Mahatma Jyotirao Fule: State Level Best Researcher Award, etc. He worked as a Guest Editor for Scopus indexed journals *Materials Today: Proceeding*, Elsevier, and *Current Pharmaceutical Biotechnology, Current Pharmaceutical Design,* and *Current Nanosciences,* Bentham Science.

www.ingramcontent.com/pod-product-compliance
Lightning Source LLC
Chambersburg PA
CBHW071319210326
41597CB00015B/1282